The Divergence
Theorem and Sets
of Finite Perimeter

PURE AND APPLIED MATHEMATICS

A Program of Monographs, Textbooks, and Lecture Notes

MONOGRAPHS AND TEXTBOOKS IN PURE AND APPLIED MATHEMATICS

Recent Titles

The Divergence Theorem and Sets of Finite Perimeter

Washek F. Pfeffer

University of California, Davis
University of Arizona, Tucson
USA

CRC Press
Taylor & Francis Group
Boca Raton London New York

CRC Press is an imprint of the
Taylor & Francis Group, an **informa** business

A CHAPMAN & HALL BOOK

CRC Press
Taylor & Francis Group
6000 Broken Sound Parkway NW, Suite 300
Boca Raton, FL 33487-2742

First issued in paperback 2019

© 2012 by Taylor & Francis Group, LLC
CRC Press is an imprint of Taylor & Francis Group, an Informa business

No claim to original U.S. Government works

ISBN-13: 978-1-4665-0719-7 (hbk)
ISBN-13: 978-0-367-38151-6 (pbk)

Library of Congress Cataloging-in-Publication Data

Pfeffer, Washek F.
  The divergence theorem and sets of finite perimeter / Washek F. Pfeffer.
      p. cm. -- (Monographs and textbooks in pure and applied mathematics)
  Includes bibliographical references and index.
  ISBN 978-1-4665-0719-7 (hardback)
  1. Divergence theorem. 2. Differential calculus. I. Title.

QA433.P493 2012
515'.4--dc23   2012005948

**Visit the Taylor & Francis Web site at
http://www.taylorandfrancis.com**

**and the CRC Press Web site at
http://www.crcpress.com**

To Lida for her love and a lifetime of companionship

Contents

Preface

The divergence theorem and the resulting integration by parts formula belong to the most frequently used tools of mathematical analysis. In its elementary form, that is for smooth vector fields defined in a neighborhood of some simple geometric object such as rectangle, cylinder, ball, etc., the divergence theorem is presented in many calculus books. Its proof is obtained by a simple application of the one-dimensional fundamental theorem of calculus and iterated Riemann integration. Appreciable difficulties arise when we consider a more general situation. Employing the Lebesgue integral is essential, but it is only the first step in a long struggle. We divide the problem into three parts.

(1) Extending the family of vector fields for which the divergence theorem holds on simple sets.

(2) Extending the family of sets for which the divergence theorem holds for Lipschitz vector fields.

(3) Proving the divergence theorem when the vector fields and sets are extended simultaneously.

Of these problems, part (2) is unquestionably the most complicated. While many mathematicians contributed to it, the Italian school represented by Caccioppoli, De Giorgi, and others obtained a complete solution by defining the sets of bounded variation (BV sets). A major contribution to part (3) is due to Federer, who proved the divergence theorem for BV sets and Lipschitz vector fields. While parts (1)–(3) can be combined, treating them separately illuminates the exposition.

We begin with sets that are locally simple — finite unions of dyadic cubes, called dyadic figures. Combining ideas of Henstock and McShane with a combinatorial argument of Jurkat, we establish the divergence theorem for very general vector fields defined on dyadic figures. The proof involves only basic properties of the Lebesgue integral and Hausdorff measures. An easy corollary of the divergence theorem is a powerful integration by parts formula. It yields results on removable sets for the Cauchy-Riemann, Laplace, and minimal surface equations.

The next goal is to move from dyadic figures to BV sets. To enhance the intuition, our starting point is the geometric definition of perimeter. The perimeter of a set is the codimension one Hausdorff measure of its essential boundary. Several properties of sets with finite perimeter are derived directly from the definition. Deeper results rest on the equivalent analytic definition.

Following the standard presentation, we say that an integrable function has bounded variation, or is a BV function, if its distributional gradient is a vector-valued measure of finite variation. The variation of a BV function is defined as the variation of its distributional gradient. A set whose indicator is a BV function is called a BV set. The variation of a BV set is the variation of its indicator.

Although we are mainly interested in BV sets, it is neither possible nor desirable to separate them from BV functions. It is often easier to prove a result about BV functions first, and state the corresponding result about BV sets as a corollary. The link between BV sets and BV functions is the coarea formula, which connects the variation of a function with that of its level sets. Our objective is to show the equivalence of the geometric and analytic definitions by equating the perimeter of a measurable set with its variation. A variety of useful results concerning BV sets follows from the interplay between the two definitions. Throughout, we derive properties of BV functions directly from the definition, without referring to corresponding properties of Sobolev spaces. Sobolev spaces are not discussed in this text, and no a priori knowledge about them is required.

Once the BV sets are defined and their main properties established, it is relatively easy to apply the divergence theorem we proved for dyadic figures to BV sets. The main tool, due to Giacomelli and Tamanini, consists of approximating arbitrary BV sets by their BV subsets with special properties. At the end, we extend the divergence theorem to a family of unbounded vector fields with controlled growth.

We pay particular attention to continuous vector fields and their weak divergence. Elaborating on ideas of Bourgain and Brezis, we characterize the distributions F for which the divergence equation $\operatorname{div} v = F$ has a continuous weak solution — a recent joint work of T. De Pauw and the author.

All of our results and proofs rely entirely on the Lebesgue integral. No exotic integrals, akin to the generalized Riemann integral of Henstock and Kurzweil, are involved. Notwithstanding, some techniques we use are inspired by investigations of these integrals. We strove to give complete and detailed proofs of all our claims. Only a few standard facts are quoted without proofs, in which case we always provide precise references. The book has three parts, roughly corresponding to parts (1)–(3) listed above. We trust that the titles of the chapters and sections are sufficiently descriptive. Results and comments we consider marginal are presented in small print. However, marginal does not mean unimportant; a useful enhancement of the main text can be found in the small print.

The first two chapters, which deal with dyadic figures, are quite elementary. Except for very basic properties of Hausdorff measures, they should be accessible to the beginning graduate students. The rest of the book presupposes the knowledge equivalent to the first year graduate course in analysis. In addition, some familiarity with Hausdorff measures and distributions is expected. Rudimentary results from functional analysis are employed in the last two chapters. Our presentation owes much to the excellent textbooks [29, 75] and monographs [1, 33], which can serve as useful references.

During the preparation of this text I largely benefited from discussions with L. Ambrosio, P. Bouafia, G.D. Chakerian, T. De Pauw, D.B. Fuchs, N. Fusco, R.J. Gardner, G. Gruenhage, Z. Nashed, M. Šilhavý, S. Solecki, V. Sverak, B.S. Thomson, and M. Torres. I am obliged to W.G. McCallum who offered me a position of Research Associate in the Mathematics Department of the University of Arizona; it gave me access to university facilities, in particular to the university library.

Editorial help provided by the publisher was invaluable. In this regard my thanks belong to K. Craig, M. Dimont, S. Kumar, S. Morkert, and R. Stern.

W.F.P.

Tucson, Arizona
February 2012

Part 1

Dyadic figures

Chapter 1

Preliminaries

We establish the notation and terminology, and present some basic facts that will be used throughout the book. Several well-known theorems are stated without proofs; however, those results for which we found no convenient references are proved in detail. In general, the reader is expected to have some prior knowledge of the concepts introduced in this chapter.

1.1. The setting

The sets of all integers and of all positive integers are denoted by $\mathbb{Z}$ and $\mathbb{N}$, respectively. Symbols $\mathbb{Q}$, $\mathbb{R}$, and $\mathbb{C}$ denote, respectively, the sets of all rational, real, and complex numbers. The sets of all positive real numbers and of all positive rationals numbers are denoted by $\mathbb{R}_+$ and $\mathbb{Q}_+$, respectively. Unless specified otherwise, by a *number* we always mean a real number. Elements of $\overline{\mathbb{R}} := \mathbb{R} \cup \{\pm\infty\}$ are called the *extended real numbers*. In $\overline{\mathbb{R}}$ we consider the usual order and topology, and define the following algebraic operations:

$$a + \infty : = +\infty + a := +\infty \text{ for } a > -\infty,$$

$$a - \infty : = -\infty + a := -\infty \text{ for } a < +\infty,$$

$$a \cdot (\pm\infty) : = \begin{cases} \pm\infty & \text{if } a > 0, \\ \mp\infty & \text{if } a < 0, \\ 0 & \text{if } a = 0. \end{cases}$$

At various places we write $P := Q$ instead of $P = Q$ to stress the fact that P is *defined* as equal to Q. Throughout, the symbol ∞ stands for $+\infty$. Unless specified otherwise, $\varepsilon \to 0$ means $\varepsilon \to 0+$.

Finite and countably infinite sets are called *countable*. We say that a family $\mathcal{E}$ of sets *covers* a set E, or is a *cover* of E, if $E \subset \bigcup \mathcal{E}$. For any pair of sets A and B, the set

$$A \triangle B := (A - B) \cup (B - A) = A \cup B - A \cap B$$

is called the *symmetric difference* of A and B.

By a function we always mean an *extended real-valued* function. A *finite function* is real-valued. If a function f equals identically to $c \in \overline{\mathbb{R}}$, we write $f \equiv c$. When no confusion can arise, the same symbol denotes a function f defined on a set A and its restriction $f \upharpoonright B$ to a set $B \subset A$. For a function f defined on a set E and $t \in \overline{\mathbb{R}}$, we let

$$\{f > t\} := \{x \in E : f(x) > t\},$$

and define the sets $\{f \geq t\}$, $\{f < t\}$, etc. similarly. The set $\{f = 0\}$ is called the *null set* of f. Further, we let

$$f^+ := \max\{f, 0\} \quad \text{and} \quad f^- := \max\{-f, 0\},$$

and note that $f = f^+ - f^-$ and $|f| = f^+ + f^-$. The value of f at $x \in E$ is denoted interchangeably by $f(x)$, $f[x]$, and $\langle f, x \rangle$.

For $m \in \mathbb{N}$, and $x := (\xi_1, \ldots, \xi_m)$ and $y := (\eta_1, \ldots, \eta_m)$ in $\mathbb{R}^m$,

$$x \cdot y := \sum_{i=1}^{m} \xi_i \eta_i \quad \text{and} \quad |x| := \sqrt{x \cdot x}.$$

In $\mathbb{R}^m$ we use exclusively the Euclidean metric induced by the norm $|x|$. The diameter, closure, interior, and boundary of a set $E \subset \mathbb{R}^m$ are denoted by $d(E)$, $\operatorname{cl} E$, $\operatorname{int} E$, and ∂E, respectively. The distance between sets $A, B \subset \mathbb{R}^m$ is denoted by $\operatorname{dist}(A, B)$, or $\operatorname{dist}(x, B)$ if $A = \{x\}$ is a singleton. Given $E \subset \mathbb{R}^m$ and $r \in \mathbb{R}_+$, we let

$$U(E, r) := \{x \in \mathbb{R}^m : \operatorname{dist}(x, E) < r\},$$
$$B(E, r) := \{x \in \mathbb{R}^m : \operatorname{dist}(x, E) \leq r\}.$$

If $E = \{x\}$ is a singleton, the sets

$$U(x, r) := U(\{x\}, r) \quad \text{and} \quad B(x, r) := B(\{x\}, r)$$

are, respectively, the *open* and *closed ball* in $\mathbb{R}^m$ of radius r centered at x. For a pair of sets $A, B \subset \mathbb{R}^m$, the symbol $A \Subset B$ indicates that $\operatorname{cl} A$ is a *compact* subset of $\operatorname{int} B$.

Given $E \subset \mathbb{R}^m$ and $s \in \mathbb{N}$, we denote by $C(E; \mathbb{R}^s)$ the linear space of all continuous maps $\phi : E \to \mathbb{R}^s$. We let $C(E) := C(E; \mathbb{R})$, and note that according to this definition, all continuous function are real-valued.

The *Urysohn function* associated with a pair A, B of closed disjoint subsets of $\mathbb{R}^m$ is a function $\mathbf{u}_{A,B} \in C(\mathbb{R}^m)$ defined by the formula

$$\mathbf{u}_{A,B}(x) := \frac{\operatorname{dist}(x, B)}{\operatorname{dist}(x, A) + \operatorname{dist}(x, B)}, \quad x \in \mathbb{R}^m. \tag{1.1.1}$$

Theorem 1.1.1 (Tietze). *Let $C \subset \mathbb{R}^m$ be a closed set. Each continuous map $\phi : C \to \mathbb{R}^s$ has a continuous extension $\psi : \mathbb{R}^m \to \mathbb{R}^s$.*

PROOF. As ϕ has a continuous extension if and only if each of its coordinates does, it suffices to show that every continuous function $f : C \to \mathbb{R}$ has a continuous extension $g : \mathbb{R}^m \to \mathbb{R}$. Now $\mathbb{R}$ is homeomorphic to the open interval $(-1, 1)$, e.g., via the strictly increasing continuous function $\varphi : x \mapsto \frac{2}{\pi} \tan^{-1} x : \mathbb{R} \to (-1, 1)$. Hence we may assume that $f : C \to (-1, 1)$. Let $\mathbf{u}_{C_{1-}, C_{1+}}$ be the Urysohn function associated with

$$C_{1\pm} := \{ x \in C : \pm f(x) \geq 1/3 \}.$$

If $g_1 := 1/3 - (2/3) \mathbf{u}_{C_{1-}, C_{1+}}$, then

$$\left| g_1(x) \right| \leq 3^{-1} \text{ for all } x \in \mathbb{R}^m,$$
$$\left| f(x) - g_1(x) \right| \leq 3^{-1} \text{ for all } x \in C.$$

Next let $f_1 = f - g_1$, and let $\mathbf{u}_{C_{2-}, C_{2+}}$ be the Urysohn function associated with

$$C_{2\pm} := \{ x \in C : \pm f_1(x) \geq 1/3^2 \}.$$

If $g_2 := 1/3^2 - (2/3^2) \mathbf{u}_{C_{2-}, C_{2+}}$, then

$$\left| g_2(x) \right| \leq 3^{-2} \text{ for all } x \in \mathbb{R}^m,$$
$$\left| f(x) - g_1(x) - g_2(x) \right| \leq 3^{-2} \text{ for all } x \in C.$$

Proceeding by recursion, we define functions $g_k \in C(\mathbb{R}^m)$ such that $|g_k| \leq 3^k$ and $\left| f - \sum_{j=1}^{k} (g_j \restriction C) \right| \leq 3^k$ for $k = 1, 2, \ldots$. It is clear that $g := \sum_{k=1}^{\infty} g_k$ belongs to $C(\mathbb{R}^m)$ and extends f. $\qquad \square$

Corollary 1.1.2. *Let $\Omega \subset \mathbb{R}^m$ be an open set, and let $C \subset \Omega$ be a closed set. Each continuous map $\phi : C \to \mathbb{R}^s$ has a continuous extension $\theta : \mathbb{R}^m \to \mathbb{R}^s$ such that $\mathrm{cl}\, \{\theta \neq 0\} \subset \Omega$.*

PROOF. By Titze's theorem ϕ has a continuous extension $\psi : C \to \mathbb{R}^s$. Let $f = \mathbf{u}_{C, \mathbb{R}^m - \Omega}$ be the Urysohn function associated with C and $\mathbb{R}^m - \Omega$, and let $D = \{ f \leq 1/2 \}$. Note $C \subset \mathbb{R}^m - D \subset \mathrm{cl}\,(\mathbb{R}^m - D) \subset \Omega$. If $\mathbf{u}_{C,D}$ is the Urysohn function associated with C and D, then $\theta = \mathbf{u}_{C,D} \psi$ is the desired extension. $\qquad \square$

If $\Omega \subset \mathbb{R}^m$ is an open set and $k \in \mathbb{N}$, then $C^k(\Omega; \mathbb{R}^s)$ denotes the linear space of all maps $\phi = (f_1, \ldots, f_s)$ from Ω to $\mathbb{R}^s$ such that each $f_i : \Omega \to \mathbb{R}$ has continuous partial derivatives of orders less than or equal to k. We let

$$C^\infty(\Omega; \mathbb{R}^s) = \bigcap_{k=1}^{\infty} C^k(\Omega; \mathbb{R}^s),$$

and refer to elements of $C^k(\Omega; \mathbb{R}^s)$ and $C^\infty(\Omega; \mathbb{R}^s)$, respectively, as C^k and C^∞ maps from Ω to $\mathbb{R}^s$. Instead of $C(E; \mathbb{R})$, $C^k(\Omega; \mathbb{R})$, and $C^\infty(\Omega; \mathbb{R})$, we write $C(E)$, $C^k(\Omega)$, and $C^\infty(\Omega)$, respectively. The elements of $C^k(\Omega)$ and $C^\infty(\Omega)$ are called, respectively, the C^k and C^∞ functions defined on Ω.

Let $E \subset \mathbb{R}^m$ and $\phi : E \to \mathbb{R}^s$. The *indicator* χ_E of E and the *zero extension* $\overline{\phi}$ of ϕ are defined by the formulae

$$\chi_E(x) := \begin{cases} 1 & \text{if } x \in E, \\ 0 & \text{if } x \in \mathbb{R}^m - E, \end{cases} \qquad \overline{\phi}(x) := \begin{cases} \phi(x) & \text{if } x \in E, \\ 0 & \text{if } x \in \mathbb{R}^m - E, \end{cases}$$

respectively. The *support* of ϕ is the set $\operatorname{spt} \phi := \operatorname{cl} \{\phi \neq 0\}$.

Let $\Omega \subset \mathbb{R}^m$ be an open set. The linear space of all $\phi \in C(\Omega; \mathbb{R}^s)$ with $\operatorname{spt} \phi \Subset \Omega$ is denoted by $C_c(\Omega; \mathbb{R}^s)$; the spaces $C_c(\Omega)$, $C_c^k(\Omega; \mathbb{R}^s)$, $C_c^\infty(\Omega; \mathbb{R}^s)$, etc., are defined similarly. We always identify $\phi \in C_c(\Omega; \mathbb{R}^s)$ with its zero extension $\overline{\phi} \in C_c(\mathbb{R}^m; \mathbb{R}^s)$. This simple convention, which will cause no confusion, legitimizes the inclusions

$$\begin{aligned} C_c(U; \mathbb{R}^s) &\subset C_c(\Omega; \mathbb{R}^s) \subset C_c(\mathbb{R}^m; \mathbb{R}^s), \\ C_c^k(U; \mathbb{R}^s) &\subset C_c^k(\Omega; \mathbb{R}^s) \subset C_c^k(\mathbb{R}^m; \mathbb{R}^s) \end{aligned} \tag{1.1.2}$$

where $U \subset \Omega$ is an open set and $k = 1, 2, \ldots, \infty$.

Throughout this book, the ambient space is $\mathbb{R}^n$ where $n \geq 1$ is a fixed integer. By $\{e_1, \ldots, e_n\}$ we denote the *standard base* in $\mathbb{R}^n$, i.e.,

$$e_i := (0, \ldots, \underbrace{1}_{i\text{-th place}}, \ldots, 0), \quad i = 1, \ldots, n.$$

The *projection* in the direction of e_i is the linear map

$$\pi_i : x \mapsto \sum_{j \neq i} (x \cdot e_j) e_j : \mathbb{R}^n \to \mathbb{R}^n.$$

As the set $\Pi_i := \pi_i(\mathbb{R}^n)$ is a linear subspace of $\mathbb{R}^n$ with bases

$$e_1, \ldots, e_{i-1}, e_{i+1}, \ldots, e_n,$$

it is isometric to $\mathbb{R}^{n-1}$. For each $x \in \Pi_i$, the set $\pi_i^{-1}(x)$ is isometric to $\mathbb{R}$. Thus whenever convenient, we tacitly identify the space Π_i with $\mathbb{R}^{n-1}$, and the set $\pi_i^{-1}(x)$ with $\mathbb{R}$. Note that if $n = 1$, then $\pi_1(x) = 0$ for each $x \in \mathbb{R}^1$, and hence $\pi_1(\mathbb{R}^1) = \{0\}$.

A *cell* in $\mathbb{R}^n$ is the set $A := \prod_{j=1}^n [a_j, b_j]$ where $a_j < b_j$ are real numbers. If $b_1 - a_1 = \cdots = b_n - a_n$, the cell A is called a *cube*. A *figure* is a finite, possibly empty, union of cells. A *k-cube* is a cube

$$\prod_{j=1}^n \left[i_j 2^{-k}, (i_j + 1) 2^{-k} \right]$$

where k and $i_1, \ldots, i_n$ are integers. The family of all k-cubes is denoted by $\mathcal{D}_k$, and the elements of the union $\mathcal{DC} := \bigcup_{k \in \mathbb{Z}} \mathcal{D}_k$ are called *dyadic cubes*. A *dyadic figure* is a finite, possibly empty, union of dyadic cubes. The family of all dyadic figures in $\mathbb{R}^n$ is denoted by $\mathcal{DF}$.

At places we employ unspecified positive constants depending on certain parameters, such as the dimension n. If κ is a constant depending *only* on parameters $p_1, \ldots, p_k$, we write $\kappa = \kappa(p_1, \ldots, p_k)$. With a few exceptions, we use no universal constants. Symbols denoting constants are tied to the context: distinct constants appearing in different contexts are often denoted by the same symbol.

1.2. Topology

All topologies considered in this book are assumed to be Hausdorff. If $\mathcal{T}$ and $\mathcal{S}$ are topologies in a set X and $\mathcal{T} \subset \mathcal{S}$, we say that $\mathcal{S}$ is *larger* than $\mathcal{T}$, or equivalently that $\mathcal{T}$ is *smaller* than $\mathcal{S}$. In a topological space $(X, \mathcal{T})$, the closure of $E \subset X$ is denoted by $\mathrm{cl}_{\mathcal{T}} E$. Unless specified otherwise, each $Y \subset X$ is given the *subspace topology*.

Let X be a topological space. A set $E \subset X$ is a G_δ *set* if it is the intersection of countably many open subsets of X. *Borel sets* in X are elements of the smallest σ-algebra in X containing all open subsets of X. A map ϕ from X to a topological space Y is called *Borel measurable*, or merely *Borel*, if

$$\phi^{-1}(B) := \{x \in X : \phi(x) \in B\}$$

is a Borel subset of X for every Borel set $B \subset Y$.

A subset E of a topological space X is called *sequentially closed* if each sequence $\{x_k\}$ in E that converges in X converges to $x \in E$. Each closed subset of X is sequentially closed but not vice versa; see Example 1.2.1 below. If the converse is true, i.e., if every sequentially closed set $E \subset X$ is closed, the space X is called *sequential*. All closed and all open subsets of a sequential space are sequential. A map ϕ from a sequential space X to any topological space Y is continuous whenever

$$\lim \phi(x_k) = \phi(\lim x_k)$$

for every convergent sequence $\{x_k\}$ in X. Each first countable space is sequential, but the converse is false; see Example 10.3.4.

Example 1.2.1. Let ω_1 be the first uncountable ordinal, and let X be the space of all ordinals smaller than or equal to ω_1 equipped with the order topology. The set $E = X - \{\omega_1\}$ is sequentially closed but not closed.

Unless specified otherwise, a *linear space* is a linear space over $\mathbb{R}$. Let X be a linear space. The zero element of X is denoted by 0, and

$$A + B := \{x + y : x \in A \text{ and } y \in B\} \quad \text{and} \quad tA := \{tx : x \in A\}$$

for $A, B \subset X$ and $t \in \mathbb{R}$. As usual

$$-A := (-1)A \quad \text{and} \quad x + A := \{x\} + A$$

for $x \in X$. A set $C \subset X$ is called

- *absorbing* if $X = \bigcup \{tC : t \in \mathbb{R}\}$,
- *symmetric* if $-C = C$,
- *convex* if $tC + (1-t)C \subset C$ for each $0 \le t \le 1$.

The *linear hull*, or *convex hull*, of a set $E \subset X$ is, respectively, the intersection of all linear subspaces of X containing E, or the intersection of all convex subsets of X containing E.

A topology in X for which the maps

$$(x, y) \mapsto x + y : X \times X \to X \quad \text{and} \quad (t, x) \mapsto tx : \mathbb{R} \times X \to X$$

are continuous is called *linear*. Since each linear topology is induced by a uniformity [28, Example 8.1.17], all topological linear spaces are completely regular [28, Theorem 8.1.21]. A *locally convex* topology is a linear topology that has a neighborhood base at zero consisting of convex sets. A linear space equipped with a linear, or locally convex, topology is called a *topological linear space*, or a *locally convex space*, respectively. In this book we encounter only locally convex spaces.

A *seminorm* in X is a functional $p : X \to \mathbb{R}$ such that

$$p(x + y) \le p(x) + p(y) \quad \text{and} \quad p(tx) = |t| p(x)$$

for all $x, y \in X$ and each $t \in \mathbb{R}$. Observe that $p(0) = 0 \le p(x)$ for each $x \in X$. A *norm* in X is a seminorm p such that $p(x) = 0$ implies $x = 0$. A family $\mathcal{P}$ of seminorms is called *separating* if $p(x) = 0$ for all $p \in \mathcal{P}$ implies $x = 0$. A separating family $\mathcal{P}$ defines a locally convex topology in X; the neighborhood base at zero is given by convex symmetric sets

$$U_{p_1,\dots,p_k;\varepsilon} := \left\{ x \in X : \max\{p_1(x), \dots, p_k(x)\} < \varepsilon \right\}$$

where $p_1, \dots, p_k$ are in $\mathcal{P}$ and $\varepsilon > 0$. Conversely, each locally convex topology in X is induced by a separating family $\mathcal{P}$ of seminorms [64, Remark 1.38, (b)]. The separating property of $\mathcal{P}$ guarantees that the topology defined by $\mathcal{P}$ is Hausdorff. A locally convex topology induced by a countable separating family $\{p_k : k \in \mathbb{N}\}$ of seminorms is metrizable; for instance, by the metric

$$\rho(x, y) := \sum_{k=1}^{\infty} 2^{-k} \frac{p_k(x-y)}{1 + p_k(x-y)}.$$

A *Fréchet space* is a completely metrizable locally convex space.

Even if a topology in X is defined by an uncountable family of seminorms, there may exist another family of seminorms in X that is countable and defines the same topology. The next example illustrates the situation.

Example 1.2.2. Let $\Omega \subset \mathbb{R}^m$ be an open set. The topology $\mathcal{T}$ of locally uniform convergence in $C(\Omega; \mathbb{R}^s)$ is defined by the seminorms

$$p_K(\phi) := \sup_{x \in K} |\phi(x)|$$

where $K \subset \Omega$ is a compact set and $\phi \in C(\Omega; \mathbb{R}^s)$. Each $x \in \Omega \cap \mathbb{Q}^m$ has an open neighborhood $U_x \Subset \Omega$. Organize $\{U_x : x \in \Omega \cap \mathbb{Q}^m\}$ into a sequence $U_1, U_2, \ldots$, and let $V_j = \bigcup_{i=1}^{j} U_i$. Since $V_j \Subset \Omega$, and since each compact set $K \subset \Omega$ is contained in some V_j, it is clear that $\mathcal{T}$ is defined by the seminorms

$$q_j(\phi) := \sup_{x \in V_j} |\phi(x)|, \quad j = 1, 2 \ldots.$$

It follows that $\mathcal{T}$ is the metrizable topology of uniform convergence on the sets V_j. Thus $\mathcal{T}$ is complete, and $\left(C(\Omega; \mathbb{R}^m), \mathcal{T}\right)$ is a Fréchet space.

Let $(X, \mathcal{T})$ be a locally convex space. A set $E \subset X$ is called *bounded* if for each convex neighborhood U of zero there is $t > 0$ such that $E \subset tU$. Every compact set $E \subset X$ is bounded. In general, $E \subset X$ is bounded if and only if $\lim t_k x_k = 0$ whenever $\{x_k\}$ is a sequence in E and $\{t_k\}$ is a sequence in $\mathbb{R}$ converging to zero [64, Theorem 1.30]. If the topology $\mathcal{T}$ is induced by a family $\mathcal{P}$ of seminorms, then $E \subset X$ is bounded if and only if for each $p \in \mathcal{P}$,

$$\sup\{p(x) : x \in E\} < \infty.$$

The *dual space* of X, abreviated as the *dual* of X, is the linear space X^* of all continuous linear functionals $x^* : X \to \mathbb{R}.$[1] To begin with, X^* is just a linear space with no topology. However, two locally convex topologies in X^* are easy to introduce:

- the *weak* topology* $\mathcal{W}^*$ defined by seminorms

$$x^* \mapsto |\langle x^*, x \rangle| : X^* \to \mathbb{R}$$

  where $x \in X$;
- the *strong topology* $\mathcal{S}^*$ defined by seminorms

$$\|x^*\|_B := \sup\left\{|\langle x^*, x \rangle| : x \in B\right\}$$

  where $x^* \in X^*$ and $B \subset X$ is a bounded set.

Since each singleton $\{x\} \subset X$ is a bounded set, the weak* topology is smaller than the strong topology.

1.3. Measures

A *measure* [2] in an arbitrary set X is a function μ defined on all subsets of X that satisfies the following conditions:

---

[1] A notable exception to the notation X^* is the space $\mathcal{D}'$ of distributions defined in Section 3.1 below.

[2] Our concept of measure is often called "outer measure", and the term "measure" is reserved for the restriction of "outer measure" to the family of all measurable sets. For our purposes, such distinction is superfluous.

(i) $\mu(\emptyset) = 0$;

(ii) $\mu(B) \leq \mu(A)$ whenever $B \subset A \subset X$;

(iii) $\mu(\bigcup_{k=1}^{\infty} A_k) \leq \sum_{k=1}^{\infty} \mu(A_k)$ whenever $A_k \subset X$ for $k = 1, 2, \ldots$.

Throughout this section, μ is a measure in a set $X \subset \mathbb{R}^m$. The *reduction* of μ to a set $Y \subset X$ is a measure $\mu \llcorner Y$ in X defined by

$$(\mu \llcorner Y)(A) := \mu(A \cap Y)$$

for each $A \subset X$. If $\mu = \mu \llcorner Y$, we say that μ *lives* in Y. A set $E \subset X$ is called μ *measurable* whenever

$$\mu = \mu \llcorner E + \mu \llcorner (X - E).$$

The *support* of μ is the set

$$\operatorname{spt} \mu := X - \bigcup \{U \subset X : U \text{ is open in } X \text{ and } \mu(U) = 0\}.$$

Since each subset of $\mathbb{R}^m$ has the Lindelöf property [28, Section 3.8 and Corollary 4.1.16], we have $\mu(X - \operatorname{spt} \mu) = 0$. In accordance with the standard terminology, the measure μ is called

- *σ-finite* if $X = \bigcup_{k=1}^{\infty} E_k$ and $\mu(E_k) < \infty$ for $k = 1, 2, \ldots$,
- *Borel* if each relatively Borel subset of X is μ measurable,
- *Borel regular* if μ is a Borel measure and each $E \subset X$ is contained in a relatively Borel subset B of X such that $\mu(B) = \mu(E)$,
- *Radon* if μ is a Borel regular measure and $\mu(K) < \infty$ for each compact set $K \subset X$,
- *metric* if $\mu(A \cup B) \geq \mu(A) + \mu(B)$ for each pair $A, B \subset X$ such that $\operatorname{dist}(A, B) > 0$.

A set $E \subset X$ is called μ *σ-finite* if the reduced measure $\mu \llcorner E$ is σ-finite. The next two theorems are proved in [29, Sections 1.1 and 1.9].

Theorem 1.3.1. *Let $\Omega \subset \mathbb{R}^m$ be an open set. Each metric measure in Ω is a Borel measure. If μ is a Borel regular measure in Ω, then $\mu \llcorner E$ is a Radon measure for each μ measurable set $E \subset \Omega$ with $\mu(E) < \infty$. If μ is a Radon measure in Ω, then the following conditions hold:*

(1) *For each set $A \subset \Omega$,*

$$\mu(A) = \inf \{\mu(U) : U \subset \Omega \text{ is open and } A \subset U\}.$$

(2) *For each μ measurable set $A \subset \Omega$,*

$$\mu(A) = \sup \{\mu(K) : K \subset A \text{ is compact}\}.$$

A set $E \subset X$ with $\mu(E) = 0$ is called μ *negligible*. Sets $A, B \subset X$ are μ *equivalent* if $\mu(A \triangle B) = 0$; they are μ *overlapping* if $\mu(A \cap B) > 0$. Maps ϕ and ψ from a set $E \subset X$ to a set Y are μ *equivalent* if the set

$$\{\phi \neq \psi\} := \{x \in E : \phi(x) \neq \psi(x)\}$$

is μ negligible. When the measure μ is clearly understood from the context, we indicate the equivalence by symbols $A \sim B$ and $\phi \sim \psi$.

Let $E \subset X$ be a μ measurable set. A map $\phi : E \to \mathbb{R}^s$ is called μ *measurable* if the set $\phi^{-1}(B)$ is μ measurable for every Borel set $B \subset \mathbb{R}^s$. The linear space of all μ measurable maps $\phi : E \to \mathbb{R}^s$ is denoted by

$$L^0(E, \mu; \mathbb{R}^s).$$

The *essential support* of $\phi \in L^0(E, \mu; \mathbb{R}^s)$ is the set

$$\operatorname{ess\,spt} \phi := \operatorname{spt} \left(\mu \llcorner \{ \phi \neq 0 \} \right).$$

Unlike the support of ϕ, the essential support of ϕ depends only on the μ equivalence class of ϕ. If $\phi \in L^0(E, \mu; \mathbb{R}^s)$, we also define

$$\operatorname{ess\,sup}_{x \in E} \left| \phi(x) \right| := \inf \left\{ \sup_{x \in E} \left| \psi(x) \right| : \psi \in L^0(E, \mu; \mathbb{R}^s) \text{ and } \psi \sim \phi \right\}.$$

For each $\phi \in L^0(E, \mu; \mathbb{R}^s)$, there is $\psi \sim \phi$ such that

$$\operatorname{ess\,spt} \phi = \operatorname{spt} \psi \quad \text{and} \quad \operatorname{ess\,sup}_{x \in E} \left| \phi(x) \right| = \sup_{x \in E} \left| \psi(x) \right|.$$

Convention 1.3.2. As is customary, we do not explicitly distinguish between an individual set $E \subset X$, or an individual map ϕ, and the μ equivalence class determined by E, or by ϕ, respectively. However, the reader should be aware of the following custom: we think of the space $L^0(E, \mu; \mathbb{R}^s)$ as consisting of equivalence classes, but when we write $\phi \in L^0(E, \mu; \mathbb{R}^s)$, we view ϕ as a specific representative of its equivalence class. In particular, when writing $\phi \in L^0(E, \mu; \mathbb{R}^s)$, we always assume that

$$\operatorname{ess\,spt} \phi = \operatorname{spt} \phi \quad \text{and} \quad \operatorname{ess\,sup}_{x \in E} \left| \phi(x) \right| = \sup_{x \in E} \left| \phi(x) \right|.$$

The next two theorems are standard tools of measure theory. Their proofs can be found in [29, Section 1.2].

Theorem 1.3.3 (Egoroff). *Let μ be a finite measure in $X \subset \mathbb{R}^m$, and let $\{\phi_k\}$ be a sequence in $L^0(X, \mu; \mathbb{R}^s)$ that converges pointwise. Given $\varepsilon > 0$, there is a μ measurable set $E \subset X$ such that $\mu(X - E) < \varepsilon$ and the sequence $\{\phi_k \restriction E\}$ converges uniformly.*

Theorem 1.3.4 (Luzin). *Let μ be a finite Borel regular measure in $X \subset \mathbb{R}^m$, and let $\phi \in L^0(X, \mu; \mathbb{R}^s)$. Given $\varepsilon > 0$, there is a compact set $K \subset X$ such that $\mu(X - K) < \varepsilon$ and the restriction $\phi \restriction K$ is continuous.*

Given a μ measurable set $E \subset X$, we let

$$\|\phi\|_{L^p(E, \mu; \mathbb{R}^s)} := \left(\int_E |\phi|^p \, d\mu \right)^{1/p} \text{ if } 1 \leq p < \infty,$$

$$\|\phi\|_{L^\infty(E, \mu; \mathbb{R}^s)} := \operatorname{ess\,sup}_{x \in E} \left| \phi(x) \right|$$

for each $\phi \in L^0(E, \mu; \mathbb{R}^s)$, and for $1 \leq p \leq \infty$, define

$$L^p(E, \mu; \mathbb{R}^s) := \{f \in L^0(E, \mu; \mathbb{R}^s) : \|f\|_{L^p(E,\mu;\mathbb{R}^s)} < \infty\}.$$

Let $1 \leq p \leq \infty$ and $1 \leq q \leq \infty$ be such that $1/p + 1/q = 1$ where we define $1/\infty := 0$. The *Hölder inequality*

$$\|fg\|_{L^1(E,\mu;\mathbb{R}^s)} \leq \|f\|_{L^p(E,\mu;\mathbb{R}^s)} \|g\|_{L^q(E,\mu;\mathbb{R}^s)} \tag{1.3.1}$$

holds for each $f, g \in L^0(E, \mu; \mathbb{R}^s)$; see [63, Theorem 3.5].

Let μ be a Borel measure in an open set $\Omega \subset \mathbb{R}^m$. For $1 \leq p \leq \infty$, we denote by $L^p_{\text{loc}}(\Omega, \mu; \mathbb{R}^s)$ the linear space of all maps $\phi \in L^0(\Omega, \mu; \mathbb{R}^s)$ such that $\phi \restriction U$ belongs to $L^p(U, \mu; \mathbb{R}^s)$ for each open set $U \Subset \Omega$. Unless stated otherwise, throughout we assume that $L^p_{\text{loc}}(\Omega, \mu; \mathbb{R}^s)$ has been equipped with the Fréchet topology defined by seminorms

$$\phi \mapsto \|\phi \restriction U\|_{L^p(U,\mu;\mathbb{R}^s)} : L^p_{\text{loc}}(\Omega, \mu; \mathbb{R}^s) \to \mathbb{R}$$

where $U \Subset \Omega$ is an open set; cf. Example 1.2.2. We write $L^p(E, \mu)$ and $L^p_{\text{loc}}(\Omega, \mu)$ instead of $L^p(E, \mu; \mathbb{R})$ and $L^p_{\text{loc}}(\Omega, \mu; \mathbb{R})$, respectively.

The following theorem is essential for establishing Theorem 2.3.7 and Proposition 7.4.3 below; in addition, it simplifies proofs of some differentiation results (Theorems 4.3.4 and 6.2.3 below). We call it the *Henstock lemma*, but the name *Saks-Henstock lemma* is also used — cf. [44] and [37].

Theorem 1.3.5. *Let μ be a Radon measure in $X \subset \mathbb{R}^m$, let $E \subset X$ be a μ measurable set with $\mu(E) < \infty$, and let $f \in L^1(E, \mu)$ be real-valued. Given $\varepsilon > 0$, there is $\delta : E \to \mathbb{R}_+$ satisfying the following condition: for every collection $\{E_1, \ldots, E_p\}$ of μ measurable μ nonoverlapping subsets of E, and for every set of points $\{x_1, \ldots, x_p\} \subset E$, the inequality*

$$\sum_{i=1}^p \left| f(x_i)\mu(E_i) - \int_{E_i} f \, d\mu \right| < \varepsilon$$

holds whenever $d(E_i \cup \{x_i\}) < \delta(x_i)$ for $i = 1, \ldots, p$.

PROOF. Choose $\varepsilon > 0$, and using the Vitali-Carathéodory theorem [63, Theorem 2.25], find functions g and h defined on E that are, respectively, upper and lower semicontinuous, and satisfy

$$g \leq f \leq h \quad \text{and} \quad \int_E (h - g) \, d\mu < \varepsilon.$$

There is $\delta : E \to \mathbb{R}_+$ such that $g(y) < f(x) + \varepsilon$ and $h(y) > f(x) - \varepsilon$ for all $x, y \in E$ with $|x - y| < \delta(x)$. If $\{E_1, \ldots, E_p\}$ and $\{x_1, \ldots, x_p\}$ satisfy the conditions of the proposition, then

$$\int_{E_i} g\, d\mu \le \int_{E_i} f\, d\mu \le \int_{E_i} h\, d\mu,$$

$$\int_{E_i} g\, d\mu - \varepsilon\mu(E_i) \le f(x_i)\mu(E_i) \le \int_{E_i} h\, d\mu + \varepsilon\mu(E_i)$$

for $i = 1, \ldots, p$. Consequently

$$\sum_{i=1}^{p}\left| f(x_i)\mu(E_i) - \int_{E_i} f\, d\mu\right| \le \sum_{i=1}^{p}\left[\int_{E_i} (h - g)\, d\mu + \varepsilon\mu(E_i)\right]$$

$$\le \int_{E} (h - g)\, d\mu + \varepsilon\mu(E)$$

$$< \varepsilon\big[1 + \mu(E)\big]. \qquad\square$$

Lebesgue measure in $\mathbb{R}^m$ is denoted by $\mathcal{L}^m$. For each subset E of the ambient space $\mathbb{R}^n$, we let

$$|E| := \mathcal{L}^n(E).$$

Sets $A, B \subset \mathbb{R}^n$ are called *overlapping* if they are $\mathcal{L}^n$ overlapping, that is to say if $|A \cap B| > 0$. Unless specified otherwise, all concepts connected with measures, such as "measurable", "negligible", etc., as well as the expressions "almost all" and "almost everywhere", refer to the measure $\mathcal{L}^n$ in $\mathbb{R}^n$. For a measurable set $E \subset \mathbb{R}^n$, we let

$$L^p(E; \mathbb{R}^s) := L^p(E, \mathcal{L}^n; \mathbb{R}^s) \quad \text{and} \quad L^p(E) := L^p(E, \mathcal{L}^n).$$

If $\Omega \subset \mathbb{R}^n$ is an open set, the meanings of $L^p_{\text{loc}}(\Omega; \mathbb{R}^s)$ and $L^p_{\text{loc}}(\Omega)$ are obvious. When no confusion is possible, we write $\int_E f(x)\, dx$ or $\int_E f$ instead of $\int_E f\, d\mathcal{L}^n$.

1.4. Hausdorff measures

We define Hausdorff measures in $\mathbb{R}^n$, and state some of their elementary properties. Select a fixed $s \ge 0$, and let

$$\Gamma(s) := \int_0^\infty t^{s-1} e^{-t}\, dt \quad \text{and} \quad \alpha(s) := \frac{\Gamma\left(\frac{1}{2}\right)^s}{\Gamma\left(\frac{s}{2} + 1\right)}.$$

Recall that $\Gamma : t \mapsto \Gamma(t)$ is the classical Euler's *gamma function* [62, Definition 8.17]. Using Fubini's theorem and induction, we obtain

$$\alpha(n) = \mathcal{L}^n\big(\{x \in \mathbb{R}^n : |x| \le 1\}\big);$$

a more advanced calculation can be found in [56, Chapter 1, Equation 1.1.7]. The function $\alpha : t \mapsto \alpha(t)$ maps $[0, \infty)$ to $[1, 5)$, has only one local maximum and one local minimum, attained at $5 < t_{\max} < 6$ and $t_{\min} = 0$, respectively; in addition $\alpha(t) \to 0$ as $t \to \infty$.

For $E \subset \mathbb{R}^n$ and $\delta > 0$, let

$$\mathcal{H}^s_\delta(E) := \inf \sum_{k=1}^{\infty} \alpha(s) \left[\frac{d(C_k)}{2} \right]^s \tag{1.4.1}$$

where the infimum is taken over all sequences $\{C_k\}$ of subsets of $\mathbb{R}^n$ such that $E \subset \bigcup_{k=1}^{\infty} C_k$ and $d(C_k) < \delta$ for $k = 1, 2, \ldots$; here we define $0^0 := 1$ and $d(\emptyset)^s := 0$. Letting

$$\mathcal{H}^s(E) := \sup_{\delta > 0} \mathcal{H}^s_\delta(E) = \lim_{\delta \to 0} \mathcal{H}^s_\delta(E),$$

the function $\mathcal{H}^s : E \mapsto \mathcal{H}^s(E)$, defined for each $E \subset \mathbb{R}^n$, is a measure in $\mathbb{R}^n$, called the *s-dimensional Hausdorff measure*. Since

$$\mathcal{H}^s(A \cup B) = \mathcal{H}^s(A) + \mathcal{H}^s(B)$$

for every pair of sets $A, B \subset \mathbb{R}^n$ with $\mathrm{dist}(A, B) > 0$, it follows from Theorem 1.3.1 that $\mathcal{H}^s$ is a Borel measure in $\mathbb{R}^n$. In addition, the measure $\mathcal{H}^s$ is Borel regular by Proposition 1.4.2 below. However, $\mathcal{H}^s$ is not a Radon measure in $\mathbb{R}^n$ when $s < n$.

It is easy to verify that $\mathcal{H}^0$ is the counting measure in $\mathbb{R}^n$. The constant $\alpha(s)/2^s$ in the definition of $\mathcal{H}^s_\delta(E)$ implies $\mathcal{H}^n = \mathcal{L}^n$. This equality follows (nontrivially) from the *isodiametric inequality*

$$\mathcal{L}^n(E) \leq \alpha(n) \left[\frac{d(E)}{2} \right]^n \tag{1.4.2}$$

which holds for every $E \subset \mathbb{R}^n$; see [29, Section 2.2]. As the diameters of sets are invariant with respect to isometric transformations, so are the Hausdorff measures. Moreover, for each $E \subset \mathbb{R}^n$ and every $t > 0$,

$$\mathcal{H}^s(tE) = t^s \, \mathcal{H}^s(E).$$

Remark 1.4.1. The value of $\mathcal{H}^s_\delta(E)$, and a fortiori that of $\mathcal{H}^s(E)$, does not change when the sequence $\{C_k\}$ in the defining equality (1.4.1) is assumed to have one of the following additional properties:

(1) Each C_k is *convex*; since the diameters of C_k and its convex hull are the same.

(2) Each C_k is *closed*; since the diameters of C_k and its closure $\mathrm{cl}\, C_k$ are the same.

(3) Each C_k is *open*; since given $\varepsilon > 0$, we can find $r_k > 0$ such that $d[U(C_k, r_k)] < \delta$ and $d[U(C_k, r_k)]^s < d(C_k)^s + \varepsilon 2^{-k}$ for $k = 1, 2, \ldots$.

(4) Each C_k is *contained in E*; since E is covered by the family $\{C_k \cap E\}$ and $d(C_k \cap E) \leq d(C_k)$ for $k = 1, 2, \ldots$.

By (4), the value $\mathcal{H}^s(E)$ depends on $\mathbb{R}^n$ only to the extent to which $\mathbb{R}^n$ defines the metric in E. In particular, if $1 \leq m < n$ is an integer, then $\mathcal{H}^s$ restricted to the subsets of $\mathbb{R}^m$ is the s-dimensional Hausdorff measure in $\mathbb{R}^m$.

Proposition 1.4.2. *Given $E \subset \mathbb{R}^n$, there is a G_δ set $B \subset \mathbb{R}^n$ such that $E \subset B$ and $\mathcal{H}^s(E) = \mathcal{H}^s(B)$. In particular, $\mathcal{H}^s$ is a Borel regular measure.*

PROOF. Assume $\mathcal{H}^s(E) < \infty$, since otherwise it suffices to let $B := \mathbb{R}^n$. Fix $k \in \mathbb{N}$. By Remark 1.4.1, (3), there are open sets $U_{k,j} \subset \mathbb{R}^n$ such that $d(U_{k,j}) < 1/k$ for $j = 1, 2, \ldots$, $U_k = \bigcup_{j=1}^\infty U_{k,j}$ contains E, and

$$\mathcal{H}^s_{1/k}(U_k) \leq \sum_{j=1}^\infty \alpha(s) \left[\frac{d(U_{k,j})}{2} \right]^s < \mathcal{H}^s(E) + \frac{1}{k}.$$

The first inequality follows directly from the definition of $\mathcal{H}^s_{1/k}$. The intersection $B = \bigcap_{k=1}^\infty U_k$ is a G_δ set containing E, and

$$\mathcal{H}^s_{1/k}(B) \leq \mathcal{H}^s_{1/k}(U_k) < \mathcal{H}^s(E) + \frac{1}{k}.$$

Letting $k \to \infty$ yields $\mathcal{H}^s(B) \leq \mathcal{H}^s(E)$. $\qquad\square$

Proposition 1.4.3. *Let $E \subset \mathbb{R}^n$, and let $0 \leq s < t$. If the measure $\mathcal{H}^s \llcorner E$ is σ-finite, then $\mathcal{H}^t(E) = 0$. Moreover, $\mathcal{H}^s \equiv 0$ for each $s > n$.*

PROOF. In proving the first claim, we may assume that $\mathcal{H}^s(E) < \infty$. Given $\delta > 0$, there is a sequence $\{C_k\}$ of subsets of $\mathbb{R}^n$ of diameters smaller than δ such that $E \subset \bigcup_{k=1}^\infty C_k$ and

$$\sum_{k=1}^\infty \alpha(s) \left[\frac{d(C_k)}{2} \right]^s < \mathcal{H}^s(E) + 1.$$

Consequently

$$\mathcal{H}^t_\delta(E) \leq \sum_{k=1}^\infty \alpha(t) \left[\frac{d(C_k)}{2} \right]^t \leq \left(\frac{\delta}{2} \right)^{t-s} \frac{\alpha(t)}{\alpha(s)} \sum_{k=1}^\infty \alpha(s) \left[\frac{d(C_k)}{2} \right]^s$$

$$\leq \left(\frac{\delta}{2} \right)^{t-s} \frac{\alpha(t)}{\alpha(s)} [\mathcal{H}^s(E) + 1],$$

and it suffices to let $\delta \to 0$.

If $s > n$, it suffices to show that $\mathcal{H}^s(Q) = 0$ for a 0-cube Q. For $k \in \mathbb{N}$, each k-cube has diameter $\delta_k := 2^{-k}\sqrt{n}$, and Q is the union of 2^{kn} such cubes. Thus

$$\mathcal{H}^s_{\delta_k}(Q) \leq \alpha(s) 2^{kn} \left(\frac{\delta_k}{2} \right)^s = \alpha(s) \left(\frac{\sqrt{n}}{2} \right)^s 2^{k(n-s)},$$

and letting $k \to \infty$ yields the desired result. $\qquad\square$

Next we relate Hausdorff measures in $\mathbb{R}^n$ to covers consisting of dyadic cubes. Recall that for $k \in \mathbb{Z}$, the family of all k-cubes is denoted by $\mathcal{D}_k$.

Proposition 1.4.4. *Let $s \geq 0$ and $E \subset \mathbb{R}^n$. Given $\varepsilon > 0$ and $p \in \mathbb{Z}$, there is a family $\mathfrak{Q} \subset \bigcup_{k \geq p} \mathfrak{D}_k$ such that E meets each $Q \in \mathfrak{Q}$, $E \subset \text{int}(\bigcup \mathfrak{Q})$, and*

$$\sum_{Q \in \mathfrak{Q}} d(Q)^s \leq \beta [\mathcal{H}^s(E) + \varepsilon]$$

where $\beta = \beta(n) > 0$.

PROOF. If $s > n$, the proposition holds with $\beta = 1$; the proof is analogous to that of the second part of Proposition 1.4.3. Hence assume $s \leq n$, and choose $\varepsilon > 0$ and $p \in \mathbb{Z}$. There is a cover $\{C_j\}$ of E such that the diameter of each C_j is smaller than $\delta = 2^{-p}$, and

$$\sum_{j=1}^{\infty} \alpha(s) \left[\frac{d(C_j)}{2} \right]^s \leq \mathcal{H}^s_\delta(E) + \varepsilon \leq \mathcal{H}^s(E) + \varepsilon.$$

Find an integer $p_j \geq p$ with $2^{-p_j-1} \leq d(C_j) < 2^{-p_j}$, and note

$$d(C_j) < d(Q)/\sqrt{n} = 2^{-p_j} \leq 2d(C_j)$$

for every p_j-cube Q. Select a p_j-cube Q with $Q \cap C_j \neq \emptyset$, and denote by $Q_{1,j}, \ldots, Q_{3^n,j}$ all p_j-cubes which meet Q, including Q itself. It follows that $C_j \subset \text{int}(\bigcup_{i=1}^{3^n} Q_{i,j})$, and hence $E \subset \text{int}(\bigcup_{j=1}^{\infty} \bigcup_{i=1}^{3^n} Q_{i,j})$. Moreover,

$$\sum_{j=1}^{\infty} \sum_{i=1}^{3^n} d(Q_{i,j})^s \leq 3^n \cdot (2\sqrt{n})^s \sum_{j=1}^{\infty} d(C_j)^s \leq 3^n \frac{(4n)^s}{\alpha(s)} [\mathcal{H}^s(E) + \varepsilon].$$

Since $0 \leq s \leq n$ implies $\alpha(s) \geq \min\{1, \alpha(n)\}$, the desired inequality holds with $\beta := (12n)^n \max\{1, 1/\alpha(n)\}$. Finally, replacing $\mathfrak{Q}$ by a smaller family $\{Q \in \mathfrak{Q} : Q \cap E \neq \emptyset\}$ completes the proof. $\qquad\square$

Additional properties of Hausdorff measures in $\mathbb{R}^n$ can be found in [30] and [46]. Hausdorff measures defined in general metric spaces are investigated in [59].

1.5. Differentiable and Lipschitz maps

Let $\Omega \subset \mathbb{R}^n$ be an open set. A map $\phi : \Omega \to \mathbb{R}^m$ is *differentiable* at $x \in \Omega$ if there is a linear map $L : \mathbb{R}^n \to \mathbb{R}^m$ such that

$$\lim_{y \to x} \frac{|\phi(y) - \phi(x) - L(y - x)|}{|y - x|} = 0.$$

If such a map L exists, it is unique. We call it the *derivative* of ϕ at x, denoted by $D\phi(x)$. If a map $\phi = (f_1, \ldots, f_n)$ from Ω to $\mathbb{R}^n$ is differentiable at $x \in \Omega$, then the *divergence* of ϕ at x is the real number

$$\operatorname{div}\phi(x) := \sum_{i=1}^{n} D_i f_i(x)$$

where $D_i := \partial/\partial\xi_i$ is the usual *partial derivative operator*.

Let $E \subset \mathbb{R}^n$ be any set, and let $\phi : E \to \mathbb{R}^m$. The *Lipschitz constant* of ϕ is the extended real number

$$\operatorname{Lip}\phi := \sup\left\{\frac{|\phi(x) - \phi(y)|}{|x - y|} : x, y \in E \text{ and } x \neq y\right\}.$$

When $\operatorname{Lip}\phi < \infty$, the map ϕ is called *Lipschitz*. If $\Omega \subset \mathbb{R}^n$ is an open set, we call a map $\phi : \Omega \to \mathbb{R}^m$ *locally Lipschitz* whenever the restriction $\phi \upharpoonright U$ is Lipschitz for each open set $U \Subset \Omega$. The linear space of all Lipschitz maps $\phi : E \to \mathbb{R}^m$ is denoted by $Lip(E; \mathbb{R}^m)$. The symbols $Lip(E)$, $Lip_c(\Omega; \mathbb{R}^m)$, $Lip_{\text{loc}}(\Omega; \mathbb{R}^m)$, etc., have the obvious meaning. For a Lipschitz map $\phi : E \to \mathbb{R}^m$ and $s \geq 0$, we obtain

$$\mathcal{H}^s\big[\phi(E)\big] \leq (\operatorname{Lip}\phi)^s \mathcal{H}^s(E). \tag{1.5.1}$$

A bijective Lipschitz map whose inverse is also Lipschitz is called a *lipeomorphism*.

Observation 1.5.1. *Let $\Omega \subset \mathbb{R}^n$ be an open set. If $\phi \in Lip_{\text{loc}}(\Omega; \mathbb{R}^m)$, then $\phi \upharpoonright K \in Lip(C; \mathbb{R}^m)$ for each compact set $K \subset \Omega$. In particular,*

$$Lip_{\text{loc}}(\Omega; \mathbb{R}^m) \cap C_c(\Omega; \mathbb{R}^m) = Lip_c(\Omega; \mathbb{R}^m).$$

PROOF. Suppose there is a compact set $K \subset \Omega$ such that ϕ is not Lipschitz in K. There are sequences $\{x_k\}$ and $\{y_k\}$ in K such that

$$\big|\phi(x_k) - \phi(y_k)\big| > k|x_k - y_k| > 0, \quad k = 1, 2, \ldots.$$

Passing to subsequences, still denoted by $\{x_k\}$ and $\{y_k\}$, we obtain the limit points $x = \lim x_k$ and $y = \lim y_k$ in K. The continuity of ϕ implies

$$\infty > \big|\phi(x) - \phi(y)\big| = \lim\big|\phi(x_k) - \phi(y_k)\big| \geq \limsup k|x_k - y_k|,$$

and consequently $x = y$. Since ϕ is Lipschitz in a neighborhood of x, there is $0 < c < \infty$ such that for all sufficiently large k,

$$c|x_k - y_k| \geq \big|\phi(x_k) - \phi(y_k)\big| > k|x_k - y_k| > 0.$$

A contradiction follows. $\square$

Proposition 1.5.2. *Let $E \subset \mathbb{R}^n$ and $\phi \in Lip(E; \mathbb{R}^m)$. There is a map $\psi \in Lip(\mathbb{R}^n; \mathbb{R}^m)$ such that $\psi(x) = \phi(x)$ for each $x \in E$,*

$$\operatorname{Lip}\psi \leq \sqrt{m}\operatorname{Lip}\phi, \quad and \quad \|\psi\|_{L^\infty(\mathbb{R}^n; \mathbb{R}^m)} \leq \|\phi\|_{L^\infty(E; \mathbb{R}^m)}.$$

PROOF. By [29, Section 3.1.1, Theorem 1], there is $\theta \in Lip(\mathbb{R}^n; \mathbb{R}^m)$ such that $\theta(x) = \phi(x)$ for each $x \in E$ and $Lip\,\theta \leq \sqrt{m}\,Lip\,\phi$. As there is nothing to prove otherwise, assume $c := \|\phi\|_{L^\infty(E;\mathbb{R}^m)}$ belongs to $\mathbb{R}_+$. Define $\gamma : \mathbb{R}^m \to \mathbb{R}^m$ by

$$\gamma(y) := \begin{cases} c\frac{y}{|y|} & \text{if } |y| > c, \\ y & \text{if } |y| \leq c. \end{cases}$$

Since $Lip\,\gamma = 1$ and $|\gamma|_{L^\infty(\mathbb{R}^m;\mathbb{R}^m)} \leq c$, the composition $\psi := \gamma \circ \theta$ is the desired extension of ϕ. $\square$

With a considerable effort, one can improve on Proposition 1.5.2 by showing that $Lip\,\psi = Lip\,\phi$. This stronger result is called *Kirschbraun's theorem* [33, Theorem 2.10.43].

The next well-known theorem has several proofs of various levels of sophistication, e.g., [33, Theorem 3.1.6], [1, Theorem 2.14], or [29, Section 6.2, Theorem 2]. For a proof with minimal prerequisites we refer to [29, Section 3.1.2].

Theorem 1.5.3 (Rademacher). *Each $\phi \in Lip(\mathbb{R}^n; \mathbb{R}^m)$ is differentiable at almost all $x \in \mathbb{R}^n$.*

Let $E \subset \mathbb{R}^n$ and $0 \leq s \leq 1$. The *s-Hölder constant* at $x \in E$ of a map $\phi : E \to \mathbb{R}^m$ is the extended real number

$$H_s\phi(x) := \limsup_{\substack{y \to x \\ y \in E}} \frac{|\phi(y) - \phi(x)|}{|y - x|^s}.$$

Clearly, $H_0\phi(x) < \infty$ if and only if ϕ is bounded in a neighborhood of x, and $H_0\phi(x) = 0$ if and only if ϕ is continuous at x. If $H_s\phi(x) < \infty$ and $0 \leq t < s$, then $H_t\phi(x) = 0$. We call

$$Lip\,\phi(x) := H_1\phi(x)$$

the *Lipschitz constant of ϕ at x*, and say that ϕ is *Lipschitz at x* whenever $Lip\,\phi(x) < \infty$. We say that ϕ is *pointwise Lipschitz* in a set $C \subset E$ if it is Lipschitz at each $x \in C$. A pointwise Lipschitz map in C need not be Lipschitz in C, even if C is compact [51, Section 1.6].

Theorem 1.5.4 (Stepanoff). *Let $\Omega \subset \mathbb{R}^n$ be an open set, and assume that $\phi : \Omega \to \mathbb{R}^m$ is pointwise Lipschitz in a set $E \subset \Omega$. Then ϕ is differentiable at almost all $x \in E$.*

For a proof of this generalization of Rademacher's theorem we refer to [33, Theorem 3.1.9], or to Section 6.2 below where a slightly more general theorem is proved in detail; see Remark 7.2.4.

Theorem 1.5.5 (Whitney). *Let $K \subset \mathbb{R}^n$ be a compact set, and let $f \in C(K)$ and $v \in C(K; \mathbb{R}^n)$ satisfy the following condition: given $\varepsilon > 0$, we can find $\delta > 0$ so that*

$$\big|f(y) - f(x) - v(x) \cdot (y - x)\big| \leq \varepsilon |y - x|$$

for all $x, y \in K$ with $|y - x| < \delta$. There is $g \in C^1(\mathbb{R}^n)$ such that

$$g(x) = f(x) \quad and \quad Dg(x) = v(x)$$

for each $x \in K$.

Theorem 1.5.5 is a special case of *Whitney's extension theorem*. Proofs of the general Whitney's result, which implies the special case, can be found in [29, Section 6.5] or in [70, Chapter 6, Section 2].

Let $\phi = (f_1, \ldots, f_n)$ be a Lipschitz map from $\mathbb{R}^n$ to $\mathbb{R}^n$. Then

$$\begin{pmatrix} Df_1 \\ \cdots \\ Df_n \end{pmatrix}$$

is an $n \times n$ matrix, whose determinant is denoted by $\det D\phi$. The *Jacobian* of ϕ is the function $J_\phi = |\det D\phi|$ defined almost everywhere in $\mathbb{R}^n$ by Rademacher's theorem. In view of (1.5.1), the inequality

$$\|J_\phi\|_{L^\infty(\mathbb{R}^n)} \leq (\operatorname{Lip} \phi)^n \tag{1.5.2}$$

is a consequence of [29, Section 3.3, Lemma 1].

The next result is called interchangeably the *area theorem* or *change of variables theorem*. It follows from [29, Section 3.3, Theorem 2].

Theorem 1.5.6. *Let $\phi : \mathbb{R}^n \to \mathbb{R}^n$ be a Lipschitz map. If $g \in L^0(\mathbb{R}^n)$ and $g \geq 0$, then $y \mapsto \sum\{g(x) : x \in \phi^{-1}(y)\}$ is a measurable function on $\mathbb{R}^n$ and*

$$\int_{\mathbb{R}^n} g(x) J_\phi(x) \, dx = \int_{\mathbb{R}^n} \sum_{x \in \phi^{-1}(y)} g(x) \, dy. \tag{1.5.3}$$

Employing Hausdorff measures and more eleborete Jacobians, formulas similar to (1.5.3) hold for Lipschitz maps $\phi : \mathbb{R}^n \to \mathbb{R}^m$ where $m \neq n$. On a few occasions when such formulas are used, we refer the reader to the appropriate sections of [29, Chapter 3].

Chapter 2

Divergence theorem for dyadic figures

Using the idea of W.B. Jurkat, we give an elementary proof of a fairly general divergence theorem for dyadic figures. While this is only a preliminary version of the divergence theorem we intend to establish, it is already a useful tool for studying removable singularities of some classical partial differential equations (Chapter 3 below).

2.1. Differentiable vector fields

If A is a figure, then for $\mathcal{H}^{n-1}$ almost all $x \in \partial A$ there is a unique *unit exterior normal* of A at x, denoted by $\nu_A(x)$. The map

$$\nu_A : x \mapsto \nu_A(x) : \partial A \to \mathbb{R}^n$$

is defined $\mathcal{H}^{n-1}$ almost everywhere, has only finitely many values, and it is $\mathcal{H}^{n-1}$ measurable. Let $E \subset \mathbb{R}^n$, and assume that $v : E \to \mathbb{R}^n$ belongs to $L^1(\partial A, \mathcal{H}^{n-1}; \mathbb{R}^n)$ for each figure $A \subset E$. A real-valued function

$$F : A \mapsto \int_{\partial A} v \cdot \nu_A \, d\mathcal{H}^{n-1} \tag{2.1.1}$$

defined on all figures $A \subset E$ is called the *flux* of v. In this context it is customary to call v a *vector field*. The term "flux" is derived from a physical example: if v is the vector field of velocities of a fluid moving in the set E, then $F(A)$ is the amount of fluid that flows out of the figure $A \subset E$ in the unit of time.

The next observation says that the flux of a vector field is an *additive function* with respect to nonoverlapping figures. Its simple verification is left to the reader.

Observation 2.1.1. *Let $E \subset \mathbb{R}^n$, and assume that $v : E \to \mathbb{R}^n$ belongs to $L^\infty(\partial A, \mathcal{H}^{n-1}; \mathbb{R}^n)$ for each figure $A \subset E$. Then*

$$\int_{\partial(A \cup B)} v \cdot \nu_{A \cup B} \, d\mathcal{H}^{n-1} = \int_{\partial A} v \cdot \nu_A \, d\mathcal{H}^{n-1} + \int_{\partial B} v \cdot \nu_B \, d\mathcal{H}^{n-1}$$

for each pair $A, B \subset E$ of nonoverlapping figures.

Proposition 2.1.2. *Let A be a cell, and let $v \in C(A; \mathbb{R}^n)$ be differentiable at each $x \in \mathrm{int}\, A$. If $\mathrm{div}\, v$ belongs to $L^1(A)$, then*

$$\int_A \mathrm{div}\, v(x)\, dx = \int_{\partial A} v \cdot \nu_A \, d\mathcal{H}^{n-1}.$$

PROOF. Let $v = (v_1, \ldots, v_n)$ and $A = \prod_{i=1}^n [a_i^-, a_i^+]$. If

$$A_i^{\pm} := \{a_i^{\pm}\} \times \pi_i(\mathrm{int}\, A),$$

then $\nu_A(x) = \pm e_i$ whenever $x \in A_i^{\pm}$, $i = 1, \ldots, n$. The boundary ∂A differs from $\bigcup_{i=1}^n (A_i^- \cup A_i^+)$ by an $\mathcal{H}^{n-1}$ negligible set. Fix i and for $x \in \mathrm{int}\, A$, write $x = (u, t)$ where $u = \pi_i(x)$ and $t = x \cdot e_i$. By Fubini's theorem and the fundamental theorem of calculus,

$$\int_A D_i v_i(x)\, dx = \int_{\pi_i(\mathrm{int}\, A)} \left(\int_{a_i^-}^{a_i^+} \frac{d}{dt} v_i(u, t)\, dt \right) du$$

$$= \int_{\pi_i(\mathrm{int}\, A)} \left[v_i(u, a_i^+) - v_i(u, a_i^-) \right] du$$

$$= \int_{A_i^+} v \cdot \nu_A \, d\mathcal{H}^{n-1} + \int_{A_i^-} v \cdot \nu_A \, d\mathcal{H}^{n-1}.$$

Summing up these equalities over $i = 1, \ldots, n$ completes the proof.   □

Corollary 2.1.3. *Let $E \subset \mathbb{R}^n$, $x \in E$, and let $\{C_k\}$ be a sequence of cubes such that $\lim d(C_k) = 0$. Assume that $C_k \subset E$ and $x \in C_k$ for $k = 1, 2, \ldots$, and that $v : E \to \mathbb{R}^m$ belongs to $L^1(\partial C, \mathcal{H}^{n-1}; \mathbb{R}^n)$ for each cube $C \subset E$.*

(1) *If $0 \le s \le 1$, then*

$$\limsup \frac{1}{d(C_k)^{n-1+s}} \int_{\partial C_k} v \cdot \nu_{C_k} \, d\mathcal{H}^{n-1} \le 2n H_s v(x).$$

(2) *If $x \in \mathrm{int}\, E$ and v is differentiable at x, then*

$$\lim \frac{1}{|C_k|} \int_{\partial C_k} v \cdot \nu_{C_k} \, d\mathcal{H}^{n-1} = \mathrm{div}\, v(x).$$

PROOF. Choose $\varepsilon > 0$ — you can never go wrong by doing so. We may assume $H_s v(x) < \infty$, and find $\delta > 0$ so that

$$|v(y) - v(x)| \le \left[H_s v(x) + \varepsilon \right] \cdot |y - x|^s$$

for each $y \in E \cap U(x,\delta)$. Denote by F the flux of v, and use Proposition 2.1.2 to show that for all sufficiently large k,

$$
\begin{aligned}
\left| F(C_k) \right| &= \left| \int_{\partial C_k} \left[v(y) - v(x) \right] \cdot \nu_{C_k}(y) \, d\mathcal{H}^{n-1}(y) \right| \\
&\leq \left[H_s v(x) + \varepsilon \right] \int_{\partial C_k} |y - x|^s \, d\mathcal{H}^{n-1}(y) \\
&\leq \left[H_s v(x) + \varepsilon \right] d(C_k)^s \, \mathcal{H}^{n-1}(\partial C_k) \\
&\leq 2n \left[H_s v(x) + \varepsilon \right] d(C_k)^{n-1+s}.
\end{aligned}
$$

If $x \in \text{int}\, E$ and v is differentiable at x, let

$$
w : y \mapsto v(x) + \left[Dv(x) \right](y - x) : \mathbb{R}^n \to \mathbb{R}^n
$$

and observe that $\text{div}\, w(y) = \text{div}\, v(x)$ for each $y \in \mathbb{R}^n$. There is $\eta > 0$ such that $U(x,\eta) \subset E$ and

$$
\left| v(y) - w(y) \right| \leq \varepsilon |y - x|
$$

for every $y \in U(x,\eta)$. As $w \in C^\infty(\mathbb{R}^n; \mathbb{R}^n)$, Proposition 2.1.2 yields

$$
\begin{aligned}
\left| F(C_k) - \text{div}\, v(x) |C_k| \right| &= \left| \int_{\partial C_k} \left[v(y) - w(y) \right] \cdot \nu_{C_k}(y) \, d\mathcal{H}^{n-1}(y) \right| \\
&\leq \varepsilon \int_{\partial C_k} |y - x| \, d\mathcal{H}^{n-1}(y) \\
&\leq \varepsilon d(C_k) \mathcal{H}^{n-1}(\partial C_k) = 2n^{3/2} \varepsilon |C_k|
\end{aligned}
$$

for all sufficiently large k. Letting $k \to \infty$, the corollary follows from the arbitrariness of ε. $\qquad \square$

We prove the divergence theorem for closed balls. While this is not essential for the logical development of our exposition, it will facilitate an early presentation of examples. If $B := B(x,r)$, then $\nu_B(y) := (y - x)/r$ is the unit exterior normal of B at $y \in \partial B$. Since the induced map $\nu_B : \partial B \to \mathbb{R}^n$ is continuous, a finite integral

$$
\int_{\partial B} v \cdot \nu_B \, d\mathcal{H}^{n-1}
$$

exists for each $v \in L^1(\partial B, \mathcal{H}^{n-1}; \mathbb{R}^n)$.

Proposition 2.1.4. *Let $B \subset \mathbb{R}^n$ be a closed ball, and let $v \in C(B; \mathbb{R}^n)$ be differentiable in each $x \in \text{int}\, B$. If $\text{div}\, v$ belongs to $L^1(B)$, then*

$$
\int_B \text{div}\, v(x) \, dx = \int_{\partial B} v \cdot \nu_B \, d\mathcal{H}^{n-1}.
$$

PROOF. The proof is similar to that of Proposition 2.1.2. In view of translation invariance, we may assume $B = B(0,r)$. Let $U := \pi_n(\text{int}\, B)$ and $g(u) = \sqrt{r^2 - |u|^2}$ for each $u \in U$. If $(\partial B)_\pm := \{x \in \partial B : \pm x \cdot e_n > 0\}$, then the bijections

$$
\phi_\pm : u \mapsto \left(u, \pm g(u) \right) : U \to (\partial B)_\pm
$$

are continuously differentiable and have the same Jacobian

$$
J = \sqrt{1 + |Dg|^2} = \frac{r}{g};
$$

see [29, Section 3.3.4, B]. Let $v_n := v \cdot e_n$ and $\nu_n := \nu_B \cdot e_n$. Observe

$$\nu_n \circ \phi_\pm = (\nu_B \circ \phi_\pm) \cdot e_n = \frac{\phi_\pm \cdot e_n}{r} = \pm \frac{g}{r}.$$

If $U_k = \pi_n\big[U(0, 1 - 2^{-k})\big]$, then the maps $\phi_\pm \upharpoonright U_k$ are Lipschitz and $U = \bigcup_{k=1}^\infty U_k$. Thus applying [29, Section 3.3.4, B] to each U_k, we obtain

$$\int_{(\partial B)_\pm} v_n \nu_n \, d\mathcal{H}^{n-1} = \int_U \big[(v_n \nu_n) \circ \phi_\pm\big] J \, d\mathcal{L}^n = \pm \int_U v_n \big[u, \pm g(u)\big] \, du.$$

Now Fubini's theorem and the fundamental theorem of calculus imply

$$\begin{aligned}
\int_B D_n v_n(x) \, dx &= \int_U \left(\int_{-g(u)}^{g(u)} \frac{d}{dt} v_n(u, t) \, dt \right) du \\
&= \int_U v_n \big[u, g(u)\big] \, du - \int_U v_n \big[u, -g(u)\big] \, du \\
&= \int_{(\partial B)_+} v_n \nu_n \, d\mathcal{H}^{n-1} + \int_{(\partial B)_-} v_n \nu_n \, d\mathcal{H}^{n-1} \\
&= \int_{\partial B} v_n \nu_n \, d\mathcal{H}^{n-1},
\end{aligned}$$

since the boundary ∂B differs from $(\partial B)_+ \cup (\partial B)_-$ by an $\mathcal{H}^{n-1}$ negligible set. The proposition follows from symmetry. $\qquad\square$

2.2. Dyadic partitions

A *partition* is a finite (possibly empty) collection

$$P := \big\{(E_1, x_1), \ldots, (E_p, x_p)\big\}$$

where $\{E_1, \ldots, E_p\}$ is a collection of nonoverlapping subsets of $\mathbb{R}^n$ such that $x_i \in E_i$ for $i = 1, \ldots, p$. The *body* of P is the union $[P] := \bigcup_{i=1}^p E_i$, and P is called a partition *in a set* $A \subset \mathbb{R}^n$ if $[P] \subset A$. Given a set $E \subset \mathbb{R}^n$ and $\delta : E \to \mathbb{R}_+$, we say that P is δ-*fine* if $x_i \in E$ and $d(E_i) < \delta(x_i)$ for $i = 1, \ldots, p$. When each set E_i is a dyadic cube, then P is called a *dyadic partition*.

If dyadic cubes A and B overlap, then either $A \subset B$ or $B \subset A$. Consequently, every family $\mathcal{C}$ of dyadic cubes has a nonoverlapping subfamily $\mathcal{Q}$ such that $\bigcup \mathcal{Q} = \bigcup \mathcal{C}$. Dyadic cubes A and B are called *adjacent* if $d(A) = d(B)$ and $A \cap B \neq \emptyset$. Every dyadic cube is adjacent to 3^n dyadic cubes, including itself. Recall that for an integer k the family of all k-cubes is denoted by $\mathcal{D}_k$.

Given a family $\mathcal{E}$ of subsets of $\mathbb{R}^n$ and $x \in \mathbb{R}^n$, we let

$$St(x, \mathcal{E}) := \{E \in \mathcal{E} : x \in E\}.$$

For each $x \in \mathbb{R}^n$ and each $k \in \mathbb{Z}$, the collection $St(x, \mathcal{D}_k)$ consists of at most 2^n k-cubes, and x belongs to the interior of $\bigcup St(x, \mathcal{D}_k)$. A *star cover* of

$E \subset \mathbb{R}^n$ is a family $\mathcal{Q}$ of dyadic cubes such that for each $x \in E$ there is $k_x \in \mathbb{Z}$ with $\mathbf{St}(x, \mathcal{D}_{k_x}) \subset \mathcal{Q}$; in this case

$$E \subset \bigcup_{x \in E} \mathrm{int}\left[\bigcup \mathbf{St}(x, \mathcal{D}_{k_x})\right] \subset \mathrm{int}\left(\bigcup \mathcal{Q}\right).$$

It follows that a star cover $\mathcal{Q}$ of a compact set $K \subset \mathbb{R}^n$ has a finite nonoverlapping subcover.

Lemma 2.2.1. *Let δ be a positive function defined on a set $E \subset \mathbb{R}^n$, and let $0 \leq t \leq n$. Given $\varepsilon > 0$, the set E has a star cover $\mathcal{C}$ which satisfies the following conditions:*

(1) *for each $C \in \mathcal{C}$ there is $x_C \in C \cap E$ such that $d(C) < \delta(x_C)$;*
(2) *$\sum_{C \in \mathcal{C}} d(C)^t \leq \kappa[\mathcal{H}^t(E) + \varepsilon]$ where $\kappa := \kappa(n) > 0$.*

Proof. To avoid trivialities, assume $E \neq \emptyset$. Denote by $\mathcal{B}$ the family of all dyadic cubes C satisfying condition (1). For $k \in \mathbb{N}$ and $x \in \mathbb{R}^n$, let

$$\mathcal{D}_{\geq k} := \bigcup\{\mathcal{D}_i : i \geq k\} \quad \text{and} \quad B_k := \{x \in \mathbb{R}^n : \mathbf{St}(x, \mathcal{D}_{\geq k}) \subset \mathcal{B}\}.$$

Clearly $\{B_k\}$ is an increasing sequence. Moreover $E \subset \bigcup_{k=1}^{\infty} B_k$, since

$$\{x \in E : \delta(x) > 2^{-k}\sqrt{n}\} \subset B_k.$$

Claim. $\mathbb{R}^n - B_k = \bigcup(\mathcal{D}_{\geq k} - \mathcal{B})$ for every $k \in \mathbb{N}$. In particular, each B_k is a Borel set.

Proof. If $x \notin B_k$, some $C_x \in \mathbf{St}(x, \mathcal{D}_{\geq k})$ does not belong to $\mathcal{B}$. Hence $x \in C_x$ and $C_x \in \mathcal{D}_{\geq k} - \mathcal{B}$. It follows that $x \in \bigcup(\mathcal{D}_{\geq k} - \mathcal{B})$. Conversely, if $x \in \bigcup(\mathcal{D}_{\geq k} - \mathcal{B})$ then $x \in D_x$ for some $D_x \in \mathcal{D}_{\geq k} - \mathcal{B}$. Thus $\mathbf{St}(x, \mathcal{D}_{\geq k}) \not\subset \mathcal{B}$, which means $x \notin B_k$.

If $E_1 := E \cap B_1$ and $E_k := E \cap (B_k - B_{k-1})$ for $k = 2, 3, \ldots$, then

$$E = \bigcup_{k=1}^{\infty} E_k \quad \text{and} \quad \mathcal{H}^t(E) = \sum_{k=1}^{\infty} \mathcal{H}^t(E_k).$$

Select $E_k \neq \emptyset$. By Proposition 1.4.4, there is a cover $\mathcal{Q}_k \subset \mathcal{D}_{\geq k}$ of E_k such that $E_k \cap Q \neq \emptyset$ for each $Q \in \mathcal{Q}_k$, and

$$\sum_{Q \in \mathcal{Q}_k} d(Q)^t \leq \beta[\mathcal{H}^t(E_k) + \varepsilon 2^{-k}]$$

where $\beta = \beta(n) > 0$. If $\mathcal{C}_k$ consists of all dyadic cubes that meet E_k and are adjacent to some $Q \in \mathcal{Q}_k$, then $\mathcal{C}_k$ is a star cover of E_k, and

$$\sum_{C \in \mathcal{C}_k} d(C)^t \leq 3^n \sum_{Q \in \mathcal{Q}_k} d(Q)^t \leq 3^n \beta[\mathcal{H}^t(E_k) + \varepsilon 2^{-k}].$$

Now $\mathcal{C}_k \subset \mathcal{D}_{\geq k}$ and $E_k \subset B_k$. Since each $C \in \mathcal{C}_k$ meets E_k, the definition of B_k implies $C \in \mathcal{B}$. Thus $\mathcal{C}_k \subset \mathcal{B}$, and we see that the family $\mathcal{C} := \bigcup_{k=1}^{\infty} \mathcal{C}_k$ is a star cover of E that satisfies condition (1). Letting $\kappa := 3^n \beta$, we obtain

$$\sum_{C \in \mathcal{C}} d(C)^t \leq \sum_{k=1}^{\infty} \sum_{C \in \mathcal{C}_k} d(C)^t$$

$$\leq \kappa \sum_{k=1}^{\infty} \left[\mathcal{H}^t(E_k) + \varepsilon 2^{-k} \right] = \kappa \left[\mathcal{H}^t(E) + \varepsilon \right]. \qquad \square$$

Proposition 2.2.2. *Let $\mathcal{E}$ be a family of disjoint subsets of a dyadic figure A, and for each $E \in \mathcal{E}$ select real numbers $0 \leq t_E \leq n$ and $\varepsilon_E > 0$. Given $\delta : A \to \mathbb{R}_+$, there is a δ-fine dyadic partition*

$$P := \left\{ (C_1, x_1), \ldots, (C_p, x_p) \right\}$$

such that $[P] = A$, and with a fixed $\kappa = \kappa(n) > 0$, the inequality

$$\sum_{x_i \in E} d(C_i)^{t_E} \leq \kappa \left[\mathcal{H}^{t_E}(E) + \varepsilon_E \right]$$

holds for each $E \in \mathcal{E}$.

PROOF. There is $k \in \mathbb{N}$ such that A is the union of k-cubes. Enlarging $\mathcal{E}$ and making δ smaller, we may assume $\bigcup \mathcal{E} = A$ and $\delta(x) < 2^{-k}\sqrt{n}$ for each $x \in A$. Let $\mathcal{C}_E$ be a star cover of $E \in \mathcal{E}$ associated with $\delta_E := \delta \restriction E$, t_E, and ε_E according to Lemma 2.2.1. For every $C \in \mathcal{C}_E$, select $x_C \in E \cap C$ with $d(C) < \delta_E(x_C)$. Since $\mathcal{C} := \bigcup_{E \in \mathcal{E}} \mathcal{C}_E$ is a star cover of the compact set A, there are nonoverlapping cubes $C_1, \ldots, C_p$ in $\mathcal{C}$ such that $A \subset \bigcup_{i=1}^{p} C_i$. It follows that

$$P := \left\{ (C_i, x_{C_i}) : |C_i \cap A| > 0 \right\}$$

is a δ-fine dyadic partition with $A \subset [P]$. As our assumption about δ implies $C_i \subset A$ whenever $|C_i \cap A| > 0$, we obtain $[P] = A$. Since $\mathcal{E}$ is a disjoint family, $\{C_i : x_{C_i} \in E\} \subset \mathcal{C}_E$ for each $E \in \mathcal{E}$. Hence with the same κ as in Lemma 2.2.1, the inequality

$$\sum_{x_{C_i} \in E} d(C_i)^{t_E} \leq \sum_{C \in \mathcal{C}_E} d(C)^{t_E} \leq \kappa \left[\mathcal{H}^{t_E}(E) + \varepsilon_E \right]$$

holds for every $E \in \mathcal{E}$. $\qquad \square$

Remark 2.2.3. Lemma 2.2.1 and Proposition 2.2.2 are due to W.B. Jurkat [42, Section 4]. The classical Cousin's lemma [17] or [51, Lemma 2.6.1], as well as its generalization obtained by E.J. Howard [41, Lemma 5], are immediate consequences of Proposition 2.2.2.

2.3. Admissible maps

Definition 2.3.1. Let $E \subset \mathbb{R}^n$ be any set. A map $\phi : E \to \mathbb{R}^m$ is called *admissible* if there are numbers $0 \le s_k < 1$, and disjoint, possibly empty, sets $E_k \subset E$ such that ϕ is pointwise Lipschitz in $E - \bigcup_{k=1}^{\infty} E_k$ and for $k = 1, 2, \ldots$, the following conditions hold:

(i) E_k is $\mathcal{H}^{n-1+s_k}$ σ-finite, and $H_{s_k} \phi(x) < \infty$ for each $x \in E_k$;

(ii) $\mathcal{H}^{n-1+s_k}(E_k) > 0$ implies $H_{s_k} \phi(x) = 0$ for each $x \in E_k$.

The family of all admissible maps from the set E to $\mathbb{R}^m$ is denoted by $Adm(E; \mathbb{R}^m)$, and we write $Adm(E)$ instead of $Adm(E; \mathbb{R})$.

Note that in Definition 2.3.1 *no topological restrictions* are placed on the exceptional sets E_k.

Remark 2.3.2. If $\mathcal{H}^{n-1+s_k}(E_k) = \infty$, then $E_k = \bigcup_{j \in \mathbb{N}} E_{k,j}$ where $E_{k,j}$ are disjoint sets with $\mathcal{H}^{n-1+s_k}(E_{k,j}) < \infty$. Thus replacing each pair (E_k, s_k) with $\mathcal{H}^{n-1+s_k}(E_k) = \infty$ by the collection $\{(E_{k,j}, s_k) : j \in \mathbb{N}\}$, condition (i) of Definition 2.3.1 can be replaced by the condition:

(i*) $\mathcal{H}^{n-1+s_k}(E_k) < \infty$, and $H_{s_k} \phi(x) < \infty$ for each $x \in E_k$.

This observation will simplify future arguments.

Remark 2.3.3. Let $E \subset \mathbb{R}^n$ and $\phi \in Adm(E; \mathbb{R}^m)$. Then $H_0 \phi(x) < \infty$ for all $x \in E$, and $H_0 \phi(x) = 0$ for all $x \in E - T$ where $T \subset E$ is $\mathcal{H}^{n-1}$ negligible. Thus ϕ is *locally bounded* in E and *continuous* in $E - T$. It follows that if E is $\mathcal{H}^{n-1}$ measurable then so is ϕ, and if E is compact then ϕ is bounded. Since each set E_k is negligible, Stepanoff's theorem implies that ϕ is differentiable at almost all $x \in \operatorname{int} E$. The restriction $\phi \upharpoonright B$ belongs to $Adm(B; \mathbb{R}^m)$ for each $B \subset A$.

Remark 2.3.4. A commonly encountered map $\phi : E \to \mathbb{R}^m$ is locally bounded in E, continuous outside an $\mathcal{H}^{n-1}$ negligible set $T \subset E$, and pointwise Lipschitz outside an $\mathcal{H}^{n-1}$ σ-finite set $S \subset E$ [53, Theorem 2.9]. Letting $s_k := 0$ for $k = 1, 2, \ldots$,

$$E_1 := T, \quad E_2 := S - T, \quad \text{and} \quad E_k = \emptyset$$

for $k = 3, 4, \ldots$, we see that ϕ is admissible. Since locally bounded and pointwise Lipschitz are extreme points of the scale represented by Hölder constants, considering admissible maps is natural.

Proposition 2.3.5. *Let $E \subset \mathbb{R}^n$. With respect to pointwise addition and multiplication, $Adm(E)$ is a commutative ring, and $Adm(A; \mathbb{R}^m)$ is a module over $Adm(E)$.*

PROOF. Choosing $\phi, \psi \in Adm(E; \mathbb{R}^m)$ and $g \in Adm(E)$, it suffices to show that $\phi + \psi$ and $g\phi$ belong to $Adm(E; \mathbb{R}^m)$. As the other proof is similar, we show only that $\theta = g\phi$ belongs to $Adm(E; \mathbb{R}^m)$. Let $\{r_k\}$ and $\{s_k\}$ be sequences in $[0, 1)$, and $\{A_k\}$ and $\{B_k\}$ be sequences of disjoint subsets of E, associated with ϕ and g, respectively, according to Definition 2.3.1. Further let $A_0 := E - \bigcup_{k \in \mathbb{N}} A_k$, $B_0 := E - \bigcup_{k \in \mathbb{N}} B_k$, and $r_0 = s_0 = 1$. Define

$$t_{i,j} := \min\{r_i, s_j\} \quad \text{and} \quad E_{i,j} := A_i \cap B_j$$

for $i, j = 0, 1, \ldots$, and observe that $0 \le t_{i,j} < 1$ whenever $(i, j) \ne (0, 0)$, and that E is the union of a disjoint collection $\{E_{i,j} : i, j = 0, 1, \ldots\}$. By Remark 2.3.3, both ϕ and g are locally bounded in E. A direct calculation shows that there are functions $a, b : E \to \mathbb{R}_+$ such that

$$H_{t_{i,j}} \theta(x) \le a(x) H_{r_i} \phi(x) + b(x) H_{s_j} g(x)$$

for each $x \in E_{i,j}$ and $i, j = 0, 1, \ldots$. Thus θ is pointwise Lipschitz in

$$E_{0,0} = E - \bigcup \{E_{i,j} : i, j = 0, 1, \ldots \text{ and } (i, j) \ne (0, 0)\},$$

and for each pair $(i, j) \ne (0, 0)$, the following conditions hold:

(i) $E_{i,j}$ is $\mathcal{H}^{n-1+t_{i,j}}$ σ-finite, and $H_{t_{i,j}} \theta(x) < \infty$ for each $x \in E_{i,j}$;
(ii) $\mathcal{H}^{n-1+t_{i,j}}(E_{i,j}) > 0$ implies $H_{t_{i,j}} \theta(x) = 0$ for each $x \in E_{i,j}$.

This verifies that θ is an admissible map.   □

Lemma 2.3.6. *Let A be a dyadic figure, let $v \in Adm(A; \mathbb{R}^n)$, and define $f : A \to \mathbb{R}$ by the formula*

$$f(x) := \begin{cases} \operatorname{div} v(x) & \text{if } x \in \operatorname{int} A \text{ and } v \text{ is differentiable at } x, \\ 0 & \text{otherwise.} \end{cases}$$

For each $\varepsilon > 0$ and each $\delta : A \to \mathbb{R}_+$, there is a δ-fine dyadic partition $P := \{(C_1, x_1), \ldots, (C_p, x_p)\}$ such that $[P] = A$ and

$$\left| \sum_{i=1}^{p} f(x_i)|C_i| - \int_{\partial A} v \cdot \nu_A \, d\mathcal{H}^{n-1} \right| < \varepsilon.$$

PROOF. By Remark 2.3.3, the flux

$$F : B \mapsto \int_{\partial B} v \cdot \nu_B \, d\mathcal{H}^{n-1}$$

is defined on the family of all figures $B \subset A$. In view of Remark 2.3.2, there are numbers $0 \le s_k < 1$ and disjoint, possibly empty, sets $E_k \subset A$ such that v is pointwise Lipschitz in $A - \bigcup_{i=1}^{\infty} E_k$, and for $k = 1, 2, \ldots$, the following conditions hold:

(i*) $\mathcal{H}^{n-1+s_k}(E_k) < \infty$, and $H_{s_k} v(x) < \infty$ for each $x \in E_k$;
(ii) $\mathcal{H}^{n-1+s_k}(E_k) > 0$ implies $H_{s_k} v(x) = 0$ for each $x \in E_k$.

By Stepanoff's theorem, $A - \bigcup_{i=1}^{\infty} E_k$ is the union of disjoint sets E_0 and $D \subset \text{int } A$ such that $\mathcal{H}^n(E_0) = 0$ and v is differentiable at each $x \in D$. Thus A is the union of disjoint sets $D, E_0, E_1, \ldots$, and we let $s_0 = 1$. The family $\{(E_k, s_k) : k = 0, 1, \ldots\}$ is the disjoint union of subfamilies

$$\{(E_k, s_k) : \mathcal{H}^{n-1+s_k}(E_k) > 0\} \quad \text{and} \quad \{(E_k, s_k) : \mathcal{H}^{n-1+s_k}(E_k) = 0\},$$

which we enumerate as $\{(E_i^+, s_i^+) : i \geq 1\}$ and $\{(E_i^0, s_i^0) : i \geq 1\}$, respectively. For $i, j \in \mathbb{N}$, let

$$E_{i,j}^0 := \{x \in E_i^0 : j - 1 \leq H_{s_i^0} v(x) < j\}$$

and define $t_i^+ := n - 1 + s_i^+$ and $t_i^0 := n - 1 + s_i^0$. Now $A - D$ is the union of disjoint sets E_i^+ and $E_{i,j}^0$. Select $c_i > \mathcal{H}^{t_i^+}(E_i^+)$ and choose $\varepsilon > 0$. By Corollary 2.1.3, there is $\gamma : A \to \mathbb{R}_+$ such that for each cube $C \subset A$, the following conditions are satisfied:

(1) $|f(x)|C| - F(C)| \leq \varepsilon|C|$ if $d(C) < \gamma(x)$ for some $x \in D \cap C$,

(2) $|F(C)| \leq \varepsilon 2^{-i} c_i^{-1} d(C)^{t_i^+}$ if $d(C) < \gamma(x)$ for some $x \in E_i^+ \cap C$,

(3) $|F(C)| \leq 2nj\, d(C)^{t_i^0}$ if $d(C) < \gamma(x)$ for some $x \in E_{i,j}^0 \cap C$.

Next choose $\delta : A \to \mathbb{R}_+$. With no loss of generality, we may assume that $\delta \leq \gamma$. According to Proposition 2.2.2, there is a δ-fine dyadic partition $P := \{(C_1, x_1), \ldots, (C_p, x_p)\}$ such that $[P] = A$ and for $\kappa = \kappa(n) > 0$,

$$\sum_{x_k \in E_i^+} d(C_k)^{t_i^+} \leq \kappa c_i \quad \text{and} \quad \sum_{x_k \in E_{i,j}^0} d(C_k)^{t_i^0} \leq \varepsilon j^{-1} 2^{-i-j}.$$

Since $f(x) = 0$ for each $x \in A - D$, these inequalities, conditions (1)–(3), and Observation 2.1.1 imply the lemma:

$$\left| \sum_{k=1}^p f(x_k)|C_k| - \int_{\partial A} v \cdot \nu_A \, d\mathcal{H}^{n-1} \right|$$

$$\leq \sum_{x_k \in D} \left| f(x_k)|C_k| - F(C_k) \right| + \sum_{x_k \in A-D} |F(C_k)|$$

$$\leq \varepsilon \sum_{x_k \in D} |C_k| + \sum_{i \geq 1} \left[\sum_{x_k \in E_i^+} |F(C_k)| + \sum_{j=1}^{\infty} \sum_{x_k \in E_{i,j}^0} |F(C_k)| \right]$$

$$\leq \varepsilon|A| + \varepsilon \sum_{i \geq 1} \left[2^{-i} c_i^{-1} \sum_{x_k \in E_i^+} d(C_k)^{t_i^+} + 2n \sum_{j=1}^{\infty} j \sum_{x_k \in E_{i,j}^0} d(C_k)^{t_i^0} \right]$$

$$\leq \varepsilon|A| + \varepsilon\kappa \sum_{i=1}^{\infty} 2^{-i} + 2n\varepsilon \sum_{i,j=1}^{\infty} 2^{-i-j} = \varepsilon(|A| + \kappa + 2n). \qquad \square$$

Theorem 2.3.7. *Let A be a dyadic figure. If $v \in Adm(A; \mathbb{R}^n)$ is such that* $\operatorname{div} v$ *belongs to $L^1(A)$, then*

$$\int_A \operatorname{div} v \, d\mathcal{L}^n = \int_{\partial A} v \cdot \nu_A \, d\mathcal{H}^{n-1}.$$

PROOF. Defining f as in Lemma 2.3.6, we have $f \in L^1(A)$ and

$$\int_A f \, d\mathcal{L}^n = \int_A \operatorname{div} v \, d\mathcal{L}^n.$$

Choose $\varepsilon > 0$, and select a function $\delta : A \to \mathbb{R}_+$ associated with ε and f according to Henstock's lemma. By Lemma 2.3.6, there is a δ-fine partition $P := \{(C_1, x_1), \ldots, (C_p, x_p)\}$ such that $[P] = A$ and

$$\left| \int_A \operatorname{div} v \, d\mathcal{L}^n - \int_{\partial A} v \cdot \nu_A \, d\mathcal{H}^{n-1} \right| \leq \left| \int_A f \, d\mathcal{L}^n - \sum_{i=1}^{p} f(x_i)|C_i| \right|$$

$$+ \left| \sum_{i=1}^{p} f(x)|C_i| - \int_{\partial A} v \cdot \nu_A \, d\mathcal{H}^{n-1} \right| < 2\varepsilon. \qquad \square$$

Remark 2.3.8. Some comments are in order.

(1) The assumptions of Theorem 2.3.7 are met if $v \in Lip(A; \mathbb{R}^n)$, since $\|\operatorname{div} v\|_{L^\infty(A)} \leq n \operatorname{Lip} v$.

(2) Let $v(0) := 0$, and $v(x) := x \cos |x|^{-n-1}$ for $x \in \mathbb{R}^n - \{0\}$. Then $v \in Adm(\mathbb{R}^n; \mathbb{R}^n)$, but $\operatorname{div} v$ does not belong to $L^1(A)$ if A is a figure containing 0. Still, the flux of v can be calculated from $\operatorname{div} v$ by an averaging process which extends the Lebesgue integral. For a deeper analysis of this phenomenon, we refer the interested reader to [49, 51]; also see Chapter 9 below.

(3) Assume $n = 1$, and let $v : \mathbb{R} \to \mathbb{R}$ be differentiable almost everywhere and such that

$$\int_a^b v' \, d\mathcal{L}^1 = v(b) - v(a)$$

for each dyadic cell $[a, b] \subset \mathbb{R}$. Since [29, Section 2.4.3] implies

$$\mathcal{H}^s \left(\{x \in \mathbb{R} : H_s v(x) > 0\} \right) = 0$$

for each $0 \leq s < 1$, condition (ii) of Definition 2.3.1 cannot be omitted. The Cantor-Vitali function (Example 9.2.5 below) and its multidimensional analogue [54] provide another rationale for the definition of admissible vector fields.

(4) It is clear that using essentially the same arguments, the divergence theorem can be established for arbitrary figures. We employed dyadic figures merely for convenience. Theorem 2.3.7 is only a preliminary

result, sufficient for proving a satisfactory integration by parts theorem stated below.

Theorem 2.3.9 (Integration by parts). *Let $\Omega \subset \mathbb{R}^n$ be an open set, and let both $v : \Omega \to \mathbb{R}^n$ and $g : \Omega \to \mathbb{R}$ be locally bounded and pointwise Lipschitz almost everywhere in Ω. Assume $\operatorname{div} v \in L^1_{\mathrm{loc}}(U)$ and $Dg \in L^1_{\mathrm{loc}}(\Omega; \mathbb{R}^n)$. If $gv \in Adm(\Omega; \mathbb{R}^n)$ and $\operatorname{spt}(gv) \Subset \Omega$, then*

$$\int_\Omega g(x) \operatorname{div} v(x)\, dx = -\int_\Omega Dg(x) \cdot v(x)\, dx.$$

PROOF. Let $w = gv$, and let $A \subset \Omega$ be a dyadic figure with $\operatorname{spt} w \subset \operatorname{int} A$. By the assumptions, v and g are measurable and bounded in A. Thus

$$\operatorname{div} w = g \operatorname{div} v + Dg \cdot v$$

belongs to $L^1(A)$. Since $w \restriction \partial A = 0$, Theorem 2.3.7 yields

$$0 = \int_A \operatorname{div} w(x)\, dx = \int_A g(x) \operatorname{div} v(x)\, dx + \int_A Dg(x) \cdot v(x)\, dx. \qquad (*)$$

Let $x \in \Omega - A$. Then $\operatorname{div} w(x) = 0$, since w vanishes in the open set $\Omega - A$. Thus $g(x) \operatorname{div} v(x) = -Dg(x) \cdot v(x)$ and either $g(x) = 0$ or $v(x) = 0$. This shows that $\operatorname{spt}(g \operatorname{div} v)$ and $\operatorname{spt}(Dg \cdot v)$ are subsets of A, and the theorem follows from $(*)$. $\qquad\square$

Remark 2.3.10. The integration by parts theorem is usually applied when both g and v are admissible, and either g or v has compact support contained in Ω; see Proposition 2.3.5.

2.4. Convergence of dyadic figures

As figures are too specialized for applications, it is desirable to extend the divergence theorem to a larger family of sets. With the sole purpose of enhancing intuition, we describe the first step of the most obvious approach to this problem. Our main results will be obtained from less obvious but more efficient ideas of R. Caccioppoli [14] and E. De Giorgi [18, 19].

Lemma 2.4.1. *Let $\{A_i\}$ be a sequence of measurable set, and let $E \subset \mathbb{R}^n$ be any set. If $\lim |E \triangle A_i| = 0$, then E is measurable and for each $f \in L^1(\mathbb{R}^n)$,*

$$\lim \int_{A_i} f(x)\, dx = \int_E f(x)\, dx.$$

PROOF. Since $\lim |E - A_i| = \lim |A_i - E| = 0$, passing to a subsequence if necessary, we may assume that $|E - A_i| \leq 2^{-i}$ and $|A_i - E| \leq 2^{-i}$ for $i = 1, 2, \ldots$. Letting

$$I := \liminf A_i = \bigcup_{j=1}^{\infty} \bigcap_{i=j}^{\infty} A_j \quad \text{and} \quad S := \limsup A_i = \bigcap_{j=1}^{\infty} \bigcup_{i=j}^{\infty} A_j,$$

we infer $|E - I| = |S - E| = 0$. As $I \subset S$ implies $|E \triangle I| = |E \triangle S| = 0$, the set E is measurable and $\lim \|\chi_E - \chi_{A_i}\|_{L^1(\mathbb{R}^n)} = 0$. Any subsequence $\{B_i\}$ of $\{A_i\}$ has a subsequence $\{C_i\}$ such that $\lim \chi_{C_i} = \chi_E$ almost everywhere. Thus for $f \in L^1(\mathbb{R}^n)$, the dominated convergence theorem yields

$$\lim \left| \int_E f(x)\, dx - \int_{C_i} f(x)\, dx \right| \leq \lim \int_{\mathbb{R}^n} \left| \chi_E(x) - \chi_{C_i}(x) \right| \cdot |f(x)|\, dx = 0.$$

The lemma follows from the arbitrariness of $\{B_i\}$. $\square$

Recall that the family of all dyadic figures in $\mathbb{R}^n$ is denoted by $\mathcal{DF}$. We say that a sequence $\{A_i\}$ in $\mathcal{DF}$ *converges* to a set $E \subset \mathbb{R}^n$ if the following conditions are satisfied:

(i) Each A_i is contained in a fixed compact set $K \subset \mathbb{R}^n$.

(ii) $\lim |A_i \bigtriangleup E| = 0$ and $\sup \mathcal{H}^{n-1}(\partial A_i) < \infty$.

By Lemma 2.4.1, the set E is measurable.

Given dyadic figures A and B, we define nonoverlapping dyadic figures

$$A \ominus B := \operatorname{cl}(A - B) \quad \text{and} \quad A \odot B = \operatorname{cl}\left[\operatorname{int}(A \cap B) \right],$$

and observe that $A = (A \ominus B) \cup (A \odot B)$.

Proposition 2.4.2. *Let F be the flux of $v \in C(\mathbb{R}^n; \mathbb{R}^n)$, and let $\{A_i\}$ be a sequence in $\mathcal{DF}$ converging to a set $E \subset \mathbb{R}^n$. There exists a finite limit*

$$\widetilde{F}(E) := \lim F(A_i),$$

which does not depend on the choice of the sequence $\{A_i\}$.

PROOF. Let $K \subset \mathbb{R}^n$ be a compact set containing all figures A_i, and let $c = \sup \mathcal{H}^{n-1}(\partial A_i)$. Choose $\varepsilon > 0$, and use the Stone-Weierstrass theorem [62, Theorem 7.32] to find a vector field $w \in C^1(\mathbb{R}^n; \mathbb{R}^n)$ such that $\|v - w\|_{L^\infty(K;\mathbb{R}^n)} \leq \varepsilon$. According to Theorem 2.3.7,

$$\left| F(A_i \ominus A_j) \right| \leq \left| \int_{\partial(A_i \ominus A_j)} (v - w) \cdot \nu_{A_i \ominus A_j}\, d\mathcal{H}^{n-1} \right|$$
$$+ \left| \int_{\partial(A_i \ominus A_j)} w \cdot \nu_{A_i \ominus A_j}\, d\mathcal{H}^{n-1} \right|$$
$$\leq \varepsilon \mathcal{H}^{n-1}\left[\partial(A_i \ominus A_j) \right] + \left| \int_{A_i \ominus A_j} \operatorname{div} w(x)\, dx \right|$$
$$\leq 2c\varepsilon + \|\operatorname{div} w\|_{L^\infty(K)} |A_i \ominus A_j|;$$

since $\partial(A_i \ominus A_j) \subset \partial A_i \cup \partial A_j$. By Observation 2.1.1,

$$\left| F(A_i) - F(A_j) \right| = \left| \left[F(A_i \ominus A_j) + F(A_i \odot Aj) \right] - \left[F(A_j \ominus A_i) + F(A_j \odot A_i) \right] \right|$$
$$\leq \left| F(A_i \ominus A_j) \right| + \left| F(A_j \ominus A_i) \right|$$
$$\leq \|\operatorname{div} w\|_{L^\infty(K)} \left(|A_i \ominus A_j| + |A_j \ominus A_i| \right) + 4c\varepsilon$$
$$= \|\operatorname{div} w\|_{L^\infty(K)} |A_i \bigtriangleup A_j| + 4c\varepsilon.$$

It follows that $\{F(A_i)\}$ is a Cauchy sequence. The value

$$\widetilde{F}(E) := \lim F(A_i)$$

does not depend on the sequence $\{A_i\}$ of dyadic figures converging to E. Indeed, if $\{B_i\}$ is another sequence of dyadic figures converging to E, then so does the interlaced sequence $\{A_1, B_1, A_2, B_2, \dots\}$, and consequently

$$\lim F(A_i) = \lim F(B_i).$$ $\square$

Denote by $\overline{\mathcal{DF}}$ the family of all sets $E \subset \mathbb{R}^n$, necessarily measurable, for which there is a sequence $\{A_i\}$ in $\mathcal{DF}$ converging to E. Clearly $\mathcal{DF} \subset \overline{\mathcal{DF}}$, and since

$$\partial(A \cup B) \cup \partial(A \ominus B) \cup \partial(A \odot B) \subset \partial A \cup \partial B,$$

it is easy to verify that the family $\overline{\mathbf{DF}}$ is closed with respect to unions, intersections, and set differences. It follows from Proposition 2.4.2 that the flux F of $v \in C(\mathbb{R}^n; \mathbb{R}^n)$ defined on figures in Section 2.1 has a unique extension $\widetilde{F} : \overline{\mathbf{DF}} \to \mathbb{R}$, still called the flux of v.

Proposition 2.4.3. *If $\widetilde{F} : \overline{\mathbf{DF}} \to \mathbb{R}$ is the flux of $v \in C(\mathbb{R}^n; \mathbb{R}^n)$, then*

$$\widetilde{F}(A \cup B) = \widetilde{F}(A) + \widetilde{F}(B)$$

for each pair of nonoverlapping sets $A, B \in \overline{\mathbf{DF}}$.

PROOF. Let $\{A_k\}$ and $\{B_k\}$ be sequences in $\mathbf{DF}$ that converge to A and B, respectively. From $(A_k - B_k) - A \subset A_k - A$ and

$$A - (A_k - B_k) = (A - A_k) \cup (A \cap B_k) \subset (A - A_k) \cup (A \cap B) \cup (B_k - B),$$

we infer $\lim |A \bigtriangleup (A_k \ominus B_k)| = 0$. As $\partial(A_k \ominus B_k) \subset \partial A_k \cup \partial B_k$, the sequences $\{A_k \ominus B_k\}$ and $\{(A_k \ominus B_k) \cup B_k\} = \{A_k \cup B_k\}$ converge to A and $A \cup B$, respectively. Thus

$$\widetilde{F}(A \cup B) = \lim F\big[(A_k \ominus B_k) \cup B_k\big]$$
$$= \lim F(A_k \ominus B_k) + \lim F(B_k) = \widetilde{F}(A) + \widetilde{F}(B). \qquad \square$$

Proposition 2.4.4. *Let $\widetilde{F} : \overline{\mathbf{DF}} \to \mathbb{R}$ be the flux of $v \in C(\mathbb{R}^n; \mathbb{R}^n)$. If v is admissible and $\operatorname{div} v$ belongs to $L^1_{\mathrm{loc}}(\mathbb{R}^n)$, then for each $E \in \overline{\mathbf{DF}}$,*

$$\widetilde{F}(E) = \int_E \operatorname{div} v(x)\, dx.$$

PROOF. If $\{A_i\}$ is a sequence of dyadic figures converging to E, then Proposition 2.4.2, Theorem 2.3.7, and Lemma 2.4.1 imply

$$\widetilde{F}(E) = \lim F(A_i) = \lim \int_{A_i} \operatorname{div} v(x)\, dx = \int_E \operatorname{div} v(x)\, dx. \qquad \square$$

Proposition 2.4.4 establishes the divergence theorem for sets in $\overline{\mathbf{DF}}$, which are the desired generalization of dyadic figures (cf. Corollary 6.7.4 below). However, the flux $\widetilde{F} : \overline{\mathbf{DF}} \to \mathbb{R}$ does not share the geometric content of the flux $F : \mathbf{DF} \to \mathbb{R}$ defined by formula (2.1.1). This will be remedied in Chapters 4–6 below, albeit with a substantial effort. We show that each set $E \in \overline{\mathbf{DF}}$ has an "essential boundary" $\partial_* E \subset \partial E$ and a "unit exterior normal" ν_E, defined $\mathcal{H}^{n-1}$ almost everywhere on $\partial_* E$, such that the flux $\widetilde{F}$ of a vector field $v \in C(\mathbb{R}^n; \mathbb{R}^n)$ is calculated by the formula

$$\widetilde{F}(E) = \int_{\partial_* E} v \cdot \nu_E \, d\mathcal{H}^{n-1}$$

analogous to (2.1.1); see formula (6.5.1) below.

Remark 2.4.5. Using the convergence of dyadic figures, it is possible to define a sequential topology $\mathcal{T}$ in $\mathbf{DF}$ that is induced by a uniformity, and show that $\overline{\mathbf{DF}}$ is the sequential completion of the space $(\mathbf{DF}, \mathcal{T})$; see Chapter 10, in particular Section 10.6. Since the flux $F : (\mathbf{DF}, \mathcal{T}) \to \mathbb{R}$ of $v \in C(\mathbb{R}^n; \mathbb{R}^n)$ is uniformly continuous by additivity, it has a unique continuous extension $\widetilde{F} : \overline{\mathbf{DF}} \to \mathbb{R}$ — a fact we proved directly in Proposition 2.4.2.

Chapter 3

Removable singularities

We will study removable singularities for the Cauchy-Riemann, Laplace, and minimal surface equations. As these equations are in the divergence form $\mathrm{div}\left[\phi(Du)\right] = 0$, the integration by parts theorem established in the previous chapter is a natural tool. We define removable sets by means of Hausdorff measures, mostly without any topological restrictions. The results are established by short and simple arguments, which rely on the relationship between weak and strong solutions of partial differential equations. A few basic facts about distributions and weak solutions are stated without proofs. We made no attempt to survey the long history concerning removable singularities.

3.1. Distributions

A *multi-index* is an n-tuple $\alpha := (\alpha_1, \ldots, \alpha_n)$ where α_i are nonnegative integers. Let $|\alpha| := \sum_{i=1}^{n} \alpha_i$ and

$$D^\alpha := D_1^{\alpha_1} \cdots D_n^{\alpha_n} = \left(\frac{\partial}{\partial \xi_1}\right)^{\alpha_1} \cdots \left(\frac{\partial}{\partial \xi_n}\right)^{\alpha_n}.$$

Note that if $|\alpha| = 0$, then $D^\alpha f = f$ for any $f : \mathbb{R}^n \to \mathbb{C}$.

Let $\Omega \subset \mathbb{R}^n$ be an open set. Employing convention (1.1.2), we say that a sequence $\{\varphi_i\}$ in $C_c^\infty(\Omega; \mathbb{C})$ *converges* to zero in the sense of *test functions* if the following conditions hold:

(i) $\{\varphi_i\}$ is a sequence in $C_c^\infty(U; \mathbb{C})$ for an open set $U \Subset \Omega$;

(ii) $\lim \|D^\alpha \varphi_i\|_{L^\infty(\Omega; \mathbb{C})} = 0$ for each multi-index α.

The complex linear space $C_c^\infty(\Omega; \mathbb{C})$ equipped with this convergence is denoted by $\mathcal{D}(\Omega; \mathbb{C})$, and the elements of $\mathcal{D}(\Omega; \mathbb{C})$ are called *test functions*. The real linear subspace of $\mathcal{D}(\Omega; \mathbb{C})$ consisting of all real-valued test functions is denoted by $\mathcal{D}(\Omega)$.

A *distribution* is a linear functional $L : \mathcal{D}(\Omega; \mathbb{C}) \to \mathbb{C}$ such that

$$\lim L(\varphi_i) = 0$$

for each sequence $\{\varphi_i\}$ in $\mathcal{D}(\Omega; \mathbb{C})$ that converges to zero in the sense of test functions. The complex linear space of all distributions is denoted by $\mathcal{D}'(\Omega; \mathbb{C})$. The real linear space $\mathcal{D}'(\Omega)$ is defined analogously.

Remark 3.1.1. In Example 3.6.5 below we define a locally convex topology $\mathcal{S}$ in the spaces $\mathcal{D}(\Omega)$ of real-valued test functions so that the space $\mathcal{D}'(\Omega)$ of distributions is the dual space of $(\mathcal{D}(\Omega), \mathcal{S})$. The reader familiar with complex locally convex spaces will recognize instantly that a similar topology can be defined in the space $\mathcal{D}(\Omega; \mathbb{C})$ of complex-valued test functions [64, Section 6.2].

Example 3.1.2. Let $f \in L^1_{\text{loc}}(\Omega; \mathbb{C})$, let μ be a Radon measure in Ω, and let $v \in L^1_{\text{loc}}(\Omega; \mathbb{R}^n)$. The linear functionals

$$L_f : \varphi \mapsto \int_\Omega f\varphi \, dx : \mathcal{D}(\Omega; \mathbb{C}) \to \mathbb{C}, \tag{1}$$

$$L_\mu : \varphi \mapsto \int_\Omega \varphi \, d\mu : \mathcal{D}(\Omega; \mathbb{C}) \to \mathbb{C}, \tag{2}$$

$$F_v : \varphi \mapsto -\int_\Omega v \cdot D\varphi \, dx : \mathcal{D}(\Omega) \to \mathbb{R} \tag{3}$$

are examples of distributions. Distribution F_v is called the *distributional divergence* of v, since if $v \in C^1(\Omega; \mathbb{R}^n)$ integration by parts yields

$$F_v(\varphi) = \int_\Omega \varphi \operatorname{div} v \, dx = L_{\operatorname{div} v}(\varphi)$$

for each test function $\varphi \in \mathcal{D}(\Omega)$.

Let α be a multi-index. If $f \in C^{|\alpha|}(\Omega; \mathbb{C})$, then repeated applications of the integration by parts theorem yield

$$\langle L_{D^\alpha f}, \varphi \rangle = \int_\Omega \varphi(x) D^\alpha f(x) \, dx$$

$$= (-1)^{|\alpha|} \int_\Omega f(x) D^\alpha \varphi(x) \, dx = (-1)^{|\alpha|} \langle L_f, D^\alpha \varphi \rangle$$

for each $\varphi \in \mathcal{D}(\Omega; \mathbb{C})$. Since for any distribution L, the linear functional

$$\varphi \mapsto (-1)^{|\alpha|} \langle L, D^\alpha \varphi \rangle : \mathcal{D}(\Omega; \mathbb{C}) \to \mathbb{C}$$

is a distribution, the previous identity suggests to define a distribution $D^\alpha L$ by the formula

$$\langle D^\alpha L, \varphi \rangle := (-1)^{|\alpha|} \langle L, D^\alpha \varphi \rangle$$

for each test function $\varphi \in \mathcal{D}(\Omega; \mathbb{C})$. Observe that

$$D^\alpha L_f = L_{D^\alpha f}$$

whenever $f \in C^{|\alpha|}(\Omega; \mathbb{C})$. Additional information about test functions and distributions can be found in many standard textbooks, for instance in [64, Chapter 6] or [27, Chapter 5].

3.2. Differential equations

A *linear differential operator* with constant coefficients is the expression

$$\Lambda := \sum_{|\alpha| \leq k} c_\alpha D^\alpha$$

where $k \in \mathbb{N}$ and $c_\alpha \in \mathbb{C}$ for each multi-index α with $|\alpha| \leq k$. In an open set $\Omega \subset \mathbb{R}^n$ we consider two types of solutions of the partial differential equation $\Lambda u = 0$.

- A *strong solution* is a complex-valued function $u \in C^k(\Omega; \mathbb{C})$ such that $\langle \Lambda u, x \rangle = 0$ for each $x \in \Omega$.
- A *weak solution* is a complex-valued function $u \in L^1(\Omega; \mathbb{C})$ such that $\Lambda L_u = 0$ where L_u is defined in Example 3.1.2, (1).

Thus $u \in L^1(\Omega; \mathbb{C})$ is a weak solution of $\Lambda u = 0$ if the equality

$$\sum_{|\alpha| \leq k} (-1)^{|\alpha|} c_\alpha \int_\Omega u(x) D^\alpha \varphi(x)\, dx = \langle \Lambda L_u, \varphi \rangle = 0$$

holds for each test function $\varphi \in \mathcal{D}(\Omega; \mathbb{C})$.

For a multi-index $\alpha = (\alpha_1, \ldots, \alpha_n)$ and $x = (\xi_1, \ldots, \xi_n)$ in $\mathbb{R}^n$, let

$$x^\alpha := \xi_1^{\alpha_1} \cdots \xi_n^{\alpha_n}$$

where $\xi^{\alpha_i} = 1$ when $\alpha_i = 0$. A complex-valued function

$$p_\Lambda : x \mapsto \sum_{|\alpha| = k} c_\alpha x^\alpha : \mathbb{R}^n \to \mathbb{C}$$

is called the *characteristic polynomial* of Λ. If $p_\Lambda(x) \neq 0$ for each x in $\mathbb{R}^n - \{0\}$, the operator Λ is called *elliptic*. Note that the ellipticity of Λ is determined by the *leading coefficients*, i.e., by c_α with $|\alpha| = k$.

Example 3.2.1. The following linear differential operators are elliptic.

(1) The *Laplace* operator

$$\triangle := D_1^2 + \cdots + D_n^2,$$

since $p_\triangle(x) = |x|^2$ for each $x \in \mathbb{R}^n$.

(2) For $n = 2$ and $i := \sqrt{-1}$, the *holomorphic* operator

$$\bar{\partial} := D_1 + i D_2,$$

since $p_{\bar{\partial}}(x) = \xi_1 + i\xi_2$ for each $x = (\xi_1, \xi_2)$ in $\mathbb{R}^2$.

Clearly, each strong solution of the equation $\Lambda u = 0$ is also a weak solution. While the converse is generally false, the next theorem, proved in [64, Corollary of Theorem 8.12], asserts that strong and weak solutions are essentially the same when Λ is an elliptic operator.

Theorem 3.2.2. *Let $\Omega \subset \mathbb{R}^n$ be an open set, let Λ be a linear differential operator with constant coefficients, and let $u_w \in L^1(\Omega; \mathbb{C})$ be a weak solution of the equation $\Lambda u = 0$. If Λ is elliptic, there is a strong solution $u_s \in C^\infty(\Omega; \mathbb{C})$ such that $u_s(x) = u_w(x)$ for almost all $x \in \Omega$.*

3.3. Holomorphic functions

Throughout this section, we assume that Ω is an open subset of the complex plane $\mathbb{C}$. A complex-valued function $f \in C^1(\Omega; \mathbb{C})$ is called *holomorphic* if the *Cauchy-Riemann equation*

$$\bar{\partial}f(z) = 0$$

is satisfied for all $z \in \Omega$, or equivalently, if the *complex derivative* $f'(z)$ exists at each $z \in \Omega$.

Making the obvious identification between $\mathbb{C}$ and $\mathbb{R}^2$, we can talk about admissibility of complex-valued functions in the sense of Definition 2.3.1. Explicitly, $f : \Omega \to \mathbb{C}$ is *admissible* if there are numbers $0 \leq s_k < 1$ and disjoint, possibly empty, sets E_k such that f is pointwise Lipschitz in $\Omega - \bigcup_{k=1}^\infty E_k$ and for each $k = 1, 2, \ldots$, the following conditions hold:

(i) E_k is $\mathcal{H}^{1+s_k}$ σ-finite, and $H_{s_k} f(z) < \infty$ for each $z \in E_k$;

(ii) $\mathcal{H}^{1+s_k}(E_k) > 0$ implies $H_{s_k} f(z) = 0$ for each $z \in E_k$.

The meaning of the symbol $Adm(\Omega; \mathbb{C})$ is clear.

The following theorem generalizes the well-known classical result of A.S. Besicovitch [2].

Theorem 3.3.1. *Let $f \in Adm(\Omega; \mathbb{C})$ have the complex derivative $f'(z)$ at almost all $z \in \Omega$. Then f can be redefined at the points of discontinuity so that it becomes holomorphic.*

PROOF. Since $f'(z)$ exists at almost all $z \in \Omega$, the Cauchy-Riemann condition $\bar{\partial}f(z) = 0$ holds for almost all $z \in \Omega$. If $z = x + iy$ and $f = u + iv$, then

$$\bar{\partial}f = \frac{\partial u}{\partial x} - \frac{\partial v}{\partial y} + i\left(\frac{\partial v}{\partial x} + \frac{\partial y}{\partial y}\right) = \operatorname{div}(u, -v) + i\operatorname{div}(v, u)$$

where the vector fields $(u, -v)$ and (v, u) belong to $Adm(\Omega; \mathbb{R}^2)$. After a simple calculation, the integration by parts theorem yields

$$\int_\Omega f\,\bar{\partial}\varphi = -\int_\Omega \varphi\,\bar{\partial}f = 0 \qquad (*)$$

for each test function $\varphi \in \mathcal{D}(\Omega; \mathbb{C})$. Indeed, $(*)$ follows from Remark 2.3.10, since $\varphi \in Adm(\Omega; \mathbb{C})$ and $\mathrm{spt}\,\varphi \Subset \Omega$. Thus f is a weak solution of $\bar\partial u = 0$. As the holomorphic operator $\bar\partial$ is elliptic, Theorem 3.2.2 implies that there is a holomorphic function $g : \Omega \to \mathbb{C}$ which is equal to f almost everywhere. In particular, $g(z) = f(z)$ for each $z \in \Omega$ at which f is continuous. $\qquad\square$

3.4. Harmonic functions

Let $\Omega \subset \mathbb{R}^n$ be an open set. A function $u \in C^2(\Omega)$ is called *harmonic* if the *Laplace equation*

$$\triangle u(x) = 0$$

is satisfied for all $x \in \Omega$. As the coefficients of the Laplace operator $\triangle$ are real numbers, in testing for real-valued weak solutions of the Laplace equation, it suffices to use only real-valued test functions.

Theorem 3.4.1. *Let $C \subset \Omega$ be an $\mathcal{H}^{n-1}$ negligible set which is relatively closed in Ω, and let $u \in Lip_{\mathrm{loc}}(\Omega)$ be differentiable at each point of $\Omega - C$. If Du belongs to $Adm(\Omega - C; \mathbb{R}^n)$ and $\triangle u(x) = 0$ for almost all $x \in \Omega$, then u is harmonic.*

Proof. Select a test function $\varphi \in \mathcal{D}(\Omega)$, and observe $u,\varphi \in Adm(\Omega)$ and $D\varphi \in Adm(\Omega; \mathbb{R}^n)$. As u is locally Lipschitz, the assumptions about C imply that the map $\phi : \Omega \to \mathbb{R}^n$ defined by the formula

$$\phi(x) := \begin{cases} Du(x) & \text{if } x \in \Omega - C, \\ 0 & \text{if } x \in C \end{cases}$$

belongs to $Adm(\Omega; \mathbb{R}^n)$. Observe that for almost all $x \in \Omega$,

$$Du(x) = \phi(x) \quad \text{and} \quad \mathrm{div}\,\phi(x) = \triangle u(x) = 0.$$

Since $\triangle\varphi = \mathrm{div}\,D\varphi$, integrating by parts twice, we obtain

$$\int_\Omega u \triangle \varphi = -\int_\Omega \phi \cdot D\varphi = \int_\Omega \varphi \triangle u = 0.$$

Thus u is a weak solution of the equation $\triangle u = 0$. As $\triangle$ is elliptic, Theorem 3.2.2 implies that there is a harmonic function h defined on Ω that equals u almost everywhere. Since u is continuous, $u \equiv h$. $\qquad\square$

3.5. The minimal surface equation

Let $\Omega \subset \mathbb{R}^n$ be an open set and $u \in Lip_{\mathrm{loc}}(\Omega)$. The set

$$G(u) := \left\{ (x,y) \in \mathbb{R}^{n+1} : y = u(x) \right\}$$

is called the *graph* of u, and the extended real number $S(u) := \mathcal{H}^n[G(u)]$ is called the *surface area* of $G(u)$. It follows from [29, Section 3.3.4, B] that

$$S(u) = \int_\Omega \sqrt{1 + |Du(x)|^2}\, dx.$$

Seeking a local minimum of the functional

$$S : u \mapsto S(u) : Lip_{loc}(\Omega) \to \overline{\mathbb{R}},$$

we assume that there is $u \in C^2(\Omega)$ such that $S(u) < \infty$ and

$$\left[\frac{d}{dt} S(u + t\varphi)\right]_{t=0} = 0$$

for each $\varphi \in \mathcal{D}(\Omega)$. Differentiating under the integral sign and integrating by parts, we obtain that for every $\varphi \in \mathcal{D}(\Omega)$,

$$0 = \left[\frac{d}{dt} S(u + t\varphi)\right]_{t=0} = \int_\Omega \frac{Du \cdot D\varphi}{\sqrt{1 + |Du|^2}} = -\int_\Omega \varphi \operatorname{div} \frac{Du}{\sqrt{1 + |Du|^2}}.$$

Consequently for all $x \in \Omega$,

$$\operatorname{div} \frac{Du(x)}{\sqrt{1 + |Du(x)|^2}} = 0. \tag{$*$}$$

Theorem 3.5.1. *Let $\Omega \subset \mathbb{R}^n$ be a bounded open set whose boundary $\partial\Omega$ is a C^2 manifold, and let $g \in C(\partial\Omega)$. There is a unique $u \in C(\operatorname{cl}\Omega)$ such that u is C^2 in Ω, $u \restriction \partial\Omega = g$, and $(*)$ holds for each $x \in \Omega$. If $v \in C(\operatorname{cl}\Omega)$ is C^2 in Ω and $v \restriction \partial\Omega = g$, then $S(u) \le S(v)$.*

Theorem 3.5.1 is proved in [36, Sections 13]. Together with the previous calculations, it explains why the equation

$$\operatorname{div} \frac{Du}{\sqrt{1 + |Du|^2}} = 0 \tag{MSE}$$

is called the *minimal surface equation*.

Unless stated otherwise, by a solution of (MSE) we mean a solution in Ω. Strong solutions of (MSE) are defined as in Section 3.2; weak solutions are defined differently, since (MSE) is a nonlinear equation.

- A *strong solution* of (MSE) is a function $u \in C^2(\Omega)$ such that equality $(*)$ holds for every $x \in \Omega$.
- A *weak solution* of (MSE) is an almost everywhere differentiable function $u : \Omega \to \mathbb{R}$ such that $Du \in L^1_{loc}(\Omega)$ and

$$\int_\Omega \frac{Du \cdot D\varphi}{\sqrt{1 + |Du|^2}} = 0$$

for each $\varphi \in C^1_c(\Omega)$.

It is clear that each strong solution of (MSE) is a weak solution. A partial converse stated below is due to E. De Giorgi [20].

Theorem 3.5.2 (De Giorgi). *If $u \in Lip_{loc}(\Omega)$ is a weak solution of (MSE), then it is a strong solution.*

We use Theorem 3.5.2 in lieu of Theorem 3.2.2, and employ the following result of L. Simon; see [68] or [36, Theorem 16.9].

Theorem 3.5.3 (Simon). *Let $C \subset \Omega$ be an $\mathcal{H}^{n-1}$ negligible set which is relatively closed in Ω. Every strong solution of (MSE) in $\Omega - C$ can be extended to a strong solution in Ω.*

Theorem 3.5.4. *Let $C \subset \Omega$ be an $\mathcal{H}^{n-1}$ negligible set which is relatively closed in Ω, and let $u : \Omega - C \to \mathbb{R}$ be differentiable at each point of $\Omega - C$. If Du belongs to $Adm(\Omega - C; \mathbb{R}^n)$ and u satisfies $(*)$ for almost all $x \in \Omega - C$, then u can be extended to Ω so that it is a strong solution of (MSE).*

PROOF. Select $\varphi \in C_c^1(\Omega - C)$, and observe that both φ and

$$x \mapsto \frac{Du(x)}{\sqrt{1 + |Du(x)|^2}} : \Omega - C \to \mathbb{R}$$

belong to $Adm(\Omega - C)$. Integrating by parts yields

$$\int_{\Omega - C} \frac{Du \cdot D\varphi}{\sqrt{1 + |Du|^2}} = -\int_{\Omega - C} \varphi \operatorname{div} \frac{Du}{\sqrt{1 + |Du|^2}} = 0,$$

and u is a weak solution of (MSE) in $\Omega - C$. As Du is admissible in $\Omega - C$ by our assumption, Du is locally bounded by Remark 2.3.3. Thus u is locally Lipschitz, and De Giorgi's theorem implies that it is a strong solution of (MSE) in $\Omega - C$. An application of Simon's theorem completes the proof. $\square$

A nontrivial strong solution of (MSE) in $\mathbb{R}^n$ is called a *Bernstein function*. The following theorem shows that in higher dimensions, solutions of (MSE) may not correspond to our intuition about minimizing the surface area [36, Section 17].

Theorem 3.5.5. *In dimension $n \leq 7$, each Bernstein function is linear, i.e., its graph is a hyperplane in $\mathbb{R}^{n+1}$. If $n > 7$, there exist nonlinear Bernstein functions.*

3.6. Injective limits

We describe a general procedure for obtaining a new locally convex topology in a locally convex space X by restricting the original topology to a suitable family of subspaces. This construction, called the internal injective limit, has many applications. In Example 3.6.5 below, we use it to define a topology $\mathcal{S}$ in the space $\mathcal{D}(\Omega)$ of test functions mentioned in Remark 3.1.1. Another application is given in Section 5.3 below.

A family $\mathcal{A}$ of sets is called *directed upward by inclusion*, or simply *directed*, if for each pair $A, B \in \mathcal{A}$ there is $C \in \mathcal{A}$ containing both A and B. For any family $\mathcal{A}$ of sets and any set E, we define the family

$$\mathcal{A} \mathbin{L} E := \{A \cap E : A \in \mathcal{A}\}.$$

In this notation, if $(X, \mathcal{T})$ is a topological space and $Y \subset X$, then $\mathcal{T} \mathbin{L} Y$ is the usual subspace topology of Y inherited from $(X, \mathcal{T})$.

Theorem 3.6.1. *Let $(X, \mathcal{T})$ be a locally convex space, and let $\mathcal{A} = \{X_\alpha : \alpha \in A\}$ be a directed family of subspaces of X such that $X = \bigcup_{\alpha \in A} X_\alpha$. There is a unique locally convex topology $\mathcal{T}_\mathcal{A}$ in X that satisfies the following conditions:*

  (i) *$\mathcal{T}_\mathcal{A} \mathbin{L} X_\alpha \subset \mathcal{T} \mathbin{L} X_\alpha$ for every $\alpha \in A$,*

  (ii) *given a locally convex space Y, a linear map $\phi : (X, \mathcal{T}_\mathcal{A}) \to Y$ is continuous whenever each restriction $\phi \restriction (X_\alpha, \mathcal{T} \mathbin{L} X_\alpha) \to Y$ is continuous.*

The topology $\mathcal{T}_\mathcal{A}$ has the following properties:

  (1) *$\mathcal{T} \subset \mathcal{T}_\mathcal{A}$ and $\mathcal{T} \mathbin{L} X_\alpha = \mathcal{T}_\mathcal{A} \mathbin{L} X_\alpha$ for every $\alpha \in A$.*

  (2) *$\mathcal{T}_\mathcal{A}$ is the largest topology among locally convex topologies $\mathcal{S}$ in X for which all inclusion maps $i_\alpha : x \mapsto x : (X_\alpha, \mathcal{T} \mathbin{L} X_\alpha) \to (X, \mathcal{S})$ are continuous.*

  (3) *$\mathcal{T}_\mathcal{A}$ has a neighborhood base $\mathcal{B} \subset \mathcal{T}_\mathcal{A}$ at zero consisting of all absorbing symmetric convex sets $U \subset X$ such that $U \cap X_\alpha \in \mathcal{T} \mathbin{L} X_\alpha$ for every $\alpha \in A$.*

PROOF. If $\mathcal{S}$ is a locally convex topology in X that satisfies conditions (i) and (ii), then a straightforward verification reveals that the identity map

$$\mathrm{id} : x \mapsto x : (X, \mathcal{T}_\mathcal{A}) \to (X, \mathcal{S})$$

is a homeomorphism. Hence $\mathcal{S} = \mathcal{T}_\mathcal{A}$, which establishes the uniqueness of $\mathcal{T}_\mathcal{A}$.

(1) By condition (ii), the map $\mathrm{id} : (X, \mathcal{T}_\mathcal{A}) \to (X, \mathcal{T})$ is continuous and so $\mathcal{T} \subset \mathcal{T}_\mathcal{A}$. Now fix $\alpha \in A$, and observe that the diagram

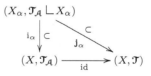

commutes. Since the maps i_α and id are continuous, so is the inclusion map j_α. Given $U \in \mathcal{T} \mathbin{L} X_\alpha$, there is $V \in \mathcal{T}$ with $V \cap X_\alpha = U$. Consequently $U = j_\alpha^{-1}(V)$ belongs to $\mathcal{T}_\mathcal{A} \mathbin{L} X_\alpha$. Now condition (i) implies $\mathcal{T} \mathbin{L} X_\alpha = \mathcal{T}_\mathcal{A} \mathbin{L} X_\alpha$.

(2) It follows from (1) that all inclusions $i_\alpha : (X_\alpha, \mathcal{T} \mathbin{L} X_\alpha) \to (X, \mathcal{T}_\mathcal{A})$ are continuous. Let $\mathcal{S}$ be a locally convex topology in X such that all inclusions $i_\alpha : (X_\alpha, \mathcal{T} \mathbin{L} X_\alpha) \to (X, \mathcal{S})$ are continuous. Then the identity map $\mathrm{id} : (X, \mathcal{T}_\mathcal{A}) \to (X, \mathcal{S})$ is continuous by condition (ii), and we conclude $\mathcal{S} \subset \mathcal{T}_\mathcal{A}$.

(3) Let $\mathcal{B}$ be the family of all absorbing, symmetric, and convex sets $U \subset X$ such that for every $\alpha \in A$ the following condition holds: $i_\alpha^{-1}(U)$ is an open neighborhood of zero in the topology $\mathcal{T} \mathbin{L} X_\alpha$, or equivalently, $U \cap X_\alpha \in \mathcal{T} \mathbin{L} X_\alpha$. If $U \in \mathcal{B}$ and $t \in \mathbb{R} - \{0\}$, then $tU \cap X_\alpha = t(U \cap X_\alpha)$ belongs to $\mathcal{T} \mathbin{L} X_\alpha$ for each $\alpha \in A$. Thus $tU \in \mathcal{B}$, and as U is convex, $\frac{1}{2}U + \frac{1}{2}U \subset U$. Since the family $\mathcal{B}$ is closed with respect to finite intersections, it is a neighborhood base at zero for a locally convex topology $\mathcal{U}$ in X, provided we show that $\mathcal{U}$ is Hausdorff. To this end, recall that in accordance with the general assumption made in Section 1.2, the topology $\mathcal{T}$ is Hausdorff. Since each symmetric convex $\mathcal{T}$ neighborhood of

zero belongs to $\mathcal{B}$, the topology $\mathcal{U}$ is also Hausdorff. From the definition of $\mathcal{B}$ it is clear that $\mathcal{U}$ is the largest among locally convex topologies $\mathcal{S}$ in X for which all inclusions

$$i_\alpha : (X_\alpha, \mathcal{T} \llcorner X_\alpha) \to (X, \mathcal{S})$$

are continuous. Thus $\mathcal{U} = \mathcal{T}_{\mathcal{A}}$ by property (2). $\qquad\square$

The topology $\mathcal{T}_{\mathcal{A}}$ defined in Theorem 3.6.1 is called the *internal injective limit* of the topologies $\mathcal{T} \llcorner X_\alpha$; see [27, Section 6.3]. The customary notation is

$$\mathcal{T}_{\mathcal{A}} = \operatorname*{inj\,lim}_{\alpha \in A}(\mathcal{T} \llcorner X_\alpha) \quad \text{and} \quad (X, \mathcal{T}_{\mathcal{A}}) = \operatorname*{inj\,lim}_{\alpha \in A}(X_\alpha, \mathcal{T} \llcorner X_\alpha).$$

Families $\mathcal{A}$ and $\mathcal{B}$ of sets are called *interlacing* if each $A \in \mathcal{A}$ is contained in some $B \in \mathcal{B}$, and each $B \in \mathcal{B}$ is contained in some $A \in \mathcal{A}$.

Proposition 3.6.2. *Let $(X, \mathcal{T})$ be a locally convex space, and let $\mathcal{A} = \{X_\alpha : \alpha \in A\}$ and $\mathcal{B} = \{X_\beta : \beta \in B\}$ be interlacing families of subspaces of X. If $\mathcal{A}$ is directed and $X = \bigcup_{\alpha \in A} X_\alpha$, then $\mathcal{B}$ is directed, $X = \bigcup_{\beta \in B} X_\beta$, and*

$$\operatorname*{inj\,lim}_{\alpha \in A}(\mathcal{T} \llcorner X_\alpha) = \operatorname*{inj\,lim}_{\beta \in B}(\mathcal{T} \llcorner X_\beta).$$

PROOF. Let $\mathcal{T}_{\mathcal{A}} = \operatorname{inj\,lim}_{\alpha \in A}(\mathcal{T} \llcorner X_\alpha)$ and $\mathcal{T}_{\mathcal{B}} = \operatorname{inj\,lim}_{\beta \in B}(\mathcal{T} \llcorner X_\beta)$. Choose $\alpha \in A$ and find $\beta \in B$ with $X_\alpha \subset X_\beta$. The diagram

$$
\begin{array}{ccc}
(X_\alpha, \mathcal{T} \llcorner X_\alpha) & \xrightarrow[\subset]{\ i_\alpha\ } & (X, \mathcal{T}_{\mathcal{B}}) \\[4pt]
\subset \downarrow & & \uparrow \subset \\[4pt]
(X_\beta, \mathcal{T} \llcorner X_\beta) & \xrightarrow[\ \mathrm{id}\ ]{} & (X_\beta, \mathcal{T}_{\mathcal{B}} \llcorner X_\beta)
\end{array}
$$

commutes, and the vertical maps are clearly continuous. By Theorem 2.6.1, condition (i), the map id is continuous as well. Thus the map i_α is continuous, and we conclude from Theorem 2.6.1, condition (ii), that $\mathrm{id} : (X, \mathcal{T}_{\mathcal{A}}) \to (X, \mathcal{T}_{\mathcal{B}})$ is continuous. The proposition follows by symmetry. $\qquad\square$

A sequence $\{x_k\}$ in a topological linear space X is called *Cauchy* if for each neighborhood U of zero, $x_k \in U$ for all but finitely many k. If each Cauchy sequence in X converges, the space X is called *sequentially complete*. A closed subspace of a sequentially complete space is sequentially complete.

Theorem 3.6.3. *Let $(X, \mathcal{T})$ be a locally convex space, and let $\mathcal{J} = \{X_j : j \in \mathbb{N}\}$ be an increasing sequence of **closed** subspaces of X such that $X = \bigcup_{j \in \mathbb{N}} X_j$. Then*

$$(X, \mathcal{S}) = \operatorname*{inj\,lim}_{j \to \infty}(X_j, \mathcal{T} \llcorner X_j)$$

has the following properties:

(1) *A sequence $\{x_k\}$ in X converges in $(X, \mathcal{S})$ if and only if $\{x_k\}$ is a sequence in some X_j that converges in $(X, \mathcal{T})$.*

(2) *$E \subset X$ is $\mathcal{S}$ bounded if and only if E is a $\mathcal{T}$ bounded subset of some X_j.*

(3) *If each $(X_j, \mathcal{T} \llcorner X_j)$ is sequential, then a linear map ϕ from $(X, \mathcal{S})$ to a locally convex space Y is continuous whenever $\lim \phi(x_k) = 0$ for each sequence $\{x_k\}$ that converges to zero in $(X, \mathcal{S})$.*

(4) *If each $(X_j, \mathcal{T} \llcorner X_j)$ is sequentially complete, then so is $(X, \mathcal{S})$.*

PROOF. (1) If $\{x_k\}$ is a sequence converging to zero in $(X, \mathcal{S})$, then $\{x_k\}$ converges to zero in $(X, \mathcal{T})$ by Theorem 3.6.1, (1). Seeking a contradiction, assume $\{x_k\}$ is not contained in any X_j, and construct recursively a subsequence $\{y_j\}$ of $\{x_k\}$ so that $y_j \notin X_j$ for $j = 1, 2, \ldots$. Since X_j is closed, there is a convex symmetric neighborhood $U_j \in \mathcal{T}$ of zero such that y_j does not belong to $W_j = X_j + U_j$; see [64, Theorem 1.10]. The intersection $W = \bigcap_{j=1}^{\infty} W_j$ is convex and symmetric. Since $X_k \subset X_j \subset W_j$ whenever $j \geq k$,

$$X_k \cap W = X_k \cap \bigcap \{W_j : j = 1, \ldots, k\} \qquad (*)$$

for $k = 1, 2, \ldots$. The finite intersection $\bigcap_{j=1}^{k} W_j$ belongs to $\mathcal{T}$, and hence $X_k \cap W$ belongs to $\mathcal{T} \llcorner X_k$. Given $x \in X$, find $k \in \mathbb{N}$ and $t \in \mathbb{R}$ so that $x \in X_k$ and $tx \in \bigcap_{j=1}^{k} W_j$. As $tx \in X_k$, equality $(*)$ implies $tx \in W$. Thus W is an absorbing set, and hence an $\mathcal{S}$ neighborhood of zero by Theorem 3.6.1, (3). Since no y_j belongs to W, this is a contradiction. The converse follows from Theorem 3.6.1, (1).

(2) If $E \subset X$ is $\mathcal{S}$ bounded, it is $\mathcal{T}$ bounded by Theorem 3.6.1, (1). Assume that E is not contained in any X_j, and construct recursively a sequence $\{x_j\}$ in E so that $x_j \notin X_j$ for $j = 1, 2, \ldots$. Since $\{\frac{1}{j} x_j\}$ converges to zero in $(X, \mathcal{S})$, part (1) of the theorem shows that all x_j belong to some X_k, a contradiction.

(3) Assume that $\lim \phi(x_k) = 0$ for each $\{x_k\}$ in X that converges to zero in $(X, \mathcal{S})$. If $\{x_k\}$ is a sequence in X_j that converges to zero in $(X_j, \mathcal{T} \llcorner X_j)$, then $\lim \phi(x_k) = 0$ by part (1) of the theorem. By our assumption, $\phi \restriction X_j : (X_j, \mathcal{T} \llcorner X_j) \to Y$ is continuous, and Theorem 3.6.1, (ii) implies that $\phi : (X, \mathcal{S}) \to Y$ is continuous.

(4) If $\{x_k\}$ is a Cauchy sequence in $(X, \mathcal{S})$, then the set $\{x_k : k \in \mathbb{N}\}$ is $\mathcal{S}$ bounded. Part (2) of the theorem and Theorem 3.6.1, (1) imply that $\{x_k\}$ is a Cauchy sequence in some $(X_j, \mathcal{T} \llcorner X_j)$. By our assumption, $\{x_k\}$ converges in $(X_j, \mathcal{T} \llcorner X_j)$, and hence in $(X, \mathcal{S})$ by part (1) of the theorem. □

Example 3.6.4. Let $\Omega \subset \mathbb{R}^n$ be an open set, and let $\mathcal{O}$ be the family of all open sets $U \Subset \Omega$. Denote by $\mathcal{T}$ the topology in $X = C_c(\Omega; \mathbb{R}^m)$ defined by the L^∞ norm, and let

$$X_U := \{v \in X : \operatorname{spt} v \subset U\}$$

for each $U \in \mathcal{O}$. There are compact sets $K_j \subset \Omega$ such that $K_j \Subset K_{j+1}$ for each $j \in \mathbb{N}$, and $\bigcup_{j \in \mathbb{N}} K_j = \Omega$. Define closed subspaces

$$X_j := \{v \in X : \operatorname{spt} v \subset K_j\}$$

of $(X, \mathcal{T})$, and note the families $\{X_U : U \in \mathcal{O}\}$ and $\{X_j : j \in \mathbb{N}\}$ are interlaced. Thus

$$\mathcal{S} = \operatorname*{inj\,lim}_{j \to \infty}(\mathcal{T} \llcorner X_j) = \operatorname*{inj\,lim}_{U \in \mathcal{O}}(\mathcal{T} \llcorner X_U)$$

by Proposition 3.6.2. In particular, $\mathcal{S}$ does not depend on the choice of $\{K_j\}$.

Although the metrizable space $(X, \mathcal{T})$ is not complete, the space $(X, \mathcal{S})$ is sequentially complete according to Theorem 3.6.3, (4). On the other hand, the space $(X, \mathcal{S})$ is not metrizable. To see this, select disjoint balls $B(x_k, r_k) \subset \Omega$, $k = 0, 1, \ldots$ so that the closure of $\{x_0, x_1, \ldots\}$ meets $\partial \Omega$. If $\varphi_k \in X$ are such that $\varphi_k(x_k) = 1$ and $\operatorname{spt} \varphi_k \subset B(x_k, r_k)$, then in $(X, \mathcal{S})$ the following is true:

$$\lim_{k \to \infty} \tfrac{1}{k} \varphi_0 = 0 \quad \text{and} \quad \lim_{j \to \infty} \left(\tfrac{1}{j} \varphi_k + \tfrac{1}{k} \varphi_0 \right) = \tfrac{1}{k} \varphi_0, \ k = 1, 2, \ldots.$$

However, there is no diagonal sequence $\{\frac{1}{j_k} \varphi_k + \frac{1}{k} \varphi_0\}$ that converges to zero in $(X, \mathcal{S})$. This shows that $\mathcal{S}$ does not have a countable neighborhood base at zero.

According to Theorem 3.6.1, (ii), the dual space $(X, \mathbf{S})^*$ consists of all linear functionals $F : X \to \mathbb{R}$ such that $F \restriction X_U$ is $\mathcal{T}$ continuous for each open set $U \Subset \Omega$. Alternatively, a linear functional $F : X \to \mathbb{R}$ belongs to $(X, \mathbf{S})^*$ if and only if

$$\|F\|_U = \sup\{F(v) : v \in X_U \text{ and } \|v\|_{L^\infty(\Omega;\mathbb{R}^m)} \leq 1\} < \infty$$

for each open set $U \Subset \Omega$.

Denote by S the collection of all sequences $s = \{s_j\}$ in $\mathbb{R}_+$ that converge to zero. Let $K_0 = \emptyset$, and for $s \in S$ and $v \in X$ define

$$p_s(v) := \sup_{j \in \mathbb{N}} \sup\left\{ s_j^{-1}|v(x)| : x \in \Omega - K_{j-1} \right\}.$$

It not difficult to verify that every p_s is a seminorm in X.

Claim. The topology $\mathbf{S}$ in X is defined by the family of seminorms $\{p_s : s \in S\}$.

Proof. Denote by $\mathcal{U}$ the topology in X defined by the seminorms p_s. Given $s \in S$, the set $V = \{v \in X : p_s(v) < 1\}$ is absorbing, symmetric, and convex. If $j \in \mathbb{N}$ and $t = \min\{s_1, \ldots, s_j\}$, then

$$V \cap X_j = X_j \cap \{v \in X : \|v\|_{L^\infty(\Omega;\mathbb{R}^m)} < t\}.$$

Thus $V \in \mathbf{S}$. Conversely, choose an $\mathbf{S}$ neighborhood W of zero and find $0 < s_1 < 1$ so that every $v \in X$ with spt $v \subset K_1$ and $\|v\|_{L^\infty(\Omega;\mathbb{R}^m)} < s_1$ is contained in W. Next find $0 < s_2 < s_1/2$ so that every $v \in X$ with spt $v \subset K_2$ and $\|v\|_{L^\infty(\Omega;\mathbb{R}^m)} < s_2$ is contained in W. Proceeding recursively, construct a decreasing sequence $s = \{s_j\}$ in $\mathbb{R}_+$ such that $\lim s_j = 0$, and every $v \in X$ with spt $v \subset K_j$ and $\|v\|_{L^\infty(\Omega;\mathbb{R}^m)} < s_j$ is contained in W. It follows that $\{v \in X : p_s(v) < 1\} \subset W$, and we conclude $\mathbf{S} = \mathcal{U}$.

If Y is a subspace of X, let $Y_U = X_U \cap Y$ for each $U \in \mathcal{O}$. In view of the claim, it is clear that the topology $\mathbf{S}_Y = \operatorname{inj lim}_{U \in \mathcal{O}}(\mathcal{T} \llcorner Y_U)$ in Y is defined by the family of seminorms $\{p_s \restriction Y : s \in S\}$. Thus $\mathbf{S}_Y = \mathbf{S} \llcorner Y$.

Example 3.6.5. Let Ω and $\mathcal{O}$ be as in Example 3.6.4. For $k \in \mathbb{N}$, let

$$p_k(\varphi) := \max_{|\alpha| \leq k} \|D^\alpha \varphi\|_{L^\infty(\Omega)}$$

where α is a multi-index. Each p_k is a norm in $\mathcal{D}(\Omega)$, and we denote by $\mathcal{T}$ the topology in $\mathcal{D}(\Omega)$ defined by the family $\{p_k : k \in \mathbb{N}\}$; see Section 1.2. Letting

$$\mathbf{S} = \operatorname{inj lim}_{U \in \mathcal{O}} \left[\mathcal{T} \llcorner \mathcal{D}(U) \right],$$

it follows from Theorem 3.6.1 that a linear functional $L : \mathcal{D}(\Omega) \to \mathbb{R}$ is a distribution if and only if it is $\mathbf{S}$ continuous. Arguing as in Example 3.6.4, we can show that the space $(\mathcal{D}(\Omega), \mathbf{S})$ is sequentially complete but not metrizable. The seminorms which define the topology $\mathbf{S}$ in $\mathcal{D}(\Omega)$ are similar in spirit to those described in Example 3.6.4, but technically more complicated [27, Section 6.3].

Part 2

Sets of finite perimeter

Chapter 4

Perimeter

Our goal is to extend the divergence theorem from dyadic figures to more general sets, called the sets of finite perimeter. These sets have two equivalent definitions: geometric, which is intuitive but difficult to work with, and analytic, which is effective but nonintuitive. Both definitions are essential, as they complement each other. In this chapter, we introduce the geometric definition of perimeter, and derive elementary properties of sets whose perimeter is finite. Some of these properties will motivate the analytic concept of variation presented in the next chapter.

4.1. Measure-theoretic concepts

We wish to define the flux of a vector field v from a set $E \subset \mathbb{R}^n$ so that it resembles the flux of v from a figure, and satisfies the divergence theorem when v and E are sufficiently regular. To this end, we need a "boundary" B and a "unit exterior normal" ν of a measurable set E such that the equality

$$\int_E \operatorname{div} v \, d\mathcal{L}^n = \int_B v \cdot \nu \, d\mathcal{H}^{n-1}$$

holds under suitable assumptions. As the left side of this equality depends only on the equivalence class of E, so must the right side. Thus ignoring temporarily the issue of "exterior normal", we will seek a definition of the "boundary" B of E which depends only on the equivalence class of E. Such a task cannot be accomplished by topological means only — in particular, we cannot let $B = \partial E$.

Fortunately, there is a suitable measure-theoretic analogue of the topological boundary. For any $E \subset \mathbb{R}^n$, the sets

$$\operatorname{ext}_* E := \left\{ x \in \mathbb{R}^n : \lim_{r \to 0} \frac{|E \cap B(x,r)|}{|B(x,r)|} = 0 \right\},$$

$$\operatorname{int}_* E := \operatorname{ext}_*(\mathbb{R}^n - E), \quad \operatorname{cl}_* E := \mathbb{R}^n - \operatorname{ext}_* E$$

$$\partial_* E := \operatorname{cl}_* E - \operatorname{int}_* E,$$

depend only on the equivalence class of E. We call them, respectively, the *essential exterior*, *essential interior*, *essential closure*, and *essential boundary* of E. A direct verification shows

$$\text{int}_* E \subset \text{cl}_* E, \partial\ {}_* E = \text{cl}_* E \cap \text{cl}_*(\mathbb{R}^n - E) = \partial_*(\mathbb{R}^n - E),$$

and from $\text{ext}_*(A \cup B) = \text{ext}_* A \cap \text{ext}_* B$, we obtain

$$\text{int}_*(A \cap B) = \text{int}_* A \cap \text{int}_* B \quad \text{and} \quad \text{cl}_*(A \cup B) = \text{cl}_* A \cup \text{cl}_* B.$$

The relationship to the corresponding topological concepts is given by

$$\text{int}\, E \subset \text{int}_* E, \quad \text{cl}_* E \subset \text{cl}\, E, \quad \text{and} \quad \partial_* E \subset \partial E.$$

Moreover, $\partial_* E = \partial E$ if and only if $\text{cl}_* E = \text{cl}\, E$ and $\text{int}_* E = \text{int}\, E$. As the converse is obvious, assume $\partial_* E = \partial E$ and observe that

$$\text{cl}\, E = \text{int}\, E \cup \partial E \subset \text{int}_* E \cup \partial_* E = \text{cl}_* E \subset \text{cl}\, E$$

implies $\text{cl}_* E = \text{cl}\, E$, and hence $\text{int}_* E = \text{int}\, E$. The obvious inclusion

$$\text{int}_* E \subset \left\{ x \in \mathbb{R}^n : \lim_{r \to 0} \frac{|E \cap B(x,r)|}{|B(x,r)|} = 1 \right\}$$

becomes equality when E is a measurable set.

For $x \in \mathbb{R}^n$, let $x' := \pi_n(x)$ and $x_n := x \cdot e_n$. Given $x \in \mathbb{R}^n$ and positive numbers r and h, we define an *open cylinder*

$$C(x; r, h) := \pi_n\big[U(x,r)\big] \times (x_n - h, x_n + h).$$

Definition 4.1.1. An open set $\Omega \subset \mathbb{R}^n$ is called a *Lipschitz domain* if for each point $x \in \partial\Omega$ there are a cylinder $C(x; r, h)$, a function $g \in Lip(\mathbb{R}^{n-1})$, and a rotation ϕ of $\mathbb{R}^n$ about x such that

$$\phi(\Omega) \cap C(x; r, h) = \big\{ y \in C(x; r, h) : g(y') < y_n \big\}.$$

If $\Omega \subset \mathbb{R}^n$ is a Lipschitz domain, then so is $\mathbb{R}^n - \text{cl}\,\Omega$. The interiors of figures are Lipschitz domains, and it follows from [66, Theorem 1.5.1] or [29, Section 6.3, Theorem 1] that so are open convex subsets of $\mathbb{R}^n$.

Proposition 4.1.2. *If Ω is a Lipschitz domain, then $\partial_*\Omega = \partial\Omega$.*

PROOF. Choose $x \in \partial\Omega$, and let $y_n := y \cdot e_n$ and $y' := \pi_n(y)$ for each $y \in \mathbb{R}^n$. As both $\partial\Omega$ and $\partial_*\Omega$ are invariant with respect to translations and rotations, we may assume that $x = 0$, and that there are a cylinder $C = C(0; r, h)$ and a Lipschitz function $g : \mathbb{R}^{n-1} \to \mathbb{R}$ such that

$$\Omega \cap C = \big\{ y \in C : g(y') < y_n \big\}.$$

Observe $g(0) = 0$, and define cones $C_\pm := \{y \in \mathbb{R}^n : (\mathrm{Lip}\, g)|y'| \leq \pm y_n\}$. Clearly $C_+ \cap C \subset \Omega \cap C$ and $C_- \cap C \cap \Omega = \emptyset$. Since for each $s > 0$,

$$\frac{|C_\pm \cap B(0, s)|}{|B(0, s)|} = c$$

where $0 < c \leq 1/2$, we see that $x \in \partial_* \Omega$. $\square$

4.2. Essential boundary

We establish some elementary facts about the essential boundaries of arbitrary subsets of $\mathbb{R}^n$.

Observation 4.2.1. *If $A, B \subset \mathbb{R}^n$, then*

$$\partial_*(A \cup B) \cup \partial_*(A \cap B) \cup \partial_*(A - B) \subset \partial_* A \cup \partial_* B.$$

PROOF. By a direct calculation,

$$\partial_*(A \cup B) = \mathrm{cl}_*(A \cup B) - \mathrm{int}_*(A \cup B) \subset \mathrm{cl}_* A \cup \mathrm{cl}_* B - \mathrm{int}_* A \cup \mathrm{int}_* B$$
$$\subset (\mathrm{cl}_* A - \mathrm{int}_* A) \cup (\mathrm{cl}_* B - \mathrm{int}_* B) = \partial_* A \cup \partial_* B,$$
$$\partial_*(A \cap B) = \mathrm{cl}_*(A \cap B) - \mathrm{int}_*(A \cap B) \subset \mathrm{cl}_* A \cap \mathrm{cl}_* B - \mathrm{int}_* A \cap \mathrm{int}_* B$$
$$\subset (\mathrm{cl}_* A - \mathrm{int}_* A) \cup (\mathrm{cl}_* B - \mathrm{int}_* B) = \partial_* A \cup \partial_* B.$$

Since $\partial_*(A - B) = \partial_*[A \cap (\mathbb{R}^n - B)]$ and $\partial_*(\mathbb{R}^n - B) = \partial_* B$, the observation follows. $\square$

Proposition 4.2.2. *If $\Omega \subset \mathbb{R}^n$ is an open set, then for each $E \subset \subset \mathbb{R}^n$,*

$$\mathrm{int}_* E \cap \Omega = \mathrm{int}_*(E \cap \Omega) \cap \Omega, \quad \mathrm{cl}_* E \cap \Omega = \mathrm{cl}_*(E \cap \Omega) \cap \Omega,$$
$$\partial_* E \cap \Omega = \partial_*(E \cap \Omega) \cap \Omega.$$

PROOF. Since a direct verification shows

$$\mathrm{int}_* E \cap \Omega \subset \mathrm{int}_* E \cap \mathrm{int}_* \Omega = \mathrm{int}_*(E \cap \Omega) \subset \mathrm{int}_* E,$$
$$\mathrm{cl}_* E \cap \Omega \subset \mathrm{cl}_*(E \cap \Omega) \subset \mathrm{cl}_* E,$$

we obtain the first two equalities by intersecting with Ω. The equality for essential boundaries follows. $\square$

The *shape* of a bounded set $C \subset \mathbb{R}^n$ is the number

$$s(C) := \begin{cases} \frac{|C|}{d(C)^n} & \text{if } d(C) > 0, \\ 0 & \text{otherwise.} \end{cases}$$

From the isodiametric inequality (1.4.2) we obtain $s(C) \leq \alpha(n)/2^n$. The bound is attained when C is a ball.

Lemma 4.2.3. *Let $\{C_k\}$ be a sequence of subsets of $\mathbb{R}^n$ and $x \in \mathbb{R}^n$. If $\lim d(C_k \cup \{x\}) = 0$ and $\inf s(C_k \cup \{x\}) > 0$, then for each $E \subset \mathbb{R}^n$,*

$$\lim \frac{|E \cap C_k|}{|C_k|} = \begin{cases} 0 & \text{if } x \in \text{ext}_* E, \\ 1 & \text{if } x \in \text{int}_* E. \end{cases}$$

PROOF. Let $d_k = d(C_k \cup \{x\})$ and $B_k = B(x, d_k)$. Observe $C_k \subset B_k$ and $\inf(|C_k|/|B_k|) \geq \inf s(C_k \cup \{x\})/\alpha(n) > 0$. If $x \in \text{ext}_* E$, then

$$0 = \lim \frac{|E \cap B_k|}{|B_k|} \geq \limsup \left(\frac{|E \cap C_k|}{|C_k|} \cdot \frac{|C_k|}{|B_k|} \right)$$

$$\geq \limsup \frac{|E \cap C_k|}{|C_k|} \inf \frac{|C_k|}{|B_k|} \geq 0$$

and hence $\lim(|E \cap C_k|/|C_k|) = 0$. If x belongs to $\text{int}_* E = \text{ext}_*(\mathbb{R}^n - E)$, then by the first part of the proof,

$$0 = \lim \frac{|(\mathbb{R}^m - E) \cap C_k|}{|C_k|} \geq \limsup \frac{|C_k| - |E \cap C_k|}{|C_k|}$$

$$= 1 - \liminf \frac{|E \cap C_k|}{|C_k|} \geq 0. \qquad \square$$

Proposition 4.2.4. *If $A, B \subset \mathbb{R}^n$, then $\text{cl}_* A \cap \text{int}_* B \subset \text{cl}_*(A \cap B)$.*

PROOF. Choose $x \in \text{cl}_* A \cap \text{int}_* B$. There is a sequence $\{r_k\}$ in $\mathbb{R}_+$ such that $\lim r_k = 0$ and

$$\lim \frac{|A \cap B(x, r_k)|}{|B(x, r_k)|} = a > 0.$$

Let $B_k := B(x, r_k)$, and observe that

$$s[(A \cap B_k) \cup \{x\}] \geq \frac{|A \cap B_k|}{(2r_k)^n} = \frac{\alpha(n)}{2^n} \cdot \frac{|A \cap B_k|}{|B_k|} > \frac{\alpha(n)}{2^n} \cdot \frac{a}{2}$$

for all sufficiently large k. Applying Lemma 4.2.3, we obtain

$$\lim \frac{|(A \cap B) \cap B_k|}{|B_k|} = \lim \frac{|B \cap (A \cap B_k)|}{|A \cap B_k|} \cdot \lim \frac{|A \cap B_k|}{|B_k|} = a.$$

Consequently $x \in \text{cl}_*(A \cap B)$. $\qquad \square$

Corollary 4.2.5. *If A and B are measurable sets, then*

$$\partial_*(A \cap B) \supset (\partial_* A \cap \text{int}_* B) \cup (\text{int}_* A \cap \partial_* B) \quad \text{and}$$

$$\partial_*(A \cap B) \subset (\partial_* A \cap \text{int}_* B) \cup (\text{int}_* A \cap \partial_* B) \cup (\partial_* A \cap \partial_* B).$$

PROOF. Proposition 4.2.4 implies

$$\partial_* A \cap \mathrm{int}_* B = (\mathrm{cl}_* A - \mathrm{int}_* A) \cap \mathrm{int}_* B$$
$$= \mathrm{cl}_* A \cap \mathrm{int}_* B - \mathrm{int}_* (A \cap B)$$
$$\subset \mathrm{cl}_* (A \cap B) - \mathrm{int}_* (A \cap B) = \partial_* (A \cap B),$$

and by symmetry, $\partial_* B \cap \mathrm{int}_* A \subset \partial_* (A \cap B)$. On the other hand,

$$\partial_* (A \cap B) \subset \mathrm{cl}_* (A \cap B) \subset \mathrm{cl}_* A \cap \mathrm{cl}_* B$$
$$= (\mathrm{int}_* A \cup \partial_* A) \cap (\mathrm{int}_* B \cup \partial_* B)$$
$$= (\partial_* A \cap \mathrm{int}_* B) \cup (\mathrm{int}_* A \cap \partial_* B) \cup (\partial_* A \cap \partial_* B),$$

since $\mathrm{int}_* (A \cap B)$ and $\partial_* (A \cap B)$ are disjoint sets.   □

4.3. Vitali's covering theorem

We prove an important combinatorial result which has many applications. In particular, it will allow us to proof some deeper properties of essential boundary. Throughout the remainder of this book, we adhere to the following convention.

Convention 4.3.1. When no attributes are added, a *ball* is always a closed ball $B(x, r)$ where $x \in \mathbb{R}^n$ and $r \in \mathbb{R}_+$. Given a ball $B = B(x, r)$, we let $B^\bullet := B(x, 5r)$.

Theorem 4.3.2 (Vitali). *Let $\mathcal{B}$ be a family of balls. If $d(B) \leq d < \infty$ for each $B \in \mathcal{B}$, then there is a disjoint family $\mathcal{C} \subset \mathcal{B}$ such that*

$$\bigcup \mathcal{B} \subset \bigcup \{C^\bullet : C \in \mathcal{C}\}.$$

PROOF. The family $\mathcal{B}$ is the union of disjoint subfamilies

$$\mathcal{B}_i := \{B \in \mathcal{B} : 2^{-i} d < d(B) \leq 2^{-i+1} d\}, \quad i = 1, 2, \ldots.$$

By Zorn's lemma [43, Chapter 0, Theorem 25], the family $\mathcal{B}_1$ contains a maximal disjoint subfamily $\mathcal{C}_1$. We use recursion to construct families $\mathcal{C}_i$, $i = 2, 3, \ldots$, so that $\mathcal{C}_i$ is a maximal disjoint subfamily of

$$\left\{ B \in \mathcal{B}_i : B \subset \mathbb{R}^n - \bigcup_{j=1}^{i-1} \left(\bigcup \mathcal{C}_i \right) = \emptyset \right\}.$$

Observe that $\mathcal{C} := \bigcup_{i=1}^{\infty} \mathcal{C}_i$ is a disjoint subfamily of $\mathcal{B}$, and choose a ball $B \in \mathcal{B}$. As $B \in \mathcal{B}_k$ for an integer $k \geq 1$, the maximality of $\mathcal{C}_k$ implies that B meets a ball $C \in \bigcup_{i=1}^{k} \mathcal{C}_i$. The inclusion $B \subset C^\bullet$ follows from the inequality $d(B) \leq 2^{-k+1} d < 2d(C)$.   □

Remark 4.3.3. As the family $\mathcal{B} = \{B(0,k) : k \in \mathbb{N}\}$ contains no non-empty disjoint subfamily, it is essential to assume that $\{d(B) : B \in \mathcal{B}\}$ is a bounded set. In the proof of Vitali's theorem, we used Zorn's lemma merely for convenience. A slightly longer constructive proof is available.

Theorem 4.3.4. *Let $\Omega \subset \mathbb{R}^n$ be an open set. If $f \in L^1_{\mathrm{loc}}(\Omega)$, then*

$$\lim_{r \to 0} \frac{1}{|B(x,r)|} \int_{B(x,r)} f(y)\,dy = f(x)$$

for almost all $x \in \Omega$.

PROOF. With no loss of generality, we may assume that Ω is bounded and f is a real-valued function. Let $F(B) := \int_B f$ for every ball $B \subset \Omega$, and denote by N the set of all $x \in \Omega$ at which the limit

$$\lim_{r \to 0} \frac{F[B(x,r)]}{|B(x,r)|}$$

either does not exist, or differs from $f(x)$. Given $x \in N$, find $\gamma_x > 0$ so that for each $\eta > 0$, there is a positive $r < \eta$ such that

$$B(x,r) \subset \Omega \quad \text{and} \quad \left| F[B(x,r)] - f(x)|B(x,r)| \right| \geq \gamma_x |B(x,r)|. \quad (*)$$

It suffices to show that each $N_k := \{x \in N : \gamma_x > 1/k\}$ is a negligible set. Fix $k \in \mathbb{N}$ and choose $\varepsilon > 0$. There is $\delta : \Omega \to \mathbb{R}_+$ such that

$$\sum_{i=1}^p \left| f(x_i)|B_i| - F(B_i) \right| < \frac{\varepsilon}{5^n k} \quad (**)$$

whenever $B_i \subset \Omega$ are disjoint balls, $x_i \in B_i$, and $d(B_i) < \delta(x_i)$ for $i = 1, \ldots, p$; see Henstock's lemma. For every $x \in N_k$ select a positive $r_x < \delta(x)$ so that $(*)$ holds for $B(x, r_x)$. By Vitali's theorem there are disjoint balls $B_i = B(x_i, r_{x_i})$ satisfying $N_k \subset \bigcup_i B_i^{\bullet}$. Combining $(*)$ and $(**)$, we obtain

$$|N_k| \leq \sum_i |B_i^{\bullet}| = 5^n \sum_i |B_i| \leq 5^n k \sum_i \left| f(x_i)|B_i| - F(B_i) \right| < \varepsilon \,,$$

and the theorem follows from the arbitrariness of ε. $\qquad \square$

4.4. Density

Let $E \subset \mathbb{R}^n$ and $x \in \mathbb{R}^n$. If the limit

$$\Theta(E, x) := \lim_{r \to 0} \frac{|E \cap B(x,r)|}{|B(x,r)|}$$

exists, it is called the *density* of E at x. Clearly,

$$\mathrm{ext}_* E = \{x \in \mathbb{R}^n : \Theta(E, x) = 0\}.$$

Observation 4.4.1. *Let $E \subset \mathbb{R}^n$ and $\theta \in \mathbb{R}$. The set*

$$E(\theta) := \{x \in \mathbb{R}^n : \Theta\,(E, x) = \theta\}$$

is Borel, and so are the sets $\mathrm{ext}_* E$, $\mathrm{int}_* E$, $\mathrm{cl}_* E$, *and* $\partial_* E$.

PROOF. Enumerate all positive rationals as $r_1, r_2, \ldots$, and note that

$$f_r : x \mapsto \frac{|E \cap B(x, r)|}{|B(x, r)|} : \mathbb{R}^n \to \mathbb{R}$$

is a continuous function for every $r > 0$. Since the functions

$$f^* := \limsup_{r \to 0} f_r = \lim_{k \to \infty} \sup\{f_{r_i} : r_i < 1/k\},$$

$$f_* := \liminf_{r \to 0} f_r = \lim_{k \to \infty} \inf\{f_{r_i} : r_i < 1/k\}$$

are Borel, so is the set $E(\theta) = \{f^* \leq \theta\} \cap \{f_* \geq \theta\}$. $\square$

Theorem 4.4.2. *If $E \subset \mathbb{R}^n$ is a measurable set, then*

$$\Theta(E, x) = \chi_E(x)$$

for almost all $x \in \mathbb{R}^n$, and the sets E, $\mathrm{int}_* E$, and $\mathrm{cl}_* E$ are equivalent.

PROOF. Since $\chi_E \in L^1_{\mathrm{loc}}(\mathbb{R}^n)$ and

$$\Theta(E, x) = \lim_{r \to 0} \frac{1}{|B(x, r)|} \int_{B(x, r)} \chi_E(y)\, dy,$$

$\Theta(E, x) = \chi_E(x)$ for almost all $x \in \mathbb{R}^n$ by Theorem 4.3.4. This implies $\mathrm{ext}_* E \sim \mathbb{R}^n - E$, and consequently $\mathrm{int}_* E \sim E$ and $\mathrm{cl}_* E \sim E$. $\square$

Lemma 4.4.3. *Let μ be a Radon measure in an open set $\Omega \subset \mathbb{R}^n$, and let $E \subset \Omega$. There is a Borel set $B \subset \Omega$ such that $E \subset B$ and*

$$\mu(E \cap C) = \mu(B \cap C)$$

for each μ measurable set $C \subset \Omega$.

PROOF. There is an increasing sequence $\{K_i\}$ of compact sets whose union equals Ω. Let $E_i := E \cap K_i$, and find a Borel set $A_i \subset \Omega$ so that $E_i \subset A_i \subset K_i$ and $\mu(A_i) = \mu(E_i)$. Letting $B_i := \bigcap_{j=i}^{\infty} A_j$, we obtain $E_i \subset B_i \subset K_i$. If $C \subset \Omega$ is a μ measurable set, then

$$\mu(E_i) = \mu(E_i \cap C) + \mu(E_i - C) \leq \mu(B_i \cap C) + \mu(B_i - C)$$
$$= \mu(B_i) \leq \mu(A_i) = \mu(E_i) < \infty.$$

This implies $\mu(E_i \cap C) = \mu(B_i \cap C)$. The union $B := \bigcup_{i=1}^{\infty} B_i$ is a Borel set containing E. Moreover $\mu(E \cap C) = \mu(B \cap C)$, since $\{E_i\}$ and $\{B_i\}$ are increasing sequences. $\square$

Corollary 4.4.4. *If $E \subset \mathbb{R}^n$, then $\Theta(E, x) = 1$ for almost all $x \in E$.*

PROOF. Let B be as in Lemma 4.4.3 applied to $\Omega = \mathbb{R}^n$ and $\mu = \mathcal{L}^n$. Then $E \subset B$, and Theorem 4.4.2 implies that $\Theta(E, x) = \Theta(B, x) = 1$ for almost all $x \in B$. $\qquad\square$

Theorem 4.4.5. *A set $E \subset \mathbb{R}^n$ is measurable if and only if $|\partial_* E| = 0$.*

PROOF. Corollary 4.4.4 implies $|E - \mathrm{cl}_* E| = 0$ and

$$|\mathrm{int}_* E - E| = \left|(\mathbb{R}^n - E) - \mathrm{cl}_*(\mathbb{R}^n - E)\right| = 0.$$

If $|\partial E| = 0$ then $\mathrm{cl}_* E \sim \mathrm{int}_* E$. Thus $E \sim \mathrm{cl}_* E$ is measurable by Observation 4.4.1. The converse follows from Theorem 4.4.2. $\qquad\square$

4.5. Definition of perimeter

Definition 4.5.1. The *perimeter* of $E \subset \mathbb{R}^n$ in an open set $\Omega \subset \mathbb{R}^n$ is the extended real number

$$\mathbf{P}(E, \Omega) := \mathcal{H}^{n-1}(\partial_* E \cap \Omega).$$

The perimeter $\mathbf{P}(E, \mathbb{R}^n)$ of E in $\mathbb{R}^n$ is called the *perimeter* of E, denoted by $\mathbf{P}(E)$.

Let $\Omega \subset \mathbb{R}^n$ be an open set. A set $E \subset \Omega$ is called

    (i) a set of *finite perimeter* in Ω if $|E| + \mathbf{P}(E, \Omega) < \infty$,

    (ii) a set of *locally finite perimeter* in Ω if $\mathbf{P}(E, U) < \infty$ for each open set $U \Subset \Omega$.

The family of all sets $E \subset \Omega$ of finite, or locally finite, perimeter in Ω is denoted by $\mathcal{P}(\Omega)$, or $\mathcal{P}_{\mathrm{loc}}(\Omega)$, respectively. Clearly

$$\mathcal{P}(\Omega) \subset \mathcal{P}_{\mathrm{loc}}(\Omega),$$

and if $E \subset \mathbb{R}^n$ belongs to $\mathcal{P}_{\mathrm{loc}}(\Omega)$, then so does $\mathbb{R}^n - E$.

    Note that in the above terminology a set $E \subset \Omega$ such that $|E| = \infty$ and $\mathbf{P}(E, \Omega) < \infty$ is called a set of locally finite, rather than finite, perimeter in Ω.

    If $\Omega \subset \mathbb{R}^n$ is an open set and $E \subset \mathbb{R}^n$, Proposition 4.2.2 implies

$$\mathbf{P}(E, \Omega) = \mathbf{P}(E \cap \Omega, \Omega) \leq \mathbf{P}(E \cap \Omega). \qquad (4.5.1)$$

It follows that when studying perimeters in an open set $\Omega \subset \mathbb{R}^n$, we may restrict our attention to subsets of Ω. The next proposition is a direct consequence of Observation 4.2.1.

Proposition 4.5.2. *If $\Omega \subset \mathbb{R}^n$ is open and $A, B \subset \Omega$, then*

$$\max\{\mathbf{P}(A \cup B, \Omega), \mathbf{P}(A \cap B, \Omega), \mathbf{P}(A - B, \Omega)\} \leq \mathbf{P}(A, \Omega) + \mathbf{P}(B, \Omega).$$

Proposition 4.5.3. *Let $\Omega \subset \mathbb{R}^n$ be an open set. Each set $E \subset \Omega$ that belongs to $\mathcal{P}_{\mathrm{loc}}(\Omega)$ is measurable.*

PROOF. Suppose that $E \subset \Omega$ is not measurable. As Ω is the union of countably many open balls $U \Subset \Omega$, there is an open ball $V \Subset \Omega$ for which $E \cap V$ is not a measurable set. Proposition 4.2.2 implies

$$\partial_* E \cap V = \partial_* (E \cap V) \cap V = \partial_* (E \cap V) - \partial V,$$

since $\partial_* (E \cap V) \subset \operatorname{cl} V$. According to Theorem 4.4.5,

$$|\partial_* E \cap V| = |\partial_* (E \cap V)| > 0$$

and $\mathbf{P}(E, V) = \infty$ by Proposition 1.4.3. $\qquad\square$

In terms of the measure $\mathcal{H}^{n-1}$, the essential boundary $\partial_* E$ of a nonmeasurable set $E \subset \mathbb{R}^n$ is enormous: it follows from [30, Thorem 5.6] and [59, Theorems 59] that $\partial_* E$ contains uncountably many disjoint subsets whose $\mathcal{H}^{n-1}$ measure is infinite.

Proposition 4.5.4. *Let $\Omega \subset \mathbb{R}^n$ be an open set. Then $E \subset \Omega$ belongs to $\mathcal{P}_{\mathrm{loc}}(\Omega)$ if and only if $\mathbf{P}(E \cap A) < \infty$ for each $A \Subset \Omega$ with $\mathbf{P}(A) < \infty$.*

PROOF. Assume $E \in \mathcal{P}_{\mathrm{loc}}(\Omega)$, and choose $A \Subset \Omega$ with $\mathbf{P}(A) < \infty$. There is a figure whose interior U satisfies $A \Subset U \Subset \Omega$. By Proposition 4.5.2 and Corollary 4.2.5,

$$\mathbf{P}(E \cap A) = \mathbf{P}(E \cap U \cap A) \leq \mathbf{P}(E \cap U) + \mathbf{P}(A)$$
$$\leq \mathbf{P}(E, U) + \mathbf{P}(U) + \mathbf{P}(A) < \infty.$$

Conversely, choose an open set $U \Subset \Omega$ and find a figure whose interior V satisfies $U \Subset V \Subset \Omega$. As inequality (4.5.1) and our assumption imply

$$\mathbf{P}(E, U) \leq \mathbf{P}(E, V) \leq \mathbf{P}(E \cap V) < \infty,$$

we see that $E \in \mathcal{P}_{\mathrm{loc}}(\Omega)$. $\qquad\square$

Lemma 4.5.5. *If μ is a Radon measure in $\mathbb{R}^n$, then each family $\mathcal{E}$ of μ nonoverlapping μ measurable sets of positive measure is countable.*

PROOF. As μ is Radon and $B(0, r)$ is compact, $\mu[B(0, r)] < \infty$ for every $r \in \mathbb{R}_+$. It follows that for $j, k = 1, 2, \ldots$, the family

$$\mathcal{E}_{j,k} = \left\{ E \in \mathcal{E} : \mu[E \cap B(0, j)] > 1/k \right\}$$

is finite. Thus $\mathcal{E} = \bigcup_{j,k=1}^{\infty} \mathcal{E}_{j,k}$ is a countable family. $\qquad\square$

Proposition 4.5.6. *Let $E \subset \mathbb{R}^n$. If $n \geq 2$, then for all but countably many balls $B \subset \mathbb{R}^n$, each pair*

$$\big[\partial_* (E \cap B), (\partial_* E \cap \operatorname{int} B) \cup (\operatorname{int}_* E \cap \partial B) \big],$$
$$\big[\partial_* (E - B), (\partial_* E - \operatorname{cl} B) \cup (\operatorname{int}_* E \cap \partial B) \big]$$

consists of $\mathcal{H}^{n-1}$ equivalent sets, and

$$\mathbf{P}(E \cap B) = \mathbf{P}(E, \operatorname{int} B) + \mathcal{H}^{n-1}(\operatorname{int}_* E \cap \partial B)$$
$$\mathbf{P}(E - B) = \mathbf{P}(E, \mathbb{R}^n - \operatorname{cl} B) + \mathcal{H}^{n-1}(\operatorname{int}_* E \cap \partial B).$$

PROOF. Let B be a ball with $\mathcal{H}^{n-1}(\partial B \cap \partial_* E) = 0$. Corollary 4.2.5 shows that the first pair consists of $\mathcal{H}^{n-1}$ equivalent sets. Replacing B by $\mathbb{R}^n - B$, we see that the same is true for the second pair. As $n \geq 2$, the boundaries of two distinct balls have $\mathcal{H}^{n-1}$ negligible intersection. The proposition follows from Lemma 4.5.5. $\square$

Remark 4.5.7. Proposition 4.5.6 holds in any dimension for concentric balls, as well as for concentric cubes.

Proposition 4.5.8. *A Lipschitz domain in $\mathbb{R}^n$ belongs to $\mathcal{P}_{\mathrm{loc}}(\mathbb{R}^n)$. It belongs to $\mathcal{P}(\mathbb{R}^n)$ whenever it is bounded.*

PROOF. If Ω is a Lipschitz domain, then $\partial_* \Omega = \partial \Omega$ by Proposition 4.1.2. According to the definition, $\partial \Omega$ is locally the graph G of a Lipschitz function g defined on a bounded open set $V \subset \mathbb{R}^{n-1}$. Thus

$$\phi : u \mapsto \big(u, g(u)\big) : V \to G$$

is a Lipschitz bijection, and inequality (1.5.1) implies $\mathcal{H}^{n-1}(G) < \infty$. Since G is a relatively open subset of $\partial \Omega$, we have $\mathcal{H}^{n-1}(K) < \infty$ for any compact set $K \subset \partial \Omega$. In particular, $\mathbf{P}(\Omega, U) < \infty$ for any $U \Subset \mathbb{R}^n$. If Ω is bounded, there is $U \Subset \mathbb{R}^n$ with $\operatorname{cl}\Omega \subset U$. $\square$

4.6. Line sections

Useful information about the perimeters of measurable sets is obtained by studying their intersections with lines. A *line* is a one-dimensional affine subspace of $\mathbb{R}^n$. Via translation and rotation, a line L can be identified with $\mathbb{R}$. This identification defines Lebesgue measure $\mathcal{L}^1$ in L, which is equal to the Hausdorff measure $\mathcal{H}^1$ restricted to subsets of L. For *distinct* points x and y in $\mathbb{R}^n$, the sets

$$(xy) := \big\{tx + (1-t)y : 0 < t < 1\big\} \quad \text{and} \quad [xy] := (xy) \cup \{x, y\}$$

are called, respectively, the *open* and *closed segments* determined by the two-point set $\{x, y\}$. Each segment is an *uncountable* set.

Lemma 4.6.1. *Let L be a line, and let X, Y, Z be disjoint sets whose union is equal to L. Suppose that Z is contained in $\operatorname{cl} X \cap \operatorname{cl} Y$ and contains no segment. If $z \in Z$ and $\delta > 0$, then there are $x \in X$ and $y \in Y$ such that $|x - y| < \delta$ and $z \in (xy)$.*

PROOF. We may assume $L = \mathbb{R}$. Choose $z \in Z$ and $\delta > 0$. The open interval $(z - \delta/2, z + \delta/2)$ contains points $x \in X$ and $y \in Y$, and no generality is lost by assuming $x < y$. If $y < z$, there is $u \in X \cup Y$ such that $z < u < z + \delta/2$, since $(z, z + \delta/2) \not\subset Z$. Either (yu) or (xu) is the desired segment. The case $z < x$ is argued similarly. $\qquad\square$

Lemma 4.6.2. *Let L be a line, and let X, Y, Z be disjoint sets whose union is equal to L. Suppose the following conditions hold:*

(a) *there are increasing sequences $\{X_k\}$ and $\{Y_k\}$ of closed subsets of L such that $X_k \cap Y_k = \emptyset$ for each $k \in \mathbb{N}$, and*

$$X \subset \bigcup\{X_k : k \in \mathbb{N}\} \quad \text{and} \quad Y \subset \bigcup\{Y_k : k \in \mathbb{N}\};$$

(b) *given $x \in X$ and $y \in Y$, the intersections*

$$(x\tilde{x}) \cap (X \cup Z) \quad \text{and} \quad (y\tilde{y}) \cap (Y \cup Z)$$

are nonempty for all $\tilde{x}, \tilde{y} \in L$ such that $\tilde{x} \neq x$ and $\tilde{y} \neq y$.

If $x \in X$ and $y \in Y$, then $(xy) \cap Z \neq \emptyset$.

PROOF. As we may assume $L = \mathbb{R}$, open segments are nonempty open intervals, and condition (b) states that each $x \in X$ is a two-sided cluster point of $X \cup Z$, and each $y \in Y$ is a two-sided cluster point of $Y \cup Z$. Choose $x_1 \in X$, $y_1 \in Y$ and, with no loss of generality, suppose $x_1 < y_1$. Working toward a contradiction, assume $(x_1, y_1) \cap Z = \emptyset$. Select $k_1 \in \mathbb{N}$ so that $x_1 \in X_{k_1}$ and $y_1 \in Y_{k_1}$, and define

$$a_1 := \sup\{x \in X_{k_1} : x < y_1\} \quad \text{and} \quad b_1 := \inf\{y \in Y_{k_1} : y > a_1\}.$$

Then $a_1 \in X_{k_1}$, $b_1 \in Y_{k_1}$, $x_1 \leq a_1 < b_1 \leq y_1$, and the intersection $(a_1, b_1) \cap (X_{k_1} \cup Y_{k_1})$ is empty. As a_1 and b_1 are two-sided cluster points of $X \cup Z$ and $Y \cup Z$, respectively, and $(a_1, b_1) \cap Z = \emptyset$, there are $x_2 \in X \cap (a_1, b_1)$ and $y_2 \in Y \cap (x_2, b_1)$. Thus we have

$$x_1 < x_2 < y_2 < y_1 \quad \text{and} \quad [x_2, y_2] \cap (X_{k_1} \cup Y_{k_1}) = \emptyset.$$

By recursion, we construct a strictly increasing sequence $\{k_i\}$ of positive integers, and points $x_i \in X_{k_i}$, $y_i \in Y_{k_i}$ such that for $i = 1, 2, \ldots$,

$$x_i < x_{i+1} < y_{i+1} < y_i \quad \text{and} \quad [x_{i+1}, y_{i+1}] \cap (X_{k_i} \cup Y_{k_i}) = \emptyset.$$

The nonempty intersection $\bigcap_{i=1}^{\infty}[x_i, y_i]$ is disjoint from $X \cup Y$ as well as from Z — a contradiction. $\qquad\square$

A *hyperplane* is an $(n-1)$-dimensional affine subspace of $\mathbb{R}^n$. To a hyperplane Π corresponds the orthogonal projection $\pi : \mathbb{R}^n \to \Pi$, and we let $L_u := \pi^{-1}(u)$ for each $u \in \Pi$. Via translation and rotation, Π can be identified with $\mathbb{R}^{n-1}$. This identification defines Lebesgue measure $\mathcal{L}^{n-1}$ in Π, which is equal to the Hausdorff measure $\mathcal{H}^{n-1}$ restricted to subsets of Π.

Theorem 4.6.3. *Let A be a measurable set, and let Π be a hyperplane. For $\mathcal{L}^{n-1}$ almost all $u \in \Pi$ the following conditions hold:*

(i) *if $z \in L_u \cap \partial_* A$ and $\delta > 0$, then there are $x \in L_u \cap \mathrm{ext}_* A$ and $y \in L_u \cap \mathrm{int}_* A$ such that $|x - y| < \delta$ and $z \in (xy)$;*

(ii) *if $x \in L_u \cap \mathrm{ext}_* A$ and $y \in L_u \cap \mathrm{int}_* A$, then $(xy) \cap \partial_* A \neq \emptyset$.*

PROOF. With no loss of generality we may assume that $\Pi = \Pi_n$. Let $\pi = \pi_n$, and $x' = x \cdot e_n$ for each $x \in \mathbb{R}^n$. Since $\mathbb{R}^n$ is the disjoint union of $\mathrm{ext}_* A$, $\mathrm{int}_* A$, and $\partial_* A$, the line L_u is the disjoint union of

$$X_u := L_u \cap \mathrm{ext}_* A, \quad Y_u := L_u \cap \mathrm{int}_* A, \quad \text{and} \quad Z_u := L_u \cap \partial_* A$$

for each $u \in \Pi$. Our proof follows those presented in [33, Theorem 4.5.11] and [29, Section 5.11, Theorem 1]: we denote by Π_0 the set of all $u \in \Pi$ such that X_u, Y_u, and Z_u satisfy the assumptions of Lemmas 4.6.1 and 4.6.2, and show that $\mathcal{L}^{n-1}(\Pi - \Pi_0) = 0$. We split the argument into four separate claims.

For $u \in \Pi$ and $r > 0$, let $B_{u,r} = \Pi \cap B(u, r)$. Denote by Π_E the set of all $u \in \Pi$ for which there exist $\delta_u > 0$ and $z_u \in L_u \cap \partial_* E$ such that

$$\{x \in L_u : 0 < |x - z_u| < \delta_u\} \subset \mathrm{ext}_* A.$$

Replacing $\mathrm{ext}_* A$ by $\mathrm{int}_* A$, a set Π_I is defined similarly. Let

$$\Pi_* := \{u \in \Pi : \mathcal{H}^1(L_u \cap \partial_* E) > 0\}$$

and $\Pi_1 := \Pi_E \cup \Pi_I \cup \Pi_*$. As $\mathcal{H}^1$ negligible sets contain no segments, the sets X_u, Y_u, and Z_u satisfy the assumptions of Lemma 4.6.1 for all $u \in \Pi - \Pi_1$.

Claim 1. $\mathcal{L}^{n-1}(\Pi_1) = 0.$

Proof. Fubini's theorem implies $\mathcal{L}^{n-1}(\Pi_*) = 0$, since $|\partial_* A| = 0$ by Theorem 4.4.5. Seeking a contradiction, we assume $\mathcal{L}^{n-1}(\Pi_E) > 0$. For each $(x, j) \in \mathbb{Q}^n \times \mathbb{N}$, denote by $\Pi_{x,j}$ the set of all $u \in \Pi_E$ such that

$$z_u \in U(x, 1/j) \quad \text{and} \quad B(x, 1/j) \subset U(z_u, \delta_u).$$

The set $N_{x,j} = \{z_u : u \in \Pi_{x,j}\}$ is negligible by Fubini's theorem, and

$$B(x, 1/j) \cap \pi^{-1}(\Pi_{x,j}) - N_{x,j} \subset \mathrm{ext}_* A. \tag{$*$}$$

We fix a pair $(x, j) \in \mathbb{Q} \times \mathbb{N}$ with $\mathcal{L}^{n-1}(\Pi_{x,j}) > 0$; such a pair exists, because

$$\Pi_E = \bigcup \{\Pi_{x,j} : (x, j) \in \mathbb{Q}^n \times \mathbb{N}\}.$$

Applying Corollary 4.4.4 to $\Pi = \mathbb{R}^{n-1}$, find a point $w \in \Pi_{x,j}$ with

$$\lim_{r \to 0} \frac{\mathcal{L}^{n-1}(B_{w,r} \cap \Pi_{x,j})}{\mathcal{L}^{n-1}(B_{w,r})} = 1.$$

Let $r > 0$ be so small that the closed cylinder

$$C_r := \{y \in \pi^{-1}(B_{w,r}) : |y' - z'_w| \leq r\}$$

is contained in $B(x, 1/j)$. Inclusion $(*)$ shows that $C_r \cap \pi^{-1}(N_{x,j})$ is contained in $\text{ext}_* A$, up to the negligible set $N_{x,j}$. Hence

$$2r\mathcal{L}^{n-1}(B_{w,r} \cap \Pi_{x,j}) = |C_r \cap \pi^{-1}(\Pi_{x,j})| \leq |C_r \cap \text{ext}_* A|.$$

As $|C_r| = 2r\mathcal{L}^{n-1}(B_{w,r})$ for each $r > 0$, we obtain

$$\lim_{r \to 0} \frac{|C_r \cap \text{ext}_* A|}{|C_r|} \geq \lim_{r \to 0} \frac{\mathcal{L}^{n-1}(B_{w,r} \cap \Pi_{x,j})}{\mathcal{L}^{n-1}(B_{w,r})} = 1.$$

The previous inequality and inclusion $B(z_w, r) \subset C_r$ yield

$$0 \geq 1 - \lim \frac{|C_r \cap \text{ext}_* A|}{|C_r|} = \lim \frac{|C_r - \text{ext}_* A|}{|C_r|}$$

$$\geq \lim_{r \to 0} \frac{|B(z_w, r) - \text{ext}_* A|}{|C_r|} = \frac{\alpha(n)}{2\alpha(n-1)} \lim_{r \to 0} \frac{|B(z_w, r) - \text{ext}_* A|}{|B(z_w, r)|}.$$

We conclude that contrary to our choice, z_w belongs to

$$\text{ext}_*(\mathbb{R}^n - \text{ext}_* A) = \text{ext}_*(\text{cl}_* A) = \text{ext}_* A \subset \mathbb{R}^n - \partial_* A.$$

This and a similar argument show that $\mathcal{L}^{n-1}(\Pi_E \cup \Pi_I) = 0$.

For each $k \in \mathbb{N}$, define closed sets

$$E_k := \left\{ x \in \mathbb{R}^n : \frac{|\text{int}_* A \cap B(x, r)|}{r^n} \leq \frac{\alpha(n-1)}{3^{n+1}} \text{ for all } 0 < r < \frac{3}{k} \right\},$$

$$I_k := \left\{ x \in \mathbb{R}^n : \frac{|\text{ext}_* A \cap B(x, r)|}{r^n} \leq \frac{\alpha(n-1)}{3^{n+1}} \text{ for all } 0 < r < \frac{3}{k} \right\},$$

and observe that the sequences $\{E_k\}$ and $\{I_k\}$ are increasing and

$$\text{ext}_* A = \text{ext}_*(\text{int}_* A) \subset \bigcup\{E_k : k \in \mathbb{N}\},$$

$$\text{int}_* A = \text{ext}_*(\text{ext}_* A) \subset \bigcup\{I_k : k \in \mathbb{N}\}. \tag{$**$}$$

Claim 2. $E_k \cap I_k = \emptyset$ for each $k \in \mathbb{N}$.

Proof. A direct calculation shows that assuming $E_k \cap I_k \neq \emptyset$ leads to an inequality $\alpha(n) \leq 2 \cdot 3^{-n-1}\alpha(n-1)$. This is impossible, since

$$\alpha(n) = |B(0, 1)| = 2 \int_{B(0,1) \cap \Pi} \sqrt{1 - |u|^2}\, du$$

$$> 2 \int_{B(0,1/3) \cap \Pi} \sqrt{1 - |u|^2}\, du > 12 \cdot 3^{-n-1}\alpha(n-1).$$

For $u \in \Pi$ and $k \in \mathbb{N}$, let $X_{u,k} := L_u \cap E_k$ and $Y_{u,k} := L_u \cap I_k$. It follows from $(**)$ and Claim 2 that for each $u \in \Pi$, the sequences $\{X_{u,k}\}$ and $\{Y_{u,k}\}$ fulfill condition (a) of Lemma 4.6.2 with respect to the sets X_u and Y_u, respectively. Denote by $S_\pm(x, j)$ the open segment determined by points x and $x \pm (3/j)e_n$. For $k, j \in \mathbb{N}$, define sets

$$E_{k,j}^{\pm} := \{x \in E_k : S_{\pm}(x,j) \subset \text{int}_* A\},$$

$$I_{k,j}^{\pm} := \{x \in I_k : S_{\pm}(x,j) \subset \text{ext}_* A\},$$

and let $E_0^{\pm} := \bigcup_{k,j \in \mathbb{N}} E_{k,j}^{\pm}$, $I_0^{\pm} := \bigcup_{k,j \in \mathbb{N}} I_{k,j}^{\pm}$, and

$$\Pi_2 = \pi(E_0^+ \cup E_0^- \cup I_0^+ \cup I_0^-).$$

Claim 3. For each $u \in \Pi - \Pi_2$, the sets X_u, Y_u, and Z_u satisfy condition (b) of Lemma 4.6.2.

Proof. Select $u \in \Pi - \Pi_2$. If $x \in X_u$, there is $k \in \mathbb{N}$ such that x belongs to $E_k - \bigcup_{j \in \mathbb{N}} (E_{k,j}^+ \cup E_{k,j}^-)$. Thus no segment $(x\tilde{x})$ with $\tilde{x} \in L_u$ and $\tilde{x} \neq x$ is contained in $\text{int}_* A$. In other words, $(x\tilde{x}) \cap (X_u \cup Z_u) \neq \emptyset$ for all $\tilde{x} \in L_u - \{x\}$. Similarly, if $y \in Y_u$ then $(y\tilde{y}) \cap (Y_u \cup Z_u) \neq \emptyset$ for all $\tilde{y} \in L_u - \{y\}$.

Claim 4. $\mathcal{L}^{n-1}(\Pi_2) = 0$.

Proof. Fix $k, j \in \mathbb{N}$, and for $p \in \mathbb{Z}$, let

$$N_p := \left\{ y \in E_{k,j}^+ : \frac{p}{j} \leq y' < \frac{p+1}{j} \right\}.$$

If $N_p \neq \emptyset$, choose $u \in \pi(N_p)$ and $0 < r < \min\{1/k, 1/j\}$. Find a point $z \in N_p \cap \pi^{-1}(B_{u,r})$ so that

$$z' > \sup\{y' : y \in N_p \cap \pi^{-1}(B_{u,r})\} - \frac{r}{2}.$$

It is easy to verify that the ball $B(z, 3r)$ contains the cylinder

$$C := \left\{ x \in \pi^{-1}[\pi(N_p) \cap B_{u,r}] : z' + r/2 < x' < z' + r \right\}.$$

Given $x \in C$, there is $y \in N_p \cap \pi^{-1}(x)$. Then $x' > z' + r/2 > y'$ by the choice of z. Since $x' < z' + r$ and $|y' - z'| < 1/j$, we obtain $0 < x' - y' < 2/j$. Hence $x \in S_+(y,j)$. As $S_+(y,j) \subset \text{int}_* A$ by the definition of $E_{k,j}^+$, we conclude that $C \subset \text{int}_* A \cap B(z, 3r)$. Consequently

$$\frac{r}{2} \mathcal{L}^{n-1}(\pi(N_p) \cap B_{u,r}) = |C| \leq |\text{int}_* E \cap B(z, 3r)| \leq \frac{\alpha(n-1)}{3^{n+1}} (3r)^n$$

where the last inequality holds because $z \in E_k$ and $0 < 3r < 3/k$. Thus

$$\limsup_{r \to 0+} \frac{\mathcal{L}^{n-1}[\pi(N_p) \cap B_{u,r}]}{\mathcal{L}^{n-1}(B_{u,r})} \leq \frac{2}{3}$$

and Corollary 4.4.4 applied to $\Pi = \mathbb{R}^{n-1}$ yields $\mathcal{L}^{n-1}[\pi(N_p)] = 0$. Since $E_{k,j}^+ = \bigcup_{p \in \mathbb{Z}} N_p$, the sets $\pi(E_{k,j}^+)$ are $\mathcal{L}^{n-1}$ negligible for all $k, j \in \mathbb{N}$, and so is $\pi(E_0^+)$. Similarly we show that $\pi(E_0^-)$ and $\pi(I_0^{\pm})$ are $\mathcal{L}^{n-1}$ negligible sets.

The theorem follows by letting $\Pi_0 = \Pi - \Pi_1 \cup \Pi_2$ and applying Lemmas 4.6.1 and 4.6.2. $\square$

Corollary 4.6.4. *Let $n = 1$. Then $A \in \mathcal{P}_{\mathrm{loc}}(\mathbb{R})$ if and only if $\partial_* A$ is a locally finite set. In this case, there is a countable collection $\mathcal{J}$ of open intervals with disjoint closures such that*

$$\mathrm{int}_* A = \bigcup \mathcal{J} \quad \text{and} \quad \partial_* A = \bigcup \{\partial J : J \in \mathcal{J}\}.$$

PROOF. Since $\mathcal{H}^0$ is the counting measure in $\mathbb{R}$, the first assertion is obvious. If $\partial_* A$ is locally finite, we can organize it into a bidirectional sequence

$$\cdots < a_{-1} < a_0 < a_1 < \cdots$$

of real numbers, which can be finite or infinite in either direction but has no cluster points in $\mathbb{R}$. Theorem 4.6.3 shows that any interval $J_k = (a_k, a_{k+1})$ is contained either in $\mathrm{ext}_* A$ or in $\mathrm{int}_* A$. Moreover, if J_k is a subset of $\mathrm{ext}_* A$, or $\mathrm{int}_* A$, then $J_{k-1} \cup J_{k+1}$ is a subset of $\mathrm{int}_* A$, or $\mathrm{ext}_* A$, respectively. □

Theorem 4.6.5. *Let $A \subset \mathbb{R}^n$ be an $\mathcal{H}^{n-1}$ measurable set, and let Π be a hyperplane. If $\mathcal{H}^{n-1} \llcorner A$ is a Radon measure, then*

$$u \mapsto \mathcal{H}^0(L_u \cap A) : \Pi \to \overline{\mathbb{R}}$$

is an $\mathcal{L}^{n-1}$ measurable function and

$$\int_\Pi \mathcal{H}^0(L_u \cap A) \, d\mathcal{L}^{n-1}(u) \le \mathcal{H}^{n-1}(A).$$

PROOF. Let $\pi : \mathbb{R}^n \to \Pi$ be the orthogonal projection. Define

$$f_E : u \mapsto \mathcal{H}^0(L_u \cap E) : \Pi \to [0, \infty]$$

for each $E \subset \mathbb{R}^n$. Since A is $\mathcal{H}^{n-1}$ measurable and $\mathcal{H}^{n-1} \llcorner A$ is a Radon measure, there are disjoint $\mathcal{H}^{n-1}$ measurable sets A_i such that $\mathcal{H}^{n-1}(A_i) < \infty$ for $i = 1, 2, \ldots$, and $A = \bigcup_{i \in \mathbb{N}} A_i$. According to Theorem 1.3.1, there are compact sets $K_{i,j} \subset A_i$ such that each set $A_i - \bigcup_{j \in \mathbb{N}} K_{i,j}$ is $\mathcal{H}^{n-1}$ negligible. If $K_p = \bigcup_{i,j \le p} K_{i,j}$ for $p = 1, 2, \ldots$, then $\{K_p\}$ is an increasing sequence of compact subsets of A, and the set

$$A - \bigcup_{p \in \mathbb{N}} K_p \subset \bigcup_{i \in \mathbb{N}} \left(A_i - \bigcup_{j \in \mathbb{N}} K_{i,j} \right)$$

is $\mathcal{H}^{n-1}$ negligible. As $\mathrm{Lip}\,\pi = 1$, the projection $N = \pi\big(A - \bigcup_{p \in \mathbb{N}} K_p\big)$ is $\mathcal{L}^{n-1}$ negligible by (1.5.1), and for each $u \in \Pi - N$,

$$f_A(u) = \lim f_{K_p}(u). \tag{$*$}$$

Let $K \subset \mathbb{R}^n$ be a compact set, and choose $j \in \mathbb{N}$ and $t > 0$. If U_j is the set of all $u \in \Pi$ such that $L_u \cap K$ can be covered by fewer than t open sets, each of diameter smaller than $1/j$, then

$$\{u \in \Pi : f_K(u) < t\} = \bigcap \{U_j : j \in \mathbb{N}\}. \tag{$**$}$$

Indeed, because $\mathcal{H}^0$ takes only integer values, equality $(**)$ follows from Remark 1.4.1, (3). Fix a point $u \in U_j$, and find open sets $V_1, \ldots, V_p$ so that $p < t$, $d(V_i) < 1/j$ for $i = 1, \ldots, p$, and $V = \bigcup_{i=1}^p V_i$ contains $L_u \cap K$. Let $B_k = \Pi \cap B(u, 1/k)$ and observe that $C_k := \pi^{-1}(B_k) \cap (K - V)$ is a compact set. Since $\bigcap_{k=1}^\infty C_k = \emptyset$, there is an integer $p \geq 1$ such that $C_p = \emptyset$. Consequently $B_p \subset U_j$, and we see that U_j is a relatively open subset of Π . This and $(**)$ show that f_K is a Borel function, and the $\mathcal{L}^{n-1}$ measurability of f_A follows from $(*)$.

For every $k \in \mathbb{N}$, there is a cover $\{B_{k,j} : j \in \mathbb{N}\}$ of A such that each $B_{k,j}$ has diameter smaller than $1/k$, and

$$\sum_{j=1}^\infty \alpha(n-1) \left[\frac{d(B_{k,j})}{2} \right]^{n-1} < \mathcal{H}_{1/k}^{n-1}(A) + \frac{1}{k}.$$

In view of Remark 1.4.1, (3), we may assume that $B_{k,j}$ are open sets. It follows that each $\pi(B_{k,j})$ is a relatively open subset of Π , and we define a Borel function

$$g_k : u \mapsto \sum_{j=1}^\infty \chi_{\pi(B_{k,j})}(u) : \Pi \to [0, \infty].$$

Choose $u \in \Pi$, and observe that $g_k(u)$ is the number of elements in the set $N_k = \{j : L_u \cap B_{k,j} \neq \emptyset\}$. Since $L_u \cap A \subset \bigcup_{j \in N_k} B_{k,j}$, we see that $\mathcal{H}_{1/k}^0(L_u \cap A) \leq g_k(u)$ for $k = 1, 2, \ldots$. Consequently

$$\mathcal{H}^0(L_u \cap A) = \lim \mathcal{H}_{1/k}^0(L_u \cap A) \leq \liminf g_k(u).$$

Fatou's lemma and the isodiametric inequality (1.4.2) imply

$$\int_\Pi \mathcal{H}^0(L_u \cap A) \, d\mathcal{L}^{n-1}(u) \leq \liminf \int_\Pi g_k \, d\mathcal{L}^{n-1}$$

$$= \liminf_{k \to \infty} \sum_{i=1}^\infty \mathcal{L}^{n-1}\left[\pi(B_{i,k}) \right]$$

$$\leq \liminf_{k \to \infty} \sum_{i=1}^\infty \alpha(n-1) \left[\frac{d(B_{i,k})}{2} \right]^{n-1}$$

$$\leq \lim_{k \to \infty} \left[\mathcal{H}_{1/k}^{n-1}(A) + \frac{1}{k} \right] = \mathcal{H}^{n-1}(A). \qquad \square$$

Let E be a subset of a line L. Identifying L with $\mathbb{R}$, the symbols

$$\text{ext}_*^L E, \quad \text{int}_*^L E, \quad \text{cl}_*^L E, \quad \text{and} \quad \partial_*^L E$$

denote, respectively, the essential exterior, essential interior, essential closure, and essential boundary of E *relative to* L.

Theorem 4.6.6. *Let $E \in \mathcal{P}_{\mathrm{loc}}(\mathbb{R}^n)$. Given a hyperplane Π, the function $u \mapsto \mathcal{H}^0(L_u \cap \partial_* E) : \Pi \to \overline{\mathbb{R}}$ is $\mathcal{L}^{n-1}$ measurable and*

$$\int_\Pi \mathcal{H}^0(L_u \cap \partial_* E)\, d\mathcal{L}^{n-1} \leq \mathbf{P}(E). \tag{4.6.1}$$

There is an $\mathcal{L}^{n-1}$ negligible set $N \subset \Pi$ such that for each $u \in \Pi - N$, the set $L_u \cap \partial_* E$ is locally countable, and

$$L_u \cap \mathrm{int}_* E = \mathrm{int}_*^{L_u}(L_u \cap E) \quad \text{and} \quad L_u \cap \partial_* E = \partial_*^{L_u}(L_u \cap E).$$

PROOF. Note that $\partial_* E$ is a Borel set by Observation 4.4.1. Since by definition, $\mathcal{H}^{n-1}(\partial_* E \cap U) = \mathbf{P}(E, U) < \infty$ for each open set $U \Subset \mathbb{R}^n$, it follows from Proposition 1.4.2 that $\mathcal{H}^{n-1} \llcorner \partial_* E$ is a Radon measure. Now the first assertion is a direct consequence of Theorems 4.4.5.

Next let $U_k = U(0, k)$, and apply Theorem 4.4.2 to the set $\partial_* E \cap U_k$. There is an $\mathcal{L}^{n-1}$ negligible set $N_k \subset \Pi$ such that $L_u \cap (\partial_* E \cap U_k)$ is a finite set for each $u \in \Pi - N_k$. The set $N = \bigcup_{k=1}^n N_k$ is still negligible, and the set $L_u \cap \partial_* E$ is locally finite for each $u \in \Pi - N$. Enlarge N so that for each $u \in \Pi - N$, the assertions of Theorem 4.6.3 hold for the set E. The theorem follows by the same argument we used in the proof of Corollary 4.6.4. $\qquad \square$

Proposition 4.6.7. *Let $E \in \mathcal{P}_{\mathrm{loc}}(\mathbb{R}^n)$, and let $\varphi \in Lip_c(\mathbb{R}^n)$ be such that $\|\varphi\|_{L^\infty(\mathbb{R}^n)} \leq 1$. For $i = 1, \ldots, n$,*

$$\int_E D_i \varphi(x)\, dx \leq \int_{\Pi_i} \mathcal{H}^0(L_u \cap \partial_* E)\, d\mathcal{L}^{n-1}(u).$$

PROOF. Since E and $\mathrm{int}_* E$ are equivalent sets by Theorem 4.4.2, Fubini's theorem shows that $L_u \cap E$ and $L_u \cap \mathrm{int}_* E$ are $\mathcal{L}^1$ equivalent for $\mathcal{L}^{n-1}$ almost all $u \in \Pi_n$. Thus with no loss of generality, we may assume that $E = \mathrm{int}_* E$. Applying Theorem 4.4.2 and Corollary 4.6.4, we find an $\mathcal{L}^{n-1}$ negligible set $N \subset \Pi_n$ such that for each point $u \in \Pi_n - N$ there is a strictly increasing finite or infinite bidirectional sequence of extended real numbers

$$\cdots < a_{u,j} < b_{u,j} < a_{u,j+1} < b_{u,j+1} < \cdots,$$

$j \in J_u$ and $J_u \subset \mathbb{Z}$, with no cluster points in $\mathbb{R}$, which satisfies

$$L_u \cap E = \{u\} \times \bigcup \{(a_{u,j} b_{u,j}) : j \in J_u\},$$
$$L_u \cap \partial_* E = \{(u, a_{u,j}), (u, b_{u,j}) : j \in J_u\} \cap \mathbb{R}^n.$$

Select $\varphi \in Lip_c(\mathbb{R}^n)$ with $\|\varphi\|_{L^\infty(\mathbb{R}^n)} \leq 1$, and for each $u \in \Pi_n$, let

$$f_u(t) := \begin{cases} \varphi(u, t) & \text{if } t \in \mathbb{R}, \\ 0 & \text{if } t = \pm\infty. \end{cases}$$

Since $f \in Lip_c(\mathbb{R})$,

$$\int_{L_u \cap E} D_n\varphi(u,t)\,dt = \sum_{j \in J_u} \int_{a_{u,j}}^{b_{u,j}} f'_u(t)\,dt$$

$$= \sum_{j \in J_u} \left[f_u(b_{u,j}) - f_u(a_{u,j}) \right]$$

$$\leq \sum_{j \in J_u} \left(\left| f_u(b_{u,j}) \right| + \left| f_u(a_{u,j}) \right| \right) \leq \mathcal{H}^0(L_u \cap \partial_* E),$$

for each $u \in \Pi_n - N$, and Fubini's theorem implies

$$\int_E D_n\varphi(x)\,dx = \int_{\Pi_n} \left(\int_{L_u \cap E} D_n\varphi(u,t)\,dt \right) d\mathcal{L}^{n-1}(u)$$

$$\leq \int_{\Pi_n} \mathcal{H}^0(L_u \cap \partial_* E)\,d\mathcal{L}^{n-1}(u).$$

The proposition follows from symmetry. $\square$

Remark 4.6.8. By Proposition 4.6.7 and Theorem 4.6.6,

$$\sup\left\{ \int_E \operatorname{div} v(x)\,dx : v \in Lip_c(\mathbb{R}^n) \text{ and } \|v\|_{L^\infty(\mathbb{R}^n;\mathbb{R}^n)} \leq 1 \right\} \leq n\mathbf{P}(E)$$

for each measurable set E. Indeed, this is true if $E \in \mathcal{P}_{\mathrm{loc}}(\mathbb{R}^n)$. Otherwise $\mathbf{P}(E) = \infty$ and there is nothing to prove. While the above inequality is useful at this stage of development, we show later that

$$\sup\left\{ \int_E \operatorname{div} v(x)\,dx : v \in Lip_c(\mathbb{R}^n) \text{ and } \|v\|_{L^\infty(\mathbb{R}^n;\mathbb{R}^n)} \leq 1 \right\} = \mathbf{P}(E)$$

for each measurable set E; see Theorem 6.5.5 below.

The $(n-1)$-dimensional *sphere* in $\mathbb{R}^n$ is the set

$$S^{n-1} := \left\{ e \in \mathbb{R}^n : |e| = 1 \right\}.$$

For each pair (e,t) in $S^{n-1} \times \mathbb{R}$ we define

$$\Pi_{e,t} := \{ x \in \mathbb{R}^n : x \cdot e = t \} \quad \text{and} \quad H_{e,t} := \{ x \in \mathbb{R}^n : x \cdot e > t \}.$$

The open *half-spaces* $H_{e,t}$ and $H_{-e,-t}$ are the connected components of $\mathbb{R}^n - \Pi_{e,t}$, and the hyperplane $\Pi_{e,t}$ is their common boundary.

Lemma 4.6.9. *Let $E \in \mathcal{P}(\mathbb{R}^n)$. For each pair (e,t) in $S^{n-1} \times \mathbb{R}$,*

$$\mathcal{L}^{n-1}(\operatorname{int}_* E \cap \Pi_{e,t}) \leq \mathcal{H}^{n-1}(\partial_* E \cap H_{\pm e,\pm t}).$$

PROOF. In view of invariance with respect to rotations and translations, it suffices to prove the lemma for $\Pi = \Pi_{e_n,0}$ and $H = H_{e_n,0}$. Since $|E| < \infty$, Fubini's theorem shows that $L_u \cap H$ meets $\operatorname{ext}_* E$ for $\mathcal{L}^{n-1}$ almost all $u \in \Pi$. By Theorem 4.6.3, (ii), the intersection $(L_u \cap H) \cap \partial_* E$ is not empty for $\mathcal{L}^{n-1}$ almost all $u \in \operatorname{int}_* E \cap \Pi$. If $\pi : H \to \Pi$ is the orthogonal projection, then $\mathrm{int}_* E \cap \Pi$ is contained in $\pi(\partial_* E \cap H)$ up to an $\mathcal{L}^{n-1}$ negligible set. As $\mathrm{Lip}\,\pi = 1$, we obtain

$$\mathcal{L}^{n-1}(\mathrm{int}_* E \cap \Pi) \leq \mathcal{L}^{n-1}\big[\pi(\partial_* E \cap H)\big] \leq \mathcal{H}^{n-1}(\partial_* E \cap H).$$

The lemma follows from symmetry.   □

A set $C \subset \mathbb{R}^n$ is called a *polytope* if it is the convex hull of a finite set $S \subset \mathbb{R}^n$ and $\mathrm{int}\,C \neq \emptyset$. Each polytope is a compact set.

Proposition 4.6.10. *Let $E \in \mathcal{P}(\mathbb{R}^n)$. If C is a polytope, then*

$$\mathbf{P}(E \cap C) \leq \mathbf{P}(E).$$

PROOF. There are (v_j, t_j) in $S^{n-1} \times \mathbb{R}$ such that $C = \bigcap_{j=1}^k H_{v_j, t_j}$. To simplify the notation, let $H_j^{\pm} := H_{\pm v_j, \pm t_j}$. By Corollary 4.2.5 and Lemma 4.6.9,

$$\begin{aligned}
\mathbf{P}(E \cap H_1^+) &\leq \mathcal{H}^{n-1}(\partial_* E \cap H_1^+) + \mathcal{H}^{n-1}(\mathrm{int}_* E \cap \partial H_1^+) \\
&\quad + \mathcal{H}^{n-1}(\partial_* E \cap \partial H_1^+) \\
&\leq \mathcal{H}^{n-1}(\partial_* E \cap H_1^+) + \mathcal{H}^{n-1}(\partial_* E \cap H_1^-) \\
&\quad + \mathcal{H}^{n-1}(\partial_* E \cap \partial H_1^+) = \mathbf{P}(E).
\end{aligned}$$

The proof is completed by induction, since

$$\mathbf{P}\big[(E \cap H_1^+) \cap H_2^+\big] \leq \mathbf{P}(E \cap H_1^+) \leq \mathbf{P}(E).\qquad\qquad □$$

Proposition 4.6.10 still holds when polytopes are replaced by arbitrary convex sets (Proposition 6.6.4 below).

Proposition 4.6.11. *Let $E \in \mathcal{P}(\mathbb{R}^n)$. If $\{C_k\}$ is a sequence of cubes such that $\lim d(C_k) = \infty$, then $\lim \mathbf{P}(E - C_k) = 0$.*

PROOF. With no loss of generality, we may assume that $C_k \subset \mathrm{int}\,C_{k+1}$ for $k = 1, 2, \ldots$. Indeed, each subsequence of $\{C_k\}$ contains a subsequence which satisfies this assumption. By Corollary 4.2.5,

$$\partial_*(E - C_k) \subset (\partial_* E - C_k) \cup (\mathrm{int}_* E \cap \partial C_k) \cup (\partial_* E \cap \partial C_k).$$

Since $\mathcal{H}^{n-1}(\partial_* E) < \infty$ and $\{\partial_* E - C_k\}$ is a decreasing sequence with empty intersection, $\lim \mathcal{H}^{n-1}(\partial_* E - C_k) = 0$. As ∂C_k are disjoint sets,

$$\sum_{k=1}^{\infty} \mathcal{H}^{n-1}(\partial_* E \cap \partial C_k) \leq \mathcal{H}^{n-1}(\partial_* E) < \infty$$

and $\lim \mathcal{H}^{n-1}(\partial_* E \cap \partial C_k) = 0$. If $C_k = \prod_{i=1}^n [a_{k,i}, b_{k,i}]$, then

$$\mathrm{int}_* E \cap \partial C_k \subset \bigcup_{i=1}^n \big[(\mathrm{int}_* E \cap \Pi_{e_i, a_{k,i}}) \cup (\mathrm{int}_* E \cap \Pi_{e_i, b_{k,i}})\big]$$

and $\lim b_{k,i} = -\lim a_{k,i} = \infty$ for $i = 1, \ldots, n$. As $\{H_{e_i,b_{k,i}}\}$ is a decreasing sequence with empty intersection, Lemma 4.6.9 implies

$$\lim \mathcal{L}^{n-1}(\text{int}_* E \cap \Pi_{e_i,b_{k,i}}) \le \lim \mathcal{H}^{n-1}(\partial_* E \cap H_{e_i,b_{k,i}}) = 0$$

for $i = 1, \ldots, n$. The proposition follows by symmetry. $\qquad\square$

The set $E = \mathbb{R}^n$ and cubes $C = [-k, k]^n$, $k = 1, 2, \ldots$, show that in Proposition 4.6.11, the assumption $E \in \mathcal{P}(\mathbb{R}^n)$ cannot be replaced by a weaker assumption $\mathbf{P}(E) < \infty$.

4.7. Lipeomorphisms

We show that a lipeomorphic image of a set of finite perimeter is again a set of finite perimeter. The proof is nontrivial and relies on topological arguments.

Proposition 4.7.1. *Let $B \subset \mathbb{R}^n$ be a closed ball, and let $\psi : B \to \mathbb{R}^n$ be a homeomorphism. Then $\text{int}\,\psi(B) = \psi(\text{int}\,B)$.*

PROOF. By the Jordan-Brouwer separation theorem [69, Chapter 4, Section 7, Theorem 15], the open set $\mathbb{R}^n - \psi(\partial B)$ has two connected components I and E such that $\partial I = \partial E = \psi(\partial B)$. By Brouwer's invariance of domain theorem [Theorem 16, ibid.], the connected set $O = \psi(\text{int}\,B)$ is open. Being connected and disjoint from $\psi(\partial B)$, the set O is contained in one of the components of $\mathbb{R}^n - \psi(\partial B)$, say $O \subset I$. Assume there is $z \in I \cap \partial O$ and find a sequence $\{x_i\}$ in $\text{int}\,B$ such that $\lim \psi(x_i) = z$. Passing to a subsequence, we may assume that $\{x_i\}$ converges to $x \in B$. Then $z = \psi(x)$, which is impossible because $z \notin O \cup \psi(\partial B)$. Thus $I \cap \partial O = \emptyset$, and O is simultaneously a relatively open and relatively closed subset of I. As $O \ne \emptyset$, we infer $O = I$. The proposition follows, since $\psi(B) = O \cup \psi(\partial B)$ and $\partial O = \psi(\partial B)$. $\qquad\square$

Throughout this section, for $r > 0$ we employ the notation

$$B_r = B(0, r) \quad \text{and} \quad U_r = U(0, r).$$

Proposition 4.7.2. *Let $B \subset \mathbb{R}^n$ be a closed ball, and let $\psi : B \to \mathbb{R}^n$ be a homeomorphism. If $K \subset \psi(\text{int}\,B)$ is a compact set and $\phi : B \to \mathbb{R}^n$ is a continuous map satisfying*

$$\sup_{x \in B} |\psi(x) - \phi(x)| < \text{dist}\,[K, \psi(\partial B)],$$

then $K \subset \phi(\text{int}\,B)$.

PROOF. If K and ϕ are as in the proposition, choose $0 < \delta < [K, \psi(\partial B)]$ so that $|\psi(x) - \phi(x)| \le \delta$ for each $x \in B$. Given $y \in K$, we show $y = \phi(x)$ for some $x \in \text{int } B$. Using translations, we may assume $y = 0$. Proposition 4.7.1 implies $B_\delta \subset \psi(\text{int } B)$. Let $\gamma := \phi \circ \left[\psi^{-1} \upharpoonright B_\delta\right]$, and observe

$$\left|x - \gamma(x)\right| = \left|\psi\left[\psi^{-1}(x)\right] - \phi\left[\psi^{-1}(x)\right]\right| \leq \delta$$

for each $x \in B_\delta$. Thus $\theta : x \mapsto x - \gamma(x)$ maps B_δ into itself. By Brouwer's fixed point theorem [69, Chapter 4, Section 7, Theorem 5], there is $z \in B_\delta$ such that $\theta(z) = z$. Hence $y = 0 = \gamma(z) = \phi\left[\psi^{-1}(z)\right]$ and $\psi^{-1}(z) \in \text{int } B$. $\square$

The following result was obtained by Z. Buczolich [11]. However, our proof follows that presented in [35, Section A.2.4]. If $A \subset \mathbb{R}^n$, then a lipeomorphism $\phi : A \to \mathbb{R}^n$ has a unique continuous extension to $\text{cl } A$. This extension is again a lipeomorphism, still denoted by ϕ.

Theorem 4.7.3. *Let $C \subset \mathbb{R}^n$, $x \in \text{cl } C$, and let $\phi : C \to \mathbb{R}^n$ be a lipeomorphism. If $\Theta(C, x) = 0$ then $\Theta\left[\phi(C), \phi(x)\right] = 0$. If C is measurable and $\Theta(C, x) = 1$, then $\Theta\left[\phi(C), \phi(x)\right] = 1$; in particular,*

$$\partial_* \phi(C) = \phi(\partial_* C).$$

PROOF. If $D = \phi(C)$, then $\text{cl } D = \phi(\text{cl } C)$ and there are $a, b \in \mathbb{R}_+$ such that

$$a|u - v| \leq |\phi(u) - \phi(v)| \leq b|u - v|$$

for all $u, v \in \text{cl } C$. Choose $x \in \text{cl } C$ and let $y = \phi(x)$. Observe that

$$D \cap B(y, r) \subset \phi\left[C \cap B(x, r/a)\right]$$

for each $r > 0$. Thus $\Theta(C, x) = 0$ implies $\Theta(D, y) = 0$, since

$$\limsup_{r \to 0} \frac{|D \cap B(y, r)|}{|B(y, r)|} \leq \left(\frac{b}{a}\right)^n \limsup_{r \to 0} \frac{|C \cap B(x, r/a)|}{|B(x, r/a)|} = 0.$$

The rest of the proof is harder. Using translations, assume $x = y = 0$ and $\Theta(C, 0) = 1$. In view of Proposition 1.5.2, the lipeomorphism $\phi : \text{cl } C \to \text{cl } D$ has a Lipschitz extension $\tilde{\phi} : \mathbb{R}^n \to \mathbb{R}^n$. Making b larger, the inequality $\left|\tilde{\phi}(u) - \tilde{\phi}(v)\right| \leq b|u - v|$ still holds for all $u, v \in \mathbb{R}^n$. For $r > 0$, define

$$\psi_r : x \mapsto r^{-1} \tilde{\phi}(rx) : \mathbb{R}^n \to \mathbb{R}^n,$$

$C_r = r^{-1} C$, and $D_r = r^{-1} D$. Clearly $\psi_r(0) = 0$, $\psi_r(C_r) = D_r$, and

$$a|u - v| \leq |\psi_r(u) - \psi_r(v)| \text{ for all } u, v \in \text{cl } C_r,$$
$$|\psi_r(u) - \psi_r(v)| \leq b|u - v| \text{ for all } u, v \in \mathbb{R}^n. \tag{$*$}$$

Claim. If $0 < c < a$, then each sequence $\{r_k\}$ in $\mathbb{R}_+$ converging to zero has a subsequence $\{s_k\}$ such that $B_c \subset \psi_{s_k}(B_1)$ for $k = 1, 2, \ldots$.

Proof. From $\Theta(C, 0) = 1$ and $|C \cap B_r|/|B_r| = |C_r \cap B_1|/|B_1|$, we obtain

$$\lim_{r \to 0} \frac{|C_r \cap B_1|}{|B_1|} = 1.$$

A sequence $\{r_k\}$ in $\mathbb{R}_+$ converging to zero has a subsequence $\{s_k\}$ such that

$$\frac{|C_{s_k} \cap B_1|}{|B_1|} > 1 - 2^{-k} \qquad (**)$$

for $k = 1, 2, \ldots$. In view of $(*)$, Ascoli's theorem [61, Chapter 9, Section 8] shows that $\{\psi_{s_k}\}$ has a subsequence, still denoted by $\{\psi_{s_k}\}$, converging uniformly on B_1 to a map $\psi : B_1 \to \mathbb{R}^n$ that satisfies

$$\psi(0) = 0 \quad \text{and} \quad |\psi(u) - \psi(v)| \leq b|u - v| \qquad (\dagger)$$

for all $u, v \in B_1$. Letting $E_j := \bigcap_{k=j}^{\infty}(C_{s_k} \cap B_1)$ and $E := \bigcup_{j=1}^{\infty} E_j$, we deduce from $(*)$ that $a|u - v| \leq |\psi_{s_k}(u) - \psi_{s_k}(v)|$ whenever $u, v \in E_j$ and $k \geq j$. Since $\{E_j\}$ is an increasing sequence,

$$a|u - v| \leq |\psi(u) - \psi(v)| \qquad (\dagger\dagger)$$

for all $u, v \in E$. Observe

$$B_1 - E = \bigcap_{j=1}^{\infty}(B_1 - E_j) = \bigcap_{j=1}^{\infty}\bigcup_{k=j}^{\infty}(B_1 - C_{s_k}).$$

The measurability of C and $(**)$ imply $|B_1 - C_{s_k}| \leq 2^{-k}|B_1|$, and so

$$|B_1 - E| \leq |B_1 - E_j| \leq 2^{-j+1}|B_1|$$

for $j = 1, 2, \ldots$. Consequently $|B_1 - E| = 0$; in particular E is a dense subset of B_1. Inequalities $(\dagger)$ and $(\dagger\dagger)$ together with the continuity of ψ show that $\psi : B_1 \to \mathbb{R}^n$ is a lipeomorphism. By Proposition 4.7.1, the set $\psi(U_1)$ is open and $\partial\psi(U_1) = \psi(\partial U_1)$. As $0 \in \psi(U_1)$ and $a = a|z - 0| \leq |\psi(z) - 0|$ for each $z \in \partial B_1$, we see that $B_c \subset U_a \subset \psi(U_1)$. Since $\lim \psi_{s_k} = \psi$ uniformly in B_1, passing to a subsequence, still denoted by $\{\psi_{s_k}\}$, we obtain

$$\sup_{z \in B_1} |\psi_{s_k}(z) - \psi(z)| < \text{dist}\,(B_c, \psi(\partial B_1))$$

for $k = 1, 2, \ldots$. The claim follows from Proposition 4.7.2.

Returning to the main proof, select $0 < c < a$ and a sequence $\{r_k\}$ in $\mathbb{R}_+$ converging to zero. Find a subsequence $\{s_k\}$ of $\{r_k\}$ according to the claim. As B_c is contained in each $\psi_{s_k}(B_1) = s_k^{-1}\phi(s_k B_1)$, there are inclusions

$$B_{cs_k} \subset \tilde{\phi}(B_{s_k}) \subset \tilde{\phi}(B_{s_k} - C) \cup \tilde{\phi}(C) = \tilde{\phi}(B_{s_k} - C) \cup D.$$

Hence $B_{cs_k} \subset \tilde{\phi}(B_{s_k} - C) \cup (D \cap B_{cs_k})$, and consequently

$$1 \leq \frac{|\tilde{\phi}(B_{s_k} - C)|}{|B_{cs_k}|} + \frac{|D \cap B_{cs_k}|}{|B_{cs_k}|}$$

for $k = 1, 2, \ldots$. This inequality and

$$\lim \frac{|\widetilde{\phi}(B_{s_k} - C)|}{|B_{cs_k}|} \leq \left(\frac{b}{c}\right)^n \lim \frac{|B_{s_k} - C|}{|B_{s_k}|}$$

$$= \left(\frac{b}{c}\right)^n \lim \left[1 - \frac{|C \cap B_{s_k}|}{|B_{s_k}|}\right] = 0$$

imply $\lim(|D \cap B_{cs_k}|/|B_{cs_k}|) = 1$. As the sequence $\{r_k\}$ has been arbitrary, the equality $\Theta(D, 0) = 1$ follows. $\square$

Corollary 4.7.4. *Let* $A \subset \mathbb{R}^n$, *and let* $\phi : A \to \mathbb{R}^m$ *be a lipeomorphism. If* A *belongs to* $\mathbf{P}(\mathbb{R}^n)$, *then so does* $\phi(A)$ *and*

$$\mathbf{P}[\phi(A)] \leq (\operatorname{Lip} \phi)^{n-1} \mathbf{P}(A).$$

If A *belongs to* $\mathbf{P}_{\mathrm{loc}}(\mathbb{R}^n)$ *then so does* $\phi(A)$.

PROOF. The claim for $A \in \mathbf{P}(\mathbb{R}^n)$ follows from Theorem 4.7.3 and inequality (1.5.1). If $A \in \mathbf{P}_{\mathrm{loc}}(\mathbb{R}^n)$, choose $x \in \operatorname{cl} A$ and let $y = \phi(x)$. For $r > 0$ and $a = 1/\operatorname{Lip}(\phi^{-1})$, Theorem 4.7.3 implies

$$\partial_* \phi(A) \cap U(y, r) = \phi(\partial_* A) \cap U(y, r) \subset \phi[\partial_* A \cap U(x, r/a)].$$

By inequality (1.5.1),

$$\mathcal{H}^{n-1}[\partial_* \phi(A) \cap U(y, r)] \leq (\operatorname{Lip} \phi)^{n-1} \mathcal{H}^{n-1}[\partial_* A \cap U(x, r/a)] < \infty.$$

As r is arbitrary, $\phi(A)$ belongs to $\mathbf{P}_{\mathrm{loc}}(\mathbb{R}^n)$. $\square$

Chapter 5

BV functions

In this chapter we present the analytic approach to the perimeter of measurable sets. We define BV functions and BV sets and establish their main properties. The emphasis is on BV functions — a connection with BV sets is provided by the coarea theorem. Sobolev's and Poincaré's inequalities are proved directly within the framework of BV functions; their validity for C^∞ functions is not presupposed. Throughout this chapter Ω denotes an open subset of $\mathbb{R}^n$.

5.1. Variation

Definition 5.1.1. Let $E \subset \mathbb{R}^n$ be such that $E \cap \Omega$ is measurable. The *variation* of E *in* Ω is the extended real number

$$\sup \left\{ \int_\Omega \chi_E \operatorname{div} v(x)\, dx : v \in Lip_c(\Omega; \mathbb{R}^n) \text{ and } \|v\|_{L^\infty(\Omega;\mathbb{R}^n)} \leq 1 \right\},$$

denoted by $\mathbf{V}(E,\Omega)$. If E is measurable, the variation $\mathbf{V}(E,\mathbb{R}^n)$ of E in $\mathbb{R}^n$ is called the *variation* of E, denoted by $\mathbf{V}(E)$.

A measurable set $E \subset \Omega$ is called

(i) a set of *bounded variation* in Ω, or a *BV set* in Ω, if

$$|E| + \mathbf{V}(E,\Omega) < \infty,$$

(ii) a set of *locally bounded variation* in Ω, or a *locally BV set* in Ω, if $\mathbf{V}(E,U) < \infty$ for each open set $U \Subset \Omega$.

The family of all BV sets in Ω, or all locally BV sets in Ω, is denoted by $\boldsymbol{BV}(\Omega)$, or $\boldsymbol{BV}_{\mathrm{loc}}(\Omega)$, respectively. Clearly $\boldsymbol{BV}(\Omega) \subset \boldsymbol{BV}_{\mathrm{loc}}(\Omega)$.

Proposition 5.1.2. *If $E \subset \Omega$ is a measurable set, then*

$$\mathbf{V}(E,\Omega) = \mathbf{V}(\Omega - E,\Omega).$$

PROOF. Choose $v \in Lip_c(\Omega; \mathbb{R}^n)$, and find a dyadic figure $A \subset \Omega$ containing spt v. In view of Remark 2.3.8, (1), the equality

$$\int_E \operatorname{div} v(x)\, dx + \int_{\Omega - E} \operatorname{div} v(x)\, dx = \int_\Omega \operatorname{div} v(x)\, dx = 0$$

follows from Theorem 2.3.7. If $\|v\|_{L^\infty(\Omega;\mathbb{R}^n)} \leq 1$, then

$$\int_{\Omega-E} \operatorname{div}\left[-v(x)\right] dx = \int_E \operatorname{div} v(x)\, dx \leq \mathbf{V}(E,\Omega).$$

From the arbitrariness of v, we obtain $\mathbf{V}(\Omega - E, \Omega) \leq \mathbf{V}(E,\Omega)$. The reverse inequality follows by symmetry. $\square$

If $E \subset \mathbb{R}^n$ is such that $E \cap \Omega$ is measurable, then

$$\mathbf{V}(E,\Omega) = \mathbf{V}(E \cap \Omega, \Omega) \leq \mathbf{V}(E \cap \Omega). \tag{5.1.1}$$

In view of the equality $\mathbf{V}(E,\Omega) = \mathbf{V}(E \cap \Omega, \Omega)$, it suffices to consider $\mathbf{V}(E,\Omega)$ only when E is a measurable subset of Ω.

Proposition 5.1.3. $\mathcal{P}(\mathbb{R}^n) \subset \mathcal{BV}(\mathbb{R}^n)$ *and* $\mathcal{P}_{\mathrm{loc}}(\mathbb{R}^n) \subset \mathcal{BV}_{\mathrm{loc}}(\mathbb{R}^n)$.

PROOF. Remark 4.6.8 implies the first inclusion. Let $E \in \mathcal{P}_{\mathrm{loc}}(\mathbb{R}^n)$, and choose an open set $U \Subset \mathbb{R}^n$. If V is the interior of a figure containing U, then $E \cap V$ belongs to $\mathcal{P}(\mathbb{R}^n)$ by Proposition 4.5.4. Thus

$$\mathbf{V}(E,U) \leq \mathbf{V}(E,V) \leq \mathbf{V}(E \cap V) < \infty,$$

and $E \in \mathcal{BV}_{\mathrm{loc}}(\mathbb{R}^n)$. $\square$

Although the variation $\mathbf{V}(E,\Omega)$ has no obvious geometric meaning at this point, our goal is to show that $\mathbf{V}(E,\Omega) = \mathbf{P}(E,\Omega)$. Since proving this equality in full generality requires a substantial effort, it is instructive to prove it first when $\Omega = \mathbb{R}^n$ and E is a dyadic figure.

Example 5.1.4. Let A be a dyadic figure. If $v \in Lip_c(\mathbb{R}^n;\mathbb{R}^n)$, then

$$\int_A \operatorname{div} v(x)\, dx = \int_{\partial A} v \cdot \nu_A\, d\mathcal{H}^{-1} \leq \|v\|_{L^\infty(\mathbb{R}^n;\mathbb{R}^n)} \mathcal{H}^{n-1}(\partial A)$$

by Theorem 2.3.7, and we infer $\mathbf{V}(A) \leq \mathbf{P}(A)$. To obtain the reverse inequality, choose $\varepsilon > 0$, and in each $(n-1)$-dimensional face A_i of A select an $(n-1)$-dimensional cell C_i such that

(i) C_i meets no $(n-2)$-dimensional face of A,
(ii) $\mathcal{H}^{n-1}\left(\partial A - \bigcup_i C_i\right) \leq \varepsilon$.

The normal ν_A restricted to $C = \bigcup_i C_i$ is a Lipschitz vector field, which can be extended to $w \in Lip(\mathbb{R}^n;\mathbb{R}^n)$ so that $\|w\|_{L^\infty(\mathbb{R}^n;\mathbb{R}^n)} \leq 1$; see Theorem 1.5.4. Select an open ball U containing A, and construct a function $\varphi \in Lip_c(\mathbb{R}^n)$ so that $0 \leq \varphi \leq 1$ and $\varphi(x) = 1$ for each $x \in U$. Then $v := \varphi w$ belongs to $Lip_c(\mathbb{R}^n;\mathbb{R}^n)$, $\|v\|_{L^\infty(\mathbb{R}^n;\mathbb{R}^n)} \leq 1$, and $v(x) = \nu_A(x)$ for each $x \in C$. Applying

Theorem 2.3.7 again, we obtain

$$\mathbf{V}(A) \geq \int_A \operatorname{div} v(x)\, dx = \int_{\partial A} v \cdot \nu_A \, d\mathcal{H}^{n-1}$$

$$= \int_C |\nu_A|^2 \, d\mathcal{H}^{n-1} + \int_{\partial A - C} v \cdot \nu_A \, d\mathcal{H}^{n-1}$$

$$\geq \mathcal{H}^{n-1}(C) - \mathcal{H}^{n-1}(\partial A - C)$$

$$\geq \mathcal{H}^{n-1}(\partial A) - 2\varepsilon = \mathbf{P}(A) - 2\varepsilon.$$

The inequality $\mathbf{V}(A) \geq \mathbf{P}(A)$ follows from the arbitrariness of ε.

Example 5.1.5. Assume $n = 1$. Let $E \subset \mathbb{R}$ be a measurable set such that $|E| > 0$ and $|\mathbb{R} - E| > 0$. By Theorem 4.4.2, there are $x \in \operatorname{int}_* E$ and $y \in \operatorname{ext}_* E$. We may assume that $x < y$. Given $0 < \eta < 1$, find $0 < \varepsilon < (y - x)/2$ so that

$$\frac{|E \cap B(x,\varepsilon)|}{|B(x,\varepsilon)|} \geq 1 - \eta \quad \text{and} \quad \frac{|E \cap B(y,\varepsilon)|}{|B(y,\varepsilon)|} \leq \eta.$$

A function $\varphi : \mathbb{R} \to [0, 1]$ defined by the formula

$$\varphi(t) := \begin{cases} \frac{1}{2\varepsilon}(t - x + \varepsilon) & \text{if } x - \varepsilon \leq t \leq x + \varepsilon, \\ \frac{1}{2\varepsilon}(-t + y + \varepsilon) & \text{if } y - \varepsilon \leq t \leq y + \varepsilon, \\ 1 & \text{if } x + \varepsilon < t < y - \varepsilon, \\ 0 & \text{otherwise} \end{cases}$$

is Lipschitz and has compact support. Since

$$\mathbf{V}(E) \geq \int_E \varphi'(t)\, dt = \frac{1}{2\varepsilon}|E \cap B(x,\varepsilon)| - \frac{1}{2\varepsilon}|E \cap B(y,\varepsilon)| \geq 1 - 2\eta,$$

the arbitrariness of η implies $\mathbf{V}(E) \geq 1$. This lower estimate cannot be improved. Indeed if $E = [0, \infty)$, then $\mathbf{V}(E) \leq \mathbf{P}(E) = 1$ by Remark 4.6.8.

We now extend the concept of variation from measurable sets to locally integrable functions.

Definition 5.1.6. The *variation* of a function $f \in L^1_{\operatorname{loc}}(\Omega)$ *in* Ω is the extended real number

$$\sup\left\{ \int_\Omega f(x) \operatorname{div} v(x)\, dx : v \in Lip_c(\Omega; \mathbb{R}^n) \text{ and } \|v\|_{L^\infty(\Omega;\mathbb{R}^n)} \leq 1 \right\},$$

denoted by $\mathbf{V}(f, \Omega)$. If $\Omega = \mathbb{R}^n$, the variation $\mathbf{V}(f, \mathbb{R}^n)$ of f in $\mathbb{R}^n$ is called the *variation* of f, denoted by $\mathbf{V}(f)$.

Let f be a function defined in Ω, and let $U \subset \Omega$ be an open set. If $f \restriction U$ belongs to $L^1_{\operatorname{loc}}(U)$, we define

$$\mathbf{V}(f, U) := \mathbf{V}(f \restriction U, U).$$

If $E \subset \mathbb{R}^n$ is such that $\Omega \cap E$ is measurable, then

$$\mathbf{V}(E, \Omega) = \mathbf{V}(\chi_E, \Omega).$$

A function $f \in L^1_{\text{loc}}(\Omega)$ is called

(i) a function of *bounded variation* in Ω, or a *BV function* in Ω, if

$$\|f\|_{L^1(\Omega)} + \mathbf{V}(f, \Omega) < \infty,$$

(ii) a function of *locally bounded variation* in Ω, or a *locally BV function* in Ω, if $\mathbf{V}(f, U) < \infty$ for each open set $U \Subset \Omega$.

The family of all BV functions in Ω, or all locally BV functions in Ω, is denoted by $BV(\Omega)$, or $BV_{\text{loc}}(\Omega)$, respectively. Note $BV(\Omega) \subset BV_{\text{loc}}(\Omega)$,

$$\boldsymbol{BV}(\Omega) = \{E \subset \Omega : \chi_E \in BV(\Omega)\},$$
$$\boldsymbol{BV}_{\text{loc}}(\Omega) = \{E \subset \Omega : \chi_E \in BV_{\text{loc}}(\Omega)\}.$$

For $f, g \in L^1_{\text{loc}}(\Omega)$ and $c \in \mathbb{R}$, a direct verification shows

$$\mathbf{V}(f + g, \Omega) \le \mathbf{V}(f, \Omega) + \mathbf{V}(g, \Omega) \quad \text{and} \quad \mathbf{V}(cf, \Omega) = |c|\mathbf{V}(f, \Omega).$$

It follows that $BV(\Omega)$ and $BV_{\text{loc}}(\Omega)$ are linear spaces, and that the functional $f \mapsto \mathbf{V}(f, \Omega)$ is a seminorm in $BV(\Omega)$. For $f \in L^1(\Omega)$, let

$$\|f\|_{BV(\Omega)} := \|f\|_{L^1(\Omega)} + \mathbf{V}(f, \Omega) \tag{5.1.2}$$

and note that $f \mapsto \|f\|_{BV(\Omega)}$ is a norm in $BV(\Omega)$.

Proposition 5.1.7. *Let $\{f_k\}$ be a sequence in $L^1_{\text{loc}}(\Omega)$ that converges in $L^1_{\text{loc}}(\Omega)$ to a function $f \in L^1_{\text{loc}}(\Omega)$. Then*

$$\mathbf{V}(f, \Omega) \le \liminf \mathbf{V}(f_k, \Omega).$$

PROOF. If $v \in C^1_c(\Omega; \mathbb{R}^n)$ and $\|v\|_{L^\infty(\Omega; \mathbb{R}^n)} \le 1$, then

$$\int_\Omega f(x) \operatorname{div} v(x) \, dx = \lim \int_\Omega f_k(x) \operatorname{div} v(x) \, dx \le \liminf \mathbf{V}(f_k, \Omega),$$

and the proposition follows from the arbitrariness of v. $\qquad\qquad\square$

5.2. Mollification

A *mollifier* is a nonnegative function $\eta \in C^\infty_c(\mathbb{R}^n)$ such that

$$\int_{\mathbb{R}^n} \eta(x) \, dx = 1, \quad \operatorname{spt} \eta \subset B(0, 1), \quad \text{and} \quad \eta(-x) = \eta(x)$$

for each $x \in \mathbb{R}^n$. For $0 < \varepsilon \le 1$, if η is a mollifier then so is

$$\eta_\varepsilon : x \mapsto \varepsilon^{-n} \eta(\varepsilon^{-1}x) : \mathbb{R}^n \to \mathbb{R}.$$

Example 5.2.1. Define a function $\eta \in C_c^\infty(\mathbb{R}^n)$ by the formula

$$\eta(x) := \begin{cases} \gamma \exp\big(|x|^2 - 1\big)^{-1} & \text{if } |x| < 1, \\ 0 & \text{if } |x| \geq 1, \end{cases}$$

where $\gamma > 0$ is such that $\int_{\mathbb{R}^n} \eta(x)\,dx = 1$. It is immediate that η is a mollifier, often called the *standard mollifier*; cf. [29, Section 4.2.1].

For Ω and $\varepsilon > 0$, we define an open set

$$\Omega_\varepsilon := \big\{x \in \Omega : B(x,\varepsilon) \subset \Omega\big\}.$$

Given $f \in L_{\mathrm{loc}}^1(\Omega)$ and a mollifier η, the convolutions

$$\eta_\varepsilon * f : x \mapsto \int_\Omega \eta_\varepsilon(x - y) f(y)\,dy : \Omega_\varepsilon \to \mathbb{R},$$

$$(D\eta_\varepsilon) * f : x \mapsto \int_\Omega (D\eta_\varepsilon)(x - y) f(y)\,dy : \Omega_\varepsilon \to \mathbb{R}^n$$

are C^∞ maps, and $\eta_\varepsilon * f = f * \eta_\varepsilon$. If $f \in L^1(\Omega)$, then its zero extension $\overline{f}$ belongs to $L^1(\mathbb{R}^n)$ and we define $\eta_\varepsilon * f(x) := \eta_\varepsilon * \overline{f}(x)$ for all $x \in \mathbb{R}^n$. For a map $\phi = (f_1, \ldots, f_m)$ in $L_{\mathrm{loc}}^1(\Omega; \mathbb{R}^m)$, we let

$$\eta_\varepsilon * \phi := (\eta_\varepsilon * f_1, \ldots, \eta_\varepsilon * f_m).$$

The next lemma is proved in [29, Sections 4.2.1 and 4.2.3]. The equality $\eta_\varepsilon * (Df) = (D\eta_\varepsilon) * f$ of assertion (4) follows from Theorem 2.3.9 and Remark 2.3.10.

Lemma 5.2.2. *Let $f \in L_{\mathrm{loc}}^1(\Omega)$ and let η be a mollifier. If $\varepsilon > 0$, then*

(1) $\|\eta_\varepsilon * f\|_{L^\infty(\Omega_\varepsilon)} \leq \|f\|_{L^\infty(\Omega)}$,

(2) $\|\eta_\varepsilon * f\|_{L^1(U)} \leq \|f\|_{L^1[B(U,\varepsilon)]}$ *for each open set $U \Subset \Omega_\varepsilon$,*

(3) $\int_\Omega (\eta_\varepsilon * f)g = \int_\Omega f(\eta_\varepsilon * g)$ *for each $g \in L^1(\Omega)$ with $\mathrm{spt}\,g \subset \Omega_\varepsilon$,*

(4) $D(\eta_\varepsilon * f) = \eta_\varepsilon * (Df) = (D\eta_\varepsilon) * f$ *when $f \in Lip_{\mathrm{loc}}(\Omega)$.*

Let $U \Subset \Omega$. If $\{\varepsilon_k\}$ is a sequence in $\mathbb{R}_+$ such that $\lim \varepsilon_k = 0$ and $U \Subset \Omega_{\varepsilon_k}$ for $k = 1, 2, \ldots,$ then

(5) $\lim \|f - \eta_{\varepsilon_k} * f\|_{L^1(U)} = 0$,

(6) $\lim \|f - \eta_{\varepsilon_k} * f\|_{L^\infty(U)} = 0$ *whenever $f \in C(\Omega)$.*

The variation $\mathbf{V}(f, \Omega)$ is usually defined by means of vector fields v which belong to $C_c^1(\Omega; \mathbb{R}^n)$, or to $C_c^\infty(\Omega; \mathbb{R}^n)$, rather than to $Lip_c(\Omega; \mathbb{R}^n)$. Utilizing mollifiers, we show that it makes no difference which space is used. In our exposition we employ all $Lip_c(\Omega; \mathbb{R}^n)$, $C_c^1(\Omega; \mathbb{R}^n)$, and $C_c^\infty(\Omega; \mathbb{R}^n)$, depending on the task at hand.

Proposition 5.2.3. *If $f \in L^1_{\text{loc}}(\Omega)$, then*

$$\mathbf{V}(f, \Omega) = \sup_v \int_\Omega f(x) \operatorname{div} v(x) \, dx$$

where $v \in C^\infty_c(\Omega; \mathbb{R}^n)$ and $\|v\|_{L^\infty(\Omega; \mathbb{R}^n)} \leq 1$.

PROOF. Denote by I the right side of the desired equality, and assume that $I < \mathbf{V}(f, \Omega)$. There is $v \in Lip_c(\Omega; \mathbb{R}^n)$ with $\|v\|_{L^\infty(\Omega; \mathbb{R}^n)} \leq 1$ and

$$I < \int_\Omega f(x) \operatorname{div} v(x) \, dx.$$

Choose a mollifier η and $\varepsilon > 0$ so that $B(\operatorname{spt} v, \varepsilon) \subset \Omega$. If $v_k := \eta_{\varepsilon/k} * v$ for $k \in \mathbb{N}$, then $v_k \in C^\infty_c(\Omega; \mathbb{R}^n)$, $\|v_k\|_{L^\infty(\Omega; \mathbb{R}^n)} \leq 1$, $\operatorname{div} v_k = \eta_{\varepsilon/k} * \operatorname{div} v$, and

$$\|\operatorname{div} v_k\|_{L^\infty(\Omega)} \leq \|\operatorname{div} v\|_{L^\infty(\Omega)} \leq n \operatorname{Lip} v,$$

$$\lim \|\operatorname{div} v_k - \operatorname{div} v\|_{L^1(\Omega)} = 0.$$

Thus $\{v_k\}$ has a subsequence $\{v_{k_j}\}$ such that $\lim \operatorname{div} v_{k_j}(x) = \operatorname{div} v(x)$ for almost all $x \in \Omega$. The dominated convergence theorem yields a contradiction:

$$\int_\Omega f(x) \operatorname{div} v(x) \, dx = \lim \int_\Omega f(x) \operatorname{div} v_{k_j}(x) \, dx \leq I.$$

The inequality $I \leq \mathbf{V}(f, \Omega)$ follows from $C^\infty_c(\Omega; \mathbb{R}^n) \subset Lip_c(\Omega; \mathbb{R}^n)$. $\qquad\square$

Remark 5.2.4. As $C^\infty_c(\Omega; \mathbb{R}^n) \subset C^1_c(\Omega; \mathbb{R}^n) \subset Lip_c(\Omega; \mathbb{R}^n)$, Proposition 5.2.3 holds when $C^\infty_c(\Omega; \mathbb{R}^n)$ is replaced by $C^1_c(\Omega; \mathbb{R}^n)$.

Example 5.2.5. Using an argument similar to that employed in Example 5.1.4, we show that $\mathbf{V}(B) = \mathbf{P}(B)$ for each ball B. It suffices to consider the ball $B := B(0, r)$. By Proposition 2.1.4, for each $v \in C^\infty_c(\mathbb{R}^n; \mathbb{R}^n)$,

$$\int_B \operatorname{div} v(x) \, dx = \int_{\partial B} v \cdot \nu_B \, d\mathcal{H}^{n-1} \leq \|v\|_{L^\infty(\mathbb{R}^n; \mathbb{R}^n)} \mathcal{H}^{n-1}(\partial B).$$

Thus $\mathbf{V}(B) \leq \mathbf{P}(B)$. For the reverse inequality, let $A := B(r + \varepsilon, 0) - B(0, r - \varepsilon)$ and $\varphi := \eta_\varepsilon * \chi_A$ where η is a mollifier and $0 < \varepsilon < r/2$. The formula

$$w(x) := \begin{cases} \varphi(x)|x|^{-1}x & \text{if } x \in \mathbb{R}^n - \{0\}, \\ 0 & \text{if } x = 0 \end{cases}$$

defines $w \in C^\infty_c(\mathbb{R}^n; \mathbb{R}^n)$ such that $\|w\|_{L^\infty(\mathbb{R}^n; \mathbb{R}^n)} \leq 1$ and $w \upharpoonright \partial B = \nu_B$. Thus

$$\mathbf{V}(B) \geq \int_B \operatorname{div} w(x) \, dx = \int_{\partial B} w \cdot \nu_B \, d\mathcal{H}^{n-1} = \mathbf{P}(B)$$

by another application of Propositions 2.1.4.

Example 5.2.6 (Caviar and Swiss cheese). Assume $n \geq 2$, and choose a countable dense set D in a closed ball $B \subset \mathbb{R}^n$. Select recursively $x_i \in D$ and $r_i > 0$ so that the balls $B_i := B(x_i, r_i)$ are disjoint, $C := \bigcup_{i=1}^\infty B_i$ is a dense subset of B, and $|C| < |B|$. According to Example 5.2.5 and Proposition 5.1.7,

$$\mathbf{V}(C) \leq \liminf_{k \to \infty} \mathbf{V}\left(\bigcup_{i=1}^k B_i\right) \leq \sum_{i=1}^\infty \mathbf{V}(B_i) < \infty.$$

Thus C is a BV set, and so is $S := B - E$. Moreover $|S| > 0$. Colloquially, the sets C and S are called "caviar" and "Swiss cheese", respectively. More complicated BV sets can be obtained by placing smaller caviar balls into each hole of a Swiss cheese, then turning each new caviar ball into a Swiss cheese, and so on. Denoting the k-th Swiss cheese by S_k, the union $\bigcup_{k=1}^{\infty} S_k$ is still a BV set, provided the diameters of the cheese holes and caviar balls are sufficiently small.

5.3. Vector valued measures

Further study of BV functions relies on measures whose values are elements of $\mathbb{R}^n$. In this section we define such measures, and establish their basic properties.

Throughout this section, we fix Ω and denote by $\mathcal{B}$ the family of all Borel sets $B \Subset \Omega$. A *division* of $B \in \mathcal{B}$ is a countable disjoint collection $\mathcal{C} \subset \mathcal{B}$ whose union equals B. An $\mathbb{R}^m$-*valued measure* (in Ω, if emphasizing is desirable) is a map $\nu : \mathcal{B} \to \mathbb{R}^m$ such that

$$\nu(B) = \sum \{\nu(C) : C \in \mathcal{C}\}$$

for each $B \in \mathcal{B}$ and each division $\mathcal{C}$ of B. An $\mathbb{R}$-valued measure is called a *signed measure*. A map $\nu = (\nu_1, \ldots, \nu_m)$ from $\mathcal{B}$ to $\mathbb{R}^m$ is an $\mathbb{R}^m$-valued measure if and only if each component ν_i is a signed measure.

Example 5.3.1. Let μ be a Radon measure. If $h = (h_1, \ldots, h_m)$ belongs to $L^1_{\text{loc}}(\Omega, \mu; \mathbb{R}^m)$, then the map

$$\mu \llcorner h : B \mapsto \left(\int_B h_1 \, d\mu, \ldots, \int_B h_m \, d\mu \right) : \mathcal{B} \to \mathbb{R}^m$$

is an $\mathbb{R}^m$-valued measure. Proposition 5.3.9 below shows that each $\mathbb{R}^m$-valued measure is of this form.

Lemma 5.3.2. *A nonnegative signed measure ν has a unique extension to a Radon measure μ.*

PROOF. Since all compact subsets of Ω belong to $\mathcal{B}$, Theorem 1.3.1 shows that there is at most one Radon measure μ with $\mu \restriction \mathcal{B} = \nu$. It remains to prove the existence. For every $E \subset \Omega$, define

$$\mu(E) := \inf_{\mathcal{A}} \sum \{\nu(A) : A \in \mathcal{A}\}$$

where $\mathcal{A} \subset \mathcal{B}$ is a countable cover of E. We show first that $\mu : E \mapsto \mu(E)$ is a measure. As ν is real-valued,

$$\nu(\emptyset) = \nu(\emptyset \cup \emptyset) = \nu(\emptyset) + \nu(\emptyset)$$

yields $\nu(\emptyset) = 0$, and hence $\mu(\emptyset) = 0$. Let $E \subset \bigcup_{i=1}^{\infty} E_i \subset \Omega$, and assume $\mu(E_i) < \infty$ for every $i \in \mathbb{N}$. Choose $\varepsilon > 0$ and select $B_{i,j} \in \mathcal{B}$ so that $E_i \subset \bigcup_{j=1}^{\infty} B_{i,j}$ and

$$\sum_{j=1}^{\infty} \nu(B_{i,j}) < \mu(E_i) + \varepsilon 2^{-i}$$

for $i = 1, 2, \dots$. From $E \subset \bigcup_{i,j=1}^{\infty} B_{i,j}$, we obtain

$$\mu(E) \leq \sum_{i,j=1}^{\infty} \nu(B_{i,j}) = \sum_{i=1}^{\infty} \sum_{j=1}^{\infty} \nu(B_{i,j}) < \sum_{i=1}^{\infty} \mu(E_i) + \varepsilon.$$

The arbitrariness of ε implies $\mu(E) \leq \sum_{i=1}^{\infty} \mu(E_i)$, and this inequality holds trivially when $\mu(E_i) = \infty$ for one or more $i \in \mathbb{N}$.

If $A, B \subset \Omega$ and $\mathrm{dist}\,(A, B) > 0$, find disjoint open sets $U, V \subset \Omega$ so that $A \subset U$ and $B \subset V$. Given $\varepsilon > 0$, there is a sequence $\{C_i\}$ in $\mathcal{B}$ such that $A \cup B \subset \sum_{i=1}^{\infty} C_i$ and

$$\mu(A \cup B) + \varepsilon > \sum_{i=1}^{\infty} \nu(C_i) \geq \sum_{i=1}^{\infty} \nu\big[C_i \cap (U \cup V)\big]$$

$$= \sum_{i=1}^{\infty} \nu(C_i \cap U) + \sum_{i=1}^{\infty} \nu(C_i \cap V)$$

$$\geq \mu(A) + \mu(B).$$

It follows that μ is a metric, and hence Borel, measure.

Given $E \subset \Omega$ with $\mu(E) < \infty$, there are sequences $\{B_{j,k}\}$ in $\mathcal{B}$ such that $E \subset \bigcup_{j=1}^{\infty} B_{j,k}$ and $\sum_{j=1}^{\infty} \nu(B_{j,k}) < \mu(E) + 1/k$. The set $B = \bigcap_{k=1}^{\infty} \bigcup_{j=1}^{\infty} B_{j,k}$ is Borel, $E \subset B \subset \bigcup_{j=1}^{\infty} B_{j,k}$, and

$$\mu(E) \leq \mu(B) \leq \sum_{j=1}^{\infty} \nu(B_{j,k}) < \mu(E) + 1/k$$

for $k = 1, 2, \dots$. This implies that μ is Borel regular. Since $\mu \upharpoonright \mathcal{B} = \nu$, we conclude that μ is a Radon measure. $\square$

Convention 5.3.3. In view of Lemma 5.3.2, we tacitly identify a nonnegative signed measure in Ω with its unique extension to a Radon measure in Ω, and denote both by the same symbol.

Corollary 5.3.4. *If μ is a Radon measure in Ω, then so is $\mu \llcorner A$ for each μ measurable set $A \subset \Omega$, and $\mu \llcorner A = \mu \llcorner \chi_A$.*

PROOF. There is an increasing sequence $\{A_k\}$ of μ measurable sets such that $A = \bigcup_{k=1}^{\infty} A_k$ and $\mu(A_k) < \infty$ for all $k \in \mathbb{N}$. Each $\mu \llcorner A_k$ is a Radon measure by Theorem 1.3.1. Given $E \subset \Omega$, there are Borel sets $B_k \subset \Omega$ such that

$E \subset B_k$ and $(\mu \mathbin{\llcorner} A_k)(E) = (\mu \mathbin{\llcorner} A_k)(B_k)$. The intersection $B = \bigcap_{k \in \mathbb{N}} B_k$ is a Borel set containing E, and

$$(\mu \mathbin{\llcorner} A)(E) = \lim(\mu \mathbin{\llcorner} A_k)(E) = \lim(\mu \mathbin{\llcorner} A_k)(B) = \lim(\mu \mathbin{\llcorner} A)(B).$$

As $(\mu \mathbin{\llcorner} A)(K) < \infty$ for every compact set $K \subset \Omega$, the previous equality implies that $\mu \mathbin{\llcorner} A$ is a Radon measure. The corollary follows from Lemma 5.3.2, since for each $C \in \mathcal{B}$,

$$(\mu \mathbin{\llcorner} A)(C) = \int_C \chi_A \, d\mu = (\mu \mathbin{\llcorner} \chi_A)(C). \qquad \square$$

Lemma 5.3.5. *Let $\{x_1, \ldots, x_p\} \subset \mathbb{R}^m$. There are a set $C \subset \mathbb{R}^m$ and constant $\beta = \beta(m) > 0$ such that*

$$\left| \sum_{x_j \in C} x_j \right| \geq \beta \sum_{j=1}^p |x_j|.$$

PROOF. For $y \in S^{m-1}$ and $B_y = B(y, 1/2)$, denote by C_y the intersection of all closed cones C in $\mathbb{R}^m$ such that the vertex of C is at the origin and $B_y \subset C$. The axis of C_y is the ray emanating from the origin and passing through y. Denote by π the orthogonal projection from C_y onto its axis, and observe that $|z| \leq 2|\pi(z)|$ for each $z \in C_y$. Thus

$$\sum_{i=1}^k |z_i| \leq 2 \sum_{i=1}^k |\pi(z_i)| = 2 \left| \sum_{i=1}^k \pi(z_i) \right| = 2 \left| \pi \left(\sum_{i=1}^k z_i \right) \right| \leq 2 \left| \sum_{i=1}^k z_i \right|$$

for each collection $\{z_1, \ldots, z_k\} \subset C_y$. Since S^{m-1} is compact, there are points $y_1, \ldots, y_q$ in S^{m-1} such that $S^{m-1} \subset \bigcup_{i=1}^q B_{y_i}$, and thus $\mathbb{R}^m = \bigcup_{i=1}^q C_{y_i}$. There is $C = C_{y_i}$ such that

$$\left| \sum_{x_j \in C} x_j \right| \geq \frac{1}{2} \sum_{x_j \in C} |x_j| \geq \frac{1}{2} \cdot \frac{1}{q} \sum_{j=1}^p |x_j|. \qquad \square$$

Let ν be an $\mathbb{R}^m$-valued measure. For $B \in \mathcal{B}$, define

$$\|\nu\|(B) := \sup_{\mathfrak{e}} \sum \left\{ |\nu(C)| : C \in \mathcal{C} \right\} \tag{5.3.1}$$

where $\mathcal{C}$ is a division of B.

Proposition 5.3.6. *If ν is an $\mathbb{R}^m$-valued measure, then*

$$\|\nu\| : B \mapsto \|\nu\|(B) : \mathcal{B} \to \mathbb{R}$$

*is a nonnegative signed measure, called the **variation** of ν.*

PROOF. Let $B \in \mathcal{B}$ be the union of a sequence $\{B_i\}$ in $\mathcal{B}$ consisting of disjoint sets. If $\mathcal{C}$ is a division of B, then $\{B_i \cap C : C \in \mathcal{C}\}$ is a division of B_i and $\{B_i \cap C : i = 1, 2, \dots\}$ is a division of $C \in \mathcal{C}$. Thus

$$\sum_{C \in \mathcal{C}} |\nu(C)| = \sum_{C \in \mathcal{C}} \left| \sum_{i=1}^{\infty} \nu(B_i \cap C) \right| \leq \sum_{C \in \mathcal{C}} \sum_{i=1}^{\infty} |\nu(B_i \cap C)|$$

$$= \sum_{i=1}^{\infty} \sum_{C \in \mathcal{C}} |\nu(B_i \cap C)| \leq \sum_{i=1}^{\infty} \|\nu\|(B_i),$$

and so $\|\nu\|(B) \leq \sum_{i=1}^{\infty} \|\nu\|(B_i)$. Next choose a division $\mathcal{C}_i$ of B_i, and observe that $\mathcal{C} = \bigcup_{i=1}^{\infty} \mathcal{C}_i$ is a division of B. Hence

$$\sum_{i=1}^{\infty} \sum_{C \in \mathcal{C}_i} |\nu(C)| \leq \|\nu\|(B).$$

From this inequality it is easy to infer $\sum_{i=1}^{\infty} \|\nu\|(B_i) \leq \|\nu\|(B)$.

It remains to show that $\|\nu\|(B) < \infty$ for each $B \in \mathcal{B}$. Seeking a contradiction, assume there is $B \in \mathcal{B}$ with $\|\nu\|(B) = \infty$. Let $\{C_j\}$ be a division of B such that

$$\sum_{j=1}^{p} |\nu(C_j)| > \frac{1}{\beta} \left(1 + |\nu(B)| \right)$$

where β is the constant from Lemma 5.3.2. After a suitable reordering,

$$\left| \nu \left(\bigcup_{j=1}^{q} C_j \right) \right| = \left| \sum_{j=1}^{q} \nu(C_j) \right| \geq \beta \sum_{j=1}^{p} |\nu(C_j)| > 1 + |\nu(B)|$$

for a positive integer $q \leq p$; see Lemma 5.3.2. Letting $C = \bigcup_{j=1}^{q} C_j$, we obtain $C \subset B$, $|\nu(C)| > 1$, and

$$|\nu(B - C)| = |\nu(B) - \nu(C)| \geq |\nu(C)| - |\nu(B)| > 1.$$

As $\|\nu\|(B) = \infty$, either $\|\nu\|(C) = \infty$ or $\|\nu\|(B - C) = \infty$. It follows there are disjoint Borel subsets B_1 and C_1 of B such that

$$\|\nu\|(B_1) = \infty \quad \text{and} \quad |\nu(C_1)| > 1.$$

Replacing B by B_1, find disjoint Borel subsets B_2 and C_2 of B_1 with

$$\|\nu\|(B_2) = \infty \quad \text{and} \quad |\nu(C_2)| > 1.$$

Proceeding recursively, we obtain a sequence $\{C_j\}$ in $\mathcal{B}$ consisting of disjoint subsets of B such that $|\nu(C_j)| > 1$ for $j = 1, 2, \dots$. Thus $\sum_{j=1}^{\infty} \nu(C_j)$ does not converge. Since $\bigcup_{i=1}^{\infty} C_j$ belongs to $\mathcal{B}$, this is a contradiction. $\qquad \square$

If ν is a signed measure in Ω, then

$$\nu_+ := \tfrac{1}{2}(\|\nu\| + \nu) \quad \text{and} \quad \nu_- = \tfrac{1}{2}(\|\nu\| - \nu) \qquad (5.3.2)$$

are nonnegative signed measures. They are called, respectively, the *positive* and *negative parts* of ν, and the identity $\nu = \nu_+ - \nu_-$ is referred to as the *Jordan decomposition* of ν.

Proposition 5.3.7. *Let μ be a Radon measure, and let $\nu = \mu \llcorner h$ for an $h \in L^1_{\mathrm{loc}}(\Omega, \mu; \mathbb{R}^m)$. Then $\|\nu\| = \mu \llcorner |h|$.*

PROOF. If $\{A_i\}$ is a division of $A \in \mathcal{B}$, then

$$\|\nu\|(A) \leq \sum_{i=1}^{\infty} \left| \int_{A_i} h \, d\mu \right| \leq \sum_{i=1}^{\infty} \int_{A_i} |h| \, d\mu = \int_A |h| \, d\mu = (\mu \llcorner |h|)(A).$$

For the reverse inequality, we fix $B \in \mathcal{B}$ and consider two cases.

Case 1. Assume $h(\Omega) = \{y_1, y_2, \dots\}$, and observe that the sets

$$B_i = \{x \in B : h(x) = y_i\}, \quad i = 1, 2, \dots,$$

form a division of B. Hence

$$\|\nu\|(B) \geq \sum_{i=1}^{\infty} |\nu(B_i)| = \sum_{i=1}^{\infty} \left| \int_{B_i} h \, d\mu \right|$$

$$= \sum_{i=1}^{\infty} |y_i| \mu(B_i) = \int_B |h| \, d\mu.$$

Case 2. For an arbitrary $h \in L^1_{\mathrm{loc}}(\Omega, \mu; \mathbb{R}^m)$, choose $\varepsilon > 0$ and construct a map $f \in L^1_{\mathrm{loc}}(\Omega, \mu; \mathbb{R}^m)$ with countably many values so that $|h(x) - f(x)| < \varepsilon$ for all $x \in \Omega$. Note that a grid in $\mathbb{R}^m$ obtained by translating the half-open cube $[0, \varepsilon\sqrt{m})^m$ facilitates the construction of f. Define an $\mathbb{R}^m$-valued measure $\lambda := \mu \llcorner f$, and find a division $\{C_i\}$ of B satisfying the inequality $\sum_{i=1}^{\infty} |\lambda(C_i)| > \|\lambda\|(B) - \varepsilon$. Since

$$\left| \sum_{i=1}^{\infty} |\nu(C_i)| - \sum_{i=1}^{\infty} |\lambda(C_i)| \right| \leq \sum_{i=1}^{\infty} |\nu(C_i) - \lambda(C_i)|$$

$$\leq \sum_{i=1}^{\infty} \int_{C_i} |h - f| \, d\mu < \varepsilon \mu\,(B),$$

applying Case 1, we obtain

$$\|\nu\|(B) \geq \sum_{i=1}^{\infty} |\nu(C_i)| > \sum_{i=1}^{\infty} |\lambda(C_i)| - \varepsilon\mu(B) > \|\lambda\|(B) - \varepsilon[1 + \mu(B)]$$

$$\geq \int_B |f| \, d\mu - \varepsilon[1 + \mu(B)] > \int_B |h| \, d\mu - \varepsilon[1 + 2\mu(B)].$$

The desired inequality follows from the arbitrariness of ε. □

Proposition 5.3.8. *Let μ be a Radon measure, and $h \in L^1_{loc}(\Omega, \mu; \mathbb{R}^m)$. If $C \subset \mathbb{R}^m$ is a closed set such that the average*

$$(h)_{B,\mu} := \frac{1}{\mu(B)} \int_B h \, d\mu$$

belongs to C for each Borel set $B \Subset \Omega$ with $\mu(B) > 0$, then $h(x)$ belongs to C for μ almost all $x \in \Omega$.

PROOF. There are countable sequences $\{B_i\}$ in $\mathbf{B}$ and $\{B(y_j, r_j)\}$ of closed balls in $\mathbb{R}^m$ such that $\Omega = \bigcup_{i=1}^\infty B_i$ and $\mathbb{R}^m - C = \bigcup_{j=1}^\infty B(y_j, r_j)$. Proceeding toward a contradiction, assume that the set

$$\{x \in \Omega : h(x) \in \mathbb{R}^m - C\} = \bigcup_{i,j=1}^\infty \{x \in B_i : h(x) \in B(y_j, r_j)\}$$

is not μ negligible. It follows that there are $B \in \mathbf{B}$ with $\mu(B) > 0$ and $B(y, r) \subset \mathbb{R}^m - C$ such that $h(x) \in B(y, r)$ for each $x \in B$. Since

$$|(h)_{B,\mu} - y| = \frac{1}{\mu(B)} \left| \int_B [h(x) - y] \, d\mu(x) \right|$$

$$\leq \frac{1}{\mu(B)} \int_B |h(x) - y| \, d\mu(x) \leq r,$$

we see that $(h)_{B,\mu}$ belongs to $B(y, r) \subset \mathbb{R}^m - C$, a contradiction. $\square$

Proposition 5.3.9. *Let ν be an $\mathbb{R}^m$-valued measure. Up to a $\|\nu\|$ equivalence, there is a unique $s \in L^1_{loc}(\Omega, \|\nu\|; \mathbb{R}^n)$ such that $\nu = \|\nu\| \llcorner s$. In addition $|s(x)| = 1$ for $\|\nu\|$ almost all $x \in \Omega$.*

PROOF. If $\nu = (\nu_1, \ldots, \nu_m)$ then $(\nu_i)_\pm \leq \|\nu_i\| \leq \|\nu\|$; see (5.3.2). By the Radon-Nikodym theorem [29, Section 1.6.2, Theorem 2], there are functions $s_{i,\pm} \in L^1_{loc}(\Omega, \|\nu\|)$ such that for each $B \in \mathbf{B}$,

$$\nu_i(B) = (\nu_i)_+(B) - (\nu_i)_-(B) = \int_B s_{i,+} \, d\|\nu\| - \int_B s_{i,-} \, d\|\nu\|;$$

see Convention 5.3.3. If $s_i = s_{i,+} - s_{i,-}$, then $s = (s_1, \ldots, s_m)$ belongs to $L^1_{loc}(\Omega, \|\nu\|; \mathbb{R}^m)$ and $\nu = \|\nu\| \llcorner s$.

To establish the uniqueness, assume $\nu = \|\nu\| \llcorner h$ where $h = (h_1, \ldots, h_m)$ is a map in $L^1_{loc}(\Omega, \|\nu\|; \mathbb{R}^m)$. Then $\int_B (s_i - h_i) \, d\|\nu\| = 0$ for $i = 1, \ldots, m$ and each $B \in \mathbf{B}$. It follows that h_i is $\|\nu\|$ equivalent to s_i for $i = 1, \ldots, m$.

Let $B \in \mathbf{B}$ and $N_k = \{x \in B : |s(x)| \leq \frac{k-1}{k}\}$ for all $k \in \mathbb{N}$. If $\{C_i\}$ is a division of N_k, then

$$\sum_{i=1}^\infty |\nu(C_i)| = \sum_{i=1}^\infty \left| \int_{C_i} s \, d\|\nu\| \right| \leq \int_{C_i} |s| \, d\|\nu\|$$

$$\leq \sum_{i=1}^\infty \frac{k-1}{k} \|\nu\|(C_i) = \frac{k-1}{k} \|\nu\|(N_k)$$

and consequently $\|\nu\|(N_k) \leq \frac{k-1}{k}\|\nu\|(N_k)$. Thus N_k is $\|\nu\|$ negligible, and so is $\bigcup_{k=1}^{\infty} N_k = \{x \in B : |s(x)| < 1\}$. As Ω is the union of countably many sets $B \in \mathcal{B}$, we see that $|s(x)| \geq 1$ for $\|\nu\|$ almost all $x \in \Omega$. On the other hand, if $B \in \mathcal{B}$ and $\|\nu\|(B) > 0$, then

$$(s)_{B,\|\nu\|} = \frac{1}{\|\nu\|(B)}\left|\int_B s\,d\|\nu\|\right| = \frac{|\nu(B)|}{\|\nu\|(B)} \leq 1.$$

Hence $|s(x)| \leq 1$ for $\|\nu\|$ almost all $x \in \Omega$ by Proposition 5.3.8. $\qquad\square$

Let ν be an $\mathbb{R}^m$-valued measure. The identity

$$\nu = \|\nu\| \, \llcorner \, s, \tag{5.3.3}$$

whose existence was established in Proposition 5.3.9, is called the *polar decomposition* of ν. Since $\|\nu\|$ is a Radon measure in Ω, we may assume that s is a Borel map. Using the polar decomposition, we define

$$\int_B v \cdot d\nu := \int_B v \cdot s\,d\mu \tag{5.3.4}$$

for $v \in L^{\infty}(\Omega, \mu; \mathbb{R}^m)$ and $B \in \mathcal{B}$.

Remark 5.3.10. Let $\nu = \|\nu\| \llcorner s$ be the polar decomposition of a signed measure ν. Since $|s(x)| = 1$ for $\|\nu\|$ almost all $x \in \Omega$ by Proposition 5.3.9,

$$\nu_{\pm} = \tfrac{1}{2}\|\nu\| \llcorner (1 \pm s) = \tfrac{1}{2}\|\nu\| \llcorner \big(|s| \pm s\big) = \|\nu\| \llcorner s^{\pm}.$$

Proposition 5.3.11. *Let ν be an $\mathbb{R}^n$-valued measure. If $\int_{\Omega} v \cdot d\nu = 0$ for each $v \in C_c^{\infty}(\Omega; \mathbb{R}^n)$, then $\nu \equiv 0$.*

PROOF. Let $\nu = \mu \llcorner s$ be the polar decomposition of ν. If v belongs to $C_c(\Omega; \mathbb{R}^n)$, construct by mollification a sequence $\{v_j\}$ in $C_c^{\infty}(\Omega; \mathbb{R}^n)$ such that $\lim \|v - v_j\|_{L^{\infty}(\Omega;\mathbb{R}^n)} = 0$ and $\operatorname{spt} v_j \subset C$ for a compact set $C \subset \Omega$ and $j = 1, 2, \ldots$. By our assumption,

$$\left|\int_{\Omega} v \cdot s\,d\mu\right| = \left|\int_{\Omega} (v - v_j) \cdot s\,d\mu\right| \leq \|v - v_j\|_{L^{\infty}(\Omega;\mathbb{R}^n)}\mu(C)$$

and hence $\int_{\Omega} v \cdot s\,d\mu = 0$.

Next choose an open set $U \Subset \Omega$ and $\varepsilon > 0$. By Luzin's theorem there is a compact set $K \subset U$ such that $s \restriction K$ is continuous and $\mu(U - K) < \varepsilon$. In view of Corollary 1.1.2, the map $s \restriction K$ extends to $v \in C_c(\Omega; \mathbb{R}^n)$ such that $\|v\|_{L^{\infty}(\Omega;\mathbb{R}^n)} \leq 1$ and $\operatorname{spt} v \subset U$. Since the first part of the proof yields $\int_U v \cdot s\,d\mu = \int_{\Omega} v \cdot s\,d\mu = \int_{\Omega} v \cdot d\nu = 0$, we obtain

$$\mu(K) = \int_K v \cdot s\,d\mu = \int_U v \cdot s\,d\mu - \int_{U-K} v \cdot s\,d\mu < \varepsilon.$$

Hence $\mu(U) \leq \mu(K) + \mu(U - K) < 2\varepsilon$, and the arbitrariness of ε implies $\mu(U) = 0$. As Ω is covered by countably many open sets $U \Subset \Omega$, we conclude $\mu \equiv 0$ and consequently $\nu \equiv 0$. $\qquad\square$

The *reduction* of an $\mathbb{R}^m$-valued measure to a Borel set $A \subset \Omega$ is an $\mathbb{R}^m$-valued measure $\nu \llcorner A$ defined by the formula

$$(\nu \llcorner A)(B) := \nu(A \cap B)$$

for each $B \in \mathcal{B}$. If $\nu = \nu \llcorner A$ we say that ν *lives in* A.

Observation 5.3.12. *Let ν be an $\mathbb{R}^m$-valued measure, and let $A \subset \Omega$ be a Borel set. If $\nu = \|\nu\| \llcorner s$ is the polar decomposition of ν, then*

$$\nu \llcorner A = (\|\nu\| \llcorner A) \llcorner s$$

is the polar decomposition of $\nu \llcorner A$. In particular $\|\nu \llcorner A\| = \|\nu\| \llcorner A$.

PROOF. Since

$$(\nu \llcorner A)(B) = \int_{A \cap B} s \, d\|\nu\| = \int_B s \, d(\|\nu\| \llcorner A)$$

for each $B \in \mathcal{B}$, we see that $\nu \llcorner A = (\|\nu\| \llcorner A) \llcorner s$. As

$$\|\nu \llcorner A\| = (\|\nu\| \llcorner A) \llcorner |s| = \|\nu\| \llcorner A$$

by Propositions 5.3.7 and 5.3.9, the observation follows from the uniqueness part of Propositions 5.3.9   □

5.4. Weak convergence

Throughout this section we again fix an open set $\Omega \Subset \mathbb{R}^n$.

Definition 5.4.1. A sequence $\{\nu_k\}$ of $\mathbb{R}^n$-valued measures *converges weakly* to an $\mathbb{R}^n$-valued measure ν if

$$\lim \int_\Omega v \cdot d\nu_k = \int_\Omega v \cdot d\nu$$

for every $v \in C_c(\Omega; \mathbb{R}^m)$.

If a sequence $\{\nu_k\}$ of $\mathbb{R}^n$-valued measures converges weakly to ν, we write

$$\text{w-lim}\, \nu_k = \nu.$$

Observation 5.4.2. *If $\{\mu_k\}$ is a weakly convergent sequence of Radon measures, then $\mu = \text{w-lim}\, \mu_k$ is a Radon measure.*

PROOF. Let $\mu = \|\mu\| \llcorner s$ be the polar decomposition of μ, and assume that $\|\mu\|(\{s < 0\}) > 0$. Find a compact set $K \subset \{s < 0\}$ with $\|\mu\|(K) > 0$, and $\varphi_j \in C_c(\Omega)$ so that $0 \leq \varphi_j \leq 1$ for $j = 1, 2, \ldots$, and $\lim \varphi_j = \chi_K$. Since $\lim \int_\Omega \varphi_j s \, d\|\mu\| = \int_K s \, d\|\mu\| < 0$, there is $\varphi = \varphi_j$ such that

$$0 > \int_\Omega \varphi s \, d\|\mu\| = \int_\Omega \varphi \, d\mu = \lim \int_\Omega \varphi \, d\mu_k \geq 0.$$

This contradiction implies $\mu = \|\mu\|$.   □

Theorem 5.4.3. *If μ and μ_j, $j = 1, 2, \ldots$, are Radon measures, then the following conditions are equivalent:*

(1) w-$\lim \mu_j = \mu$;

(2) $\mu(U) \leq \liminf \mu_j(U)$ *and* $\limsup \mu_j(K) \leq \mu(K)$ *for each open set $U \subset \Omega$ and each compact set $K \subset \Omega$;*

(3) $\lim \mu_j(B) = \mu(B)$ *for each Borel set $B \Subset \Omega$ such that $\mu(\partial B) = 0$.*

PROOF. $(1) \Rightarrow (2)$ Choose an open set $U \subset \Omega$ and a compact set $K \subset U$, and find $\varphi \in C_c(\Omega)$ with $\chi_K \leq \varphi \leq \chi_U$. Then

$$\mu(K) \leq \int_\Omega \varphi \, d\mu = \lim \int_\Omega \varphi \, d\mu_j \leq \liminf \mu_j(U),$$

$$\limsup \mu_j(K) \leq \lim \int_\Omega \varphi \, d\mu_j = \int_\Omega \varphi \, d\mu \leq \mu(U).$$

Since U and K are arbitrary and μ is a Radon measure, condition (2) follows from Theorem 1.3.1.

$(2) \Rightarrow (3)$ If $B \Subset \Omega$ is a Borel set and $\mu(\partial E) = 0$, then

$$\mu(E) = \mu(\text{int } E) \leq \liminf \mu_j(\text{int } E)$$
$$\leq \limsup \mu_j(\text{cl } E) \leq \mu(\text{cl } E) = \mu(E).$$

$(3) \Rightarrow (1)$ From Lemma 4.5.5 we deduce that only countably many hyperplanes perpendicular to the vectors $e_1, \ldots, e_k$ have positive μ measure. It follows that given $\rho > 0$, there is a positive $r < \rho$ such that $\mu(\partial C) = 0$ for each half-open cube

$$C = \prod_{i=1}^n [j_i r, (j_i + 1)r) \tag{$*$}$$

where $j_1, \ldots, j_n$ are integers. Given $\varphi \in C_c(\Omega)$ and $\varepsilon > 0$, find $\delta > 0$ so that $\left|\varphi(x) - \varphi(y)\right| \leq \varepsilon$ for all $x, y \in \Omega$ with $|x - y| < \delta$. There is a positive $r < \delta / \sqrt{n}$ and disjoint half-open cubes $C_1, \ldots, C_p$, defined in $(*)$, such that spt $\varphi \subset \bigcup_{j=1}^p C_j \subset \Omega$ and $\mu(\partial C_j) = 0$ for $j = 1, \ldots, p$. If

$$s_j = \inf_{x \in C_j} \varphi(x) \quad \text{and} \quad S_j = \sup_{x \in C_j} \varphi(x),$$

then $0 \leq S_j - s_j \leq \varepsilon$ and

$$s_j \mu_k(C_j) \leq \int_{C_j} \varphi \, d\mu_k \leq S_j \mu_k(C_j),$$

$$s_j \mu(C_j) \leq \int_{C_j} \varphi \, d\mu \leq S_j \mu(C_j).$$

With $D = \bigcup_{j=1}^{p} C_j$ and $\beta = \max\{|s_j| + |S_j| : j = 1, \ldots, p\}$, we obtain

$$\left| \int_\Omega \varphi \, d\mu_k - \int_\Omega \varphi \, d\mu \right| \leq \sum_{j=1}^{p} \left| \int_{C_j} \varphi \, d\mu_k - \int_{C_j} \varphi \, d\mu \right|$$

$$\leq \sum_{j=1}^{p} \left([S_j - s_j] \mu_k(C_j) + (|s_j| + |S_j|) |\mu_k(C_j) - \mu(C_j)| \right)$$

$$\leq \varepsilon \mu_k(D) + \beta |\mu_k(D) - \mu(D)|.$$

As $\mu(\partial D) = 0$, letting $k \to \infty$ yields

$$\left| \lim \int_\Omega \varphi \, d\mu_k - \int_\Omega \varphi \, d\mu \right| \leq \varepsilon \mu(D).$$

The proposition follows from the arbitrariness of ε. $\square$

We show that $\mathbb{R}^n$-valued measures possess a compactness property with respect to weak convergence. The argument relies on the *Riesz representation theorem*, stated below without proof. The reader interested in proving the Riesz theorem is referred to [29, Section 1.8].

Recall from (1.1.2) that for an open set $U \subset \Omega$ there is a legitimate inclusion $C_c(U; \mathbb{R}^m) \subset C_c(\Omega; \mathbb{R}^m)$.

Theorem 5.4.4 (Riesz). *Let $F : C_c(\Omega; \mathbb{R}^m) \to \mathbb{R}$ be a linear functional such that for each open set $U \Subset \Omega$,*

$$\sup\{F(v) : v \in C_c(U; \mathbb{R}^m) \text{ and } \|v\|_{L^\infty(U;\mathbb{R}^m)} \leq 1\} < \infty.$$

There is a unique $\mathbb{R}^m$-valued measure ν in Ω satisfying

$$F(v) = \int_\Omega v \cdot d\nu$$

for all $v \in C_c(\Omega; \mathbb{R}^m)$. In addition, for every open set $U \Subset \Omega$,

$$\|\nu\|(U) = \sup\{F(v) : v \in C_c(U; \mathbb{R}^m) \text{ and } \|v\|_{L^\infty(\Omega;\mathbb{R}^m)} \leq 1\}.$$

Remark 5.4.5. Let $\mathcal{S}$ be the injective limit topology in $C_c(\Omega; \mathbb{R}^m)$ defined in Example 3.6.4. Denote by $M(\Omega; \mathbb{R}^m)$ the linear space of all $\mathbb{R}^m$-valued measures in Ω , and observe that for each $\nu \in M(\Omega; \mathbb{R}^m)$, the linear functional

$$F_\nu : v \mapsto \int_\Omega v \cdot d\nu : \left(C_c(\Omega, \mathbb{R}^m), \mathcal{S} \right) \to \mathbb{R}$$

is continuous. With this notation, the Riesz theorem can be reformulated as follows.

Theorem. The map $\nu \mapsto F_\nu$ is a linear isomorphism from $M(\Omega; \mathbb{R}^m)$ onto the dual space $\left(C_c(\Omega; \mathbb{R}^m), \mathcal{S} \right)^*$, and $\|F_\nu\|_U = \|\nu\|(U)$ for each open set $U \Subset \Omega$.

Corollary 5.4.6. *If ν is an $\mathbb{R}^m$ valued measure, then*

$$\|\nu\|(U) = \sup\left\{ \int_\Omega v \cdot d\nu : v \in C_c(U; \mathbb{R}^m) \text{ and } \|v\|_{L^\infty(\Omega;\mathbb{R}^m)} \leq 1 \right\}$$

for every open set $U \subset \Omega$.

Proof. Since it is clear that the linear functional

$$F : v \mapsto \int_{\Omega} v \cdot d\nu : C_c(\Omega; \mathbb{R}^m) \to \mathbb{R}$$

satisfies the conditions of the Riesz theorem, the desired equality holds for each open set $V \Subset \Omega$. If U is any open subset of Ω, then

$$\|\nu\|(U) = \sup\{\|\nu\|(V) : V \text{ is open and } V \Subset U\}$$

$$= \sup\left\{ \int_{\Omega} v \cdot d\nu : v \in C_c(V; \mathbb{R}^m),\ \|v\|_{L^{\infty}(\Omega;\mathbb{R}^m)} \leq 1,\ V \Subset U \right\}$$

$$= \sup\left\{ \int_{\Omega} v \cdot d\nu : v \in C_c(U; \mathbb{R}^m) \text{ and } \|v\|_{L^{\infty}(\Omega;\mathbb{R}^m)} \leq 1 \right\}. \qquad \square$$

Proposition 5.4.7. *Let $\{\nu_k\}$ be a sequence of $\mathbb{R}^m$-valued measures that converges weakly to an $\mathbb{R}^m$-valued measure ν. Then*

$$\|\nu\|(U) \leq \liminf \|\nu_k\|(U)$$

for each open set $U \subset \Omega$.

Proof. Let $U \subset \Omega$ be an open set, and let $v \in C_c(U; \mathbb{R}^m)$ be such that $\|v\|_{L^{\infty}(\Omega;\mathbb{R}^m)} \leq 1$. Using our assumption and the polar decompositions of ν_k, we obtain

$$\int_{\Omega} v \cdot d\nu = \lim \int_{\Omega} v \cdot d\nu_k \leq \liminf \|\nu_k\|(U).$$

As v is arbitrary, the proposition follows from Corollary 5.4.6. $\qquad \square$

Theorem 5.4.8. *Let $\{\nu_k\}$ be a sequence of $\mathbb{R}^m$-valued measures. If*

$$\limsup \|\nu_k\|(U) < \infty$$

for each open set $U \Subset \Omega$, then $\{\nu_k\}$ has a subsequence which converges weakly to an $\mathbb{R}^m$-valued measure.

Proof. There is a sequence $\{K_i\}$ of compact sets such that $\Omega = \bigcup_{i=1}^{\infty} K_i$ and $K_i \Subset K_{i+1}$ for $i = 1, 2, \ldots$. Construct recursively subsequences $s_i = \{\nu_{i,j}\}$ of $\{\nu_k\}$ so that $\sup_j \|\nu_{i,j}\|(K_i) < \infty$ and s_{i+1} is a subsequence of s_i. Then $\{\nu_{j,j}\}$ is a subsequence of $\{\nu_k\}$ and $\sup_j \|\nu_{j,j}\|(K_i) < \infty$ for $i = 1, 2, \ldots$. Since every Borel set $B \Subset \Omega$ is contained in some K_j, we obtain that $\sup \|\nu_{j,j}\|(B) < \infty$ for each Borel set $B \Subset \Omega$. To simplify the notation, let $\mu_k := \nu_{k,k}$ for $k = 1, 2, \ldots$.

Topologize $C_c(\Omega; \mathbb{R}^m)$ and each $C_i := \{v \in C_c(\Omega; \mathbb{R}^m) : \operatorname{spt} v \subset K_i\}$ by the L^{∞} norm. Choose $D = \{v_p \in C_c(\Omega; \mathbb{R}^m) : p \in \mathbb{N}\}$ so that $D \cap C_i$ is dense in C_i for $i = 1, 2, \ldots$. By the previous paragraph,

$$\sup_{k \in \mathbb{N}} \left| \int_{\Omega} v_p \cdot d\mu_k \right| \leq \sup_{k \in \mathbb{N}} \int_{\Omega} |v_p|\, d\|\mu_k\|$$

$$\leq \|v_p\|_{L^{\infty}(\Omega;\mathbb{R}^m)} \sup_{k \in \mathbb{N}} \|\mu_k\|(\operatorname{spt} v_p) < \infty$$

for every $p \in \mathbb{N}$. Thus each sequence $\{\int_\Omega v_p \cdot d\mu_k\}_{k=1}^\infty$ has a convergent subsequence. Using a diagonal argument similar to that employed in the previous paragraph, construct a subsequence $\{\mu_{k_j}\}$ of $\{\mu_k\}$ so that for every $p \in \mathbb{N}$ there exists a finite limit

$$G(v_p) := \lim_{j\to\infty} \int_\Omega v_p \cdot d\mu_{k_j}.$$

Denote by H_i the linear hull of $D \cap C_i$, and note that $G \upharpoonright (D \cap C_i)$ has a unique linear extension $G_i : H_i \to \mathbb{R}$. Observe that

$$t_i := \sup\{\|\mu_k\|(K_i) : k \in \mathbb{N}\} < \infty$$

and $G_i(v) \le t_i \|v\|_{L^\infty(\Omega;\mathbb{R}^m)}$ for every v in H_i. Since H_i is a dense subspace of C_i, the linear functional G_i has a unique continuous extension $F_i : C_i \to \mathbb{R}$, necessarily linear and satisfying $F_i(v) \le t_i \|v\|_{L^\infty(\Omega;\mathbb{R}^m)}$ for each $v \in C_i$. As the uniqueness implies $F_{i+1} \upharpoonright C_i = F_i$, there is a linear functional F defined on $C_c(\Omega;\mathbb{R}^m) = \bigcup_{i=1}^\infty C_i$ whose restriction to C_i equals F_i for $i = 1, 2, \dots$. If $U \Subset \Omega$, then U is contained in some K_i. Thus $F \upharpoonright U = F_i$, and consequently

$$\sup\{F(v) : v \in C_c(U;\mathbb{R}^m) \text{ and } \|v\|_{L^\infty(U;\mathbb{R}^m)} \le 1\} \le t_i < \infty.$$

By the Riesz theorem, there is an $\mathbb{R}^m$-valued measure μ with $F(\varphi) = \int_\Omega v \cdot d\mu$ for every $v \in C_c(\Omega;\mathbb{R}^m)$. Choose $v \in C_c(\Omega;\mathbb{R}^m)$ and $\varepsilon > 0$. As $v \in C_i$ for some $i \in \mathbb{N}$, there is $v_p \in D \cap C_i$ with $\|v - v_p\|_{L^\infty(\Omega;\mathbb{R}^m)} < \varepsilon$. Note

$$\left| \int_\Omega v_p \cdot d\mu_{k_j} - \int_\Omega v_p \cdot d\mu \right| = \left| \int_\Omega v_p \cdot d\mu_{k_j} - G(v_p) \right| < \varepsilon$$

for all sufficiently large integers j. Since

$$\left| \int_\Omega v \cdot d\mu_{k_j} - \int_\Omega v_p \cdot d\mu_{k_j} \right| \le \int_{K_i} |v - v_p| \, d\|\mu_{k_j}\| \le \varepsilon t_i, \quad j = 1, 2, \dots,$$

$$\left| \int_\Omega v_p \cdot d\mu - \int_\Omega v \cdot d\mu \right| \le \int_{K_i} |v_p - v| \, d\|\mu\| \le \varepsilon \|\mu\|(K_i),$$

we obtain that for all sufficiently large integers j,

$$\left| \int_\Omega v \cdot d\mu_{k_j} - \int_\Omega v \cdot d\mu \right| \le \varepsilon [t_i + 1 + \|\mu\|(K_i)].$$

Hence $\int_\Omega v \cdot d\mu = \lim \int_\Omega v \cdot d\mu_{k_j}$ for each $v \in C_c(\Omega;\mathbb{R}^m)$. $\quad\square$

Proposition 5.4.9. *Let ν and ν_k be $\mathbb{R}^m$-valued measures, and let μ be a Radon measure in Ω. Assume that*

$$\text{w-}\lim \nu_k = \nu \quad \text{and} \quad \text{w-}\lim \|\nu_k\| = \mu.$$

Then $\|\nu\| \le \mu$, and for each Borel set $B \Subset \Omega$ with $\mu(\partial B) = 0$,

$$\lim \nu_k(B) = \nu(B).$$

PROOF. If $U \subset \Omega$ is an open set, find open sets $U_1 \Subset U_2 \Subset \cdots \Subset U$ so that $U = \bigcup_{i=1}^{\infty} U_i$. There are $\varphi_i \in C_c(\Omega)$ such that $\chi_{U_i} \leq \varphi_i \leq \chi_U$ for $i = 1, 2, \ldots$. By Proposition 5.4.7,

$$\|\nu\|(U_i) \leq \liminf_{k \to \infty} \|\nu_k\|(U_i) \leq \liminf_{k \to \infty} \int_\Omega \varphi_i \, d\|\nu_k\| = \liminf_{k \to \infty} \int_\Omega \varphi_i \, d\mu \leq \mu(U)$$

and hence $\|\nu\|(U) = \lim \|\nu\|(U_i) \leq \mu(U)$. As $\|\nu\|$ and μ are Radon measures, the inequality $\|\nu\| \leq \mu$ follows.

Choose a Borel set $B \Subset \Omega$ with $\mu(\partial B) = 0$. If

$$\nu_k = (\nu_k^1, \ldots, \nu_k^m) \quad \text{and} \quad \nu = (\nu^1, \ldots, \nu^m),$$

then w-$\lim \nu_k^i = \nu^i$ for $i = 1, \ldots, m$. Let $\nu_k^1 = \nu_{k+}^1 - \nu_{k-}^1$ be the Jordan decomposition of ν_k^1; see (5.3.3). As $\nu_{k\pm}^1 \leq \|\nu_k^1\| \leq \|\nu_k\|$, Theorem 5.4.8 shows that there is a subsequence $\{\nu_{s_1(k)}\}$ of $\{\nu_k\}$ such that the sequences $\{\nu_{s_1(k)\pm}^1\}$ converge weakly to Radon measures $\mu_\pm$. Clearly

$$\mu_+ - \mu_- = \text{w-}\lim(\nu_{s_1(k)+}^1 - \nu_{s_1(k)-}^1) = \text{w-}\lim \nu_{s_1(k)}^1 = \nu^1,$$

although this need not be the Jordan decomposition of ν^1. In view of Observation 5.4.2, the inequalities $\nu_{s_1(k)\pm}^1 \leq \|\nu_{s_1(k)}^1\|$, $k = 1, 2, \ldots$, yield $\mu_\pm \leq \mu$. In particular $\mu_\pm(\partial B) = 0$, and so $\mu_\pm(B) = \lim \nu_{s_1(k)\pm}^1(B)$ by Theorem 5.4.3. Hence

$$\nu^1(B) = \mu_+(B) - \mu_-(B) = \lim \left[\nu_{s_1(k)+}^1(B) - \nu_{s_1(k)-}^1(B)\right] = \lim \nu_{s_1(k)}^1(B).$$

Next find a subsequence $\{\nu_{s_2(k)}\}$ of $\{\nu_{s_1(k)}\}$ with $\nu^2(B) = \lim \nu_{s_2(k)}^2(B)$. Proceeding recursively, it is obvious that $\{\nu_{s_m(k)}\}$ is a subsequence of $\{\nu_k\}$ such that $\nu(B) = \lim \nu_{s_m(k)}(B)$. However, more is true: each subsequence $\{\nu_{s(k)}\}$ of $\{\nu_k\}$ contains a subsequence $\{\nu_{s(k_j)}\}$ such that $\nu(B) = \lim \nu_{s(k_j)}(B)$. This implies $\nu(B) = \lim \nu_k(B)$. $\qquad \square$

Example 5.4.10. Let $h_k(t) := \sin 2^k t$ for $t \in \mathbb{R}$ and $k \in \mathbb{N}$, and define $\nu_k := \mathcal{L}^1 \lfloor h_k$. If $f \in C_c(\mathbb{R})$, choose $\varepsilon > 0$ and find an open set $U \Subset \mathbb{R}$ containing spt f. There is $\varphi \in C_c^1(U)$ with $\|f - \varphi\|_{L^\infty(\mathbb{R})} < \varepsilon$. Thus

$$\left|\int_{\mathbb{R}} f \, d\nu_k\right| \leq \int_U |f(t) - \varphi(t)| \, dt + 2^{-k} \left|\int_{\mathbb{R}} \varphi'(t) \cos 2^k t \, dt\right|$$
$$\leq \varepsilon |U| + 2^{-k} \|\varphi'\|_{L^1(\mathbb{R})},$$

and hence w-$\lim \nu_k = 0$. As a direct verification shows $\|\nu_k\| = (2/\pi)\mathcal{L}^1$ for $k = 1, 2, \ldots$, we see that $\| \text{w-}\lim \nu_k\| < \text{w-}\lim \|\nu_k\|$.

5.5. Properties of BV functions

Theorem 5.5.1. *Let $f \in BV_{\mathrm{loc}}(\Omega)$. There is a unique $\mathbb{R}^n$-valued measure Df in Ω such that for each $v \in Lip_c(\Omega; \mathbb{R}^n)$,*

$$\int_\Omega f(x) \operatorname{div} v(x)\, dx = -\int_\Omega v \cdot d(Df). \tag{5.5.1}$$

If $U \subset \Omega$ is an open set, then $\|Df\|(U) = \mathbf{V}(f, U)$.

PROOF. For every open set $U \subset \Omega$, equip the linear spaces

$$C_U := C_c(U; \mathbb{R}^n) \quad \text{and} \quad Lip_U := Lip_c(U; \mathbb{R}^n)$$

with the L^∞ norm. Select open sets $\Omega_1 \Subset \Omega_2 \Subset \cdots$ whose union is Ω. The linear functional

$$G : v \mapsto -\int_\Omega f(x) \operatorname{div} v(x)\, dx : Lip_\Omega \to \mathbb{R} \tag{*}$$

satisfies $G(v) \leq \mathbf{V}(f, \Omega_i)\|v\|_{L^\infty(\Omega_i; \mathbb{R}^n)}$ for every $v \in Lip_{\Omega_i}$ and $i = 1, 2, \dots$. As Lip_{Ω_i} is a dense subspace of C_{Ω_i}, the restriction $G \restriction Lip_{\Omega_i}$ has a unique continuous extension $G_i : C_{\Omega_i} \to \mathbb{R}$, necessarily linear and satisfying

$$G_i(v) \leq \mathbf{V}(f, \Omega_i)\|v\|_{L^\infty(\Omega_i; \mathbb{R}^n)}$$

for each $v \in C_{\Omega_i}$. By the uniqueness $G_{i+1} \restriction C_{\Omega_i} = G_i$. It follows that G extends to a linear functional F defined on $C_\Omega = \bigcup_{i=1}^\infty C_{\Omega_i}$. As each open set $U \Subset \Omega$ is contained in some Ω_i, the functional F satisfies the assumptions of the Riesz theorem. Thus there is an $\mathbb{R}^n$-valued measure Df with

$$F(v) = \int_\Omega v \cdot d(Df)$$

for each $v \in C_\Omega$, and (*) implies (5.5.1). For an open set $U \subset \Omega$,

$$\|Df\|(U) = \sup\{F(v) : v \in C_U \text{ and } \|v\|_{L^\infty(U; \mathbb{R}^n)} \leq 1\}$$

by the Riesz theorem and Corollary 5.4.6. Since Lip_U is dense in C_U,

$$\|Df\|(U) = \sup\{F(v) : v \in Lip_U \text{ and } \|v\|_{L^\infty(U; \mathbb{R}^n)} \leq 1\} = \mathbf{V}(f, U).$$

Finally, if ν is an $\mathbb{R}^n$-valued measure satisfying

$$\int_\Omega f(x) \operatorname{div} v(x)\, dx = -\int_\Omega v \cdot d\nu$$

for each $v \in Lip_\Omega$, then arguing as above, we verify that $F(v) = \int_\Omega v \cdot d\nu$ for every $v \in C_\Omega$. The equality $\nu = Df$ follows from the uniqueness assertion of the Riesz theorem. $\qquad\square$

Remark 5.5.2. In the proofs of Theorems 5.4.8 and 5.5.1, extending the linear functionals can be simplified by employing the injective limit topology defined in Example 3.6.4; see Remark 5.4.5.

The $\mathbb{R}^n$-valued measure Df of Theorem 5.5.1 is called the *variational measure* of $f \in BV_{\mathrm{loc}}(\Omega)$. The components of Df are denoted by $D_1 f, \ldots, D_n f$. Theorem 5.5.1 shows that Df is the *distributional gradient* of f, and that $D_1 f, \ldots, D_n f$ are the *distributional partial derivatives* of f; see Section 3.1, in particular Example 3.1.2.

Note that while the variation $\mathbf{V}(f, \Omega)$ is defined for each $f \in L^1_{\mathrm{loc}}(\Omega)$, the symbol $\|Df\|(\Omega)$ has meaning only when $f \in BV_{\mathrm{loc}}(\Omega)$.

Proposition 5.5.3. *Let $f \in L^1_{\mathrm{loc}}(\Omega)$, and let ν be an $\mathbb{R}^n$-valued measure in Ω such that*

$$\int_\Omega f(x)\,\mathrm{div}\,v(x)\,dx = -\int_\Omega v \cdot d\nu$$

for each $v \in C^1_c(\Omega; \mathbb{R}^n)$. Then $f \in BV_{\mathrm{loc}}(\Omega)$ and $\nu = Df$.

PROOF. Proposition 5.2.3 implies that $\mathbf{V}(f, U) \le \|\nu\|(U) < \infty$ for each open set $U \Subset \Omega$. Hence $f \in BV_{\mathrm{loc}}(\Omega)$. Since $C^1_c(U; \mathbb{R}^n)$ is a dense subspace of $Lip_c(\Omega; \mathbb{R}^n)$ equipped with the L^∞ norm, an argument similar to that employed in the proof of Theorem 5.5.1 shows that the equality

$$\int_\Omega f(x)\,\mathrm{div}\,v(x)\,dx = -\int_\Omega v \cdot d\nu$$

is valid for every $v \in Lip_c(\Omega; \mathbb{R}^n)$. The proposition follows from the uniqueness assertion of Theorem 5.5.1. $\square$

Let $f : \Omega \to \mathbb{R}$ be pointwise Lipschitz and $Df \in L^1_{\mathrm{loc}}(\Omega; \mathbb{R}^n)$. Given $v \in Lip_c(\Omega; \mathbb{R}^n)$, integration by parts (Theorem 2.3.9) yields

$$\int_\Omega f(x)\,\mathrm{div}\,v(x)\,dx = -\int_\Omega Df(x) \cdot v(x)\,dx.$$

Thus $f \in BV_{\mathrm{loc}}(\Omega)$, and Propositions 5.5.3 and 5.3.7 imply

$$Df = \mathcal{L}^n \llcorner Df \quad \text{and} \quad \|Df\| = \mathcal{L}^n \llcorner |Df|. \tag{5.5.2}$$

This standard notation is ambiguous: the same symbol Df denotes both the classical gradient of f as well as the distributional gradient of f. The symbols $D_1 f, \ldots, D_n f$ have similar double meaning. Fortunately, the correct interpretation will be always clear from the context.

Proposition 5.5.4. *Let $\{f_k\}$ be a sequence in $BV_{\mathrm{loc}}(\Omega)$ converging in $L^1_{\mathrm{loc}}(\Omega)$ to $f \in BV_{\mathrm{loc}}(\Omega)$. If $\limsup \|Df_k\|(U) < \infty$ for each $U \Subset \Omega$ then*

$$\text{w-}\lim Df_k = Df.$$

If in addition $\lim \|Df_k\|(\Omega) = \|Df\|(\Omega) < \infty$, then

$$\limsup \|Df_k\|(C) \le \|Df\|(C)$$

for each relatively closed set $C \subset \Omega$, and $\text{w-}\lim \|Df_k\| = \|Df\|$.

Proof. For each $v \in Lip_c(\Omega; \mathbb{R}^n)$, Theorem 5.5.1 implies

$$\lim \int_\Omega v \cdot d(Df_k) = -\lim \int_\Omega f_k \operatorname{div} v = -\int_\Omega f \operatorname{div} v = \int_\Omega v \cdot d(Df).$$

Since $Lip_c(\Omega, \mathbb{R}^n)$ is a dense subspace of $C_c(\Omega; \mathbb{R}^n)$ equipped with the L^∞ norm, it is easy to infer that w-$\lim Df_k = Df$. Proposition 5.1.7 shows that $\|Df\|(U) \leq \liminf \|Df_k\|(U)$ for each open set $U \subset \Omega$. Thus under the additional assumption $\lim \|Df_k\|(\Omega) = \|Df\|(\Omega) < \infty$, we obtain

$$\limsup \|Df_k\|(C) = \lim \|Df_k\|(\Omega) - \liminf \|Df_k\|(\Omega - C)$$
$$\leq \|Df\|(\Omega) - \|Df\|(\Omega - C) = \|Df\|(C)$$

for every relatively closed set $C \subset \Omega$. An application of Theorem 5.4.3 completes the proof. $\square$

Example 5.5.5. Let $\Omega = (0, \pi)$ and $g_k(t) = 2^{-k} \cos 2^k t$ for $t \in \Omega$ and $k \in \mathbb{N}$. Each g_k belongs to $BV(\Omega)$ and $\lim \|g_k\|_{L^\infty(\Omega)} = 0$. Moreover, $\|Dg_k\|(\Omega) = 2$ for $k = 1, 2, \ldots$ by Example 5.4.10. Proposition 5.5.4 implies w-$\lim Dg_k = 0$.

Lemma 5.5.6. *Let $f \in L^1_{loc}(\Omega)$ and $U \Subset \Omega$. If η is a mollifier, then*

$$\|D(\eta_\varepsilon * f)\|(U) \leq \mathbf{V}(f, \Omega)$$

for each $\varepsilon > 0$ with $B(U, \varepsilon) \subset \Omega$. Moreover,

$$\|D(\eta_\varepsilon * f)\|(\mathbb{R}^n) \leq \mathbf{V}(f)$$

for each $\varepsilon > 0$ when $\Omega = \mathbb{R}^n$.

Proof. Choose $\varepsilon > 0$ with $B(U, \varepsilon) \subset \Omega$, and select $v \in C^1_c(U; \mathbb{R}^n)$ so that $\|v\|_{L^\infty(U; \mathbb{R}^n)} \leq 1$. Integration by parts yields

$$-\int_U D(\eta_\varepsilon * f) \cdot v = \int_U (\eta_\varepsilon * f) \operatorname{div} v = \int_\Omega (\eta_\varepsilon * f) \operatorname{div} v$$
$$= \int_\Omega f \operatorname{div}(\eta_\varepsilon * v) \leq \mathbf{V}(f, \Omega).$$

As v is arbitrary, $\|D(\eta_\varepsilon * f)\|(U) \leq \mathbf{V}(f, \Omega)$. If $\Omega = \mathbb{R}^n$, then

$$\|D(\eta_\varepsilon * f)\|(\mathbb{R}^n) = \sup\left\{\|D(\eta_\varepsilon * f)\|(V) : V \Subset \mathbb{R}^n \text{ is open}\right\} \leq \mathbf{V}(f)$$

for every $\varepsilon > 0$ by the first part of the proof. $\square$

Proposition 5.5.7. *If $f, g \in L^1_{loc}(\Omega)$, then*

$$\mathbf{V}(fg, \Omega) \leq \|f\|_{L^\infty(\Omega)} \mathbf{V}(g, \Omega) + \|g\|_{L^\infty(\Omega)} \mathbf{V}(f, \Omega).$$

In particular, $BV(\Omega) \cap L^\infty(\Omega)$ and $BV_{loc}(\Omega) \cap L^\infty_{loc}(\Omega)$ are algebras.

PROOF. Choose an open set $V \Subset \Omega$ and a mollifier η. For a small $\varepsilon > 0$,

$$f_k := \eta_{1/k} * f \quad \text{and} \quad g_k := \eta_{1/k} * g$$

belong to $L^1[U(V, \varepsilon)]$ for each $k \in \mathbb{N}$, and $\lim \|f_k g_k - fg\|_{L^1(V)} = 0$. Thus by Proposition 5.1.7 and Lemma 5.5.6,

$$\mathbf{V}(fg, V) \le \liminf \big\| D(f_k g_k) \big\|(V) \le \liminf \int_V |f_k Dg_k + g_k Df_k|$$

$$\le \liminf \big[\|f_k\|_{L^\infty(V)} \|Dg_k\|(V) + \|g_k\|_{L^\infty(V)} \|Df_k\|(V) \big]$$

$$\le \|f\|_{L^\infty(\Omega)} \mathbf{V}(g, \Omega) + \|g\|_{L^\infty(\Omega)} \mathbf{V}(f, \Omega).$$

The proposition follows, since V is arbitrary and

$$\mathbf{V}(fg, \Omega) = \sup \big\{ \mathbf{V}(fg, V) : V \Subset \Omega \text{ is open} \big\}. \qquad \square$$

Example 5.5.8. Assume $n = 2$, and for $t_k = k^{-3/2}$, define

$$U_k := \big\{ x \in \mathbb{R}^2 : t_{2k+1} < |x| < t_{2k} \big\}.$$

Following Example 5.2.5, observe $\mathbf{V}(U_k) = 2\pi(t_{2k+1} + t_{2k})$. This and Proposition 5.1.7 imply that $U = \bigcup_{k=1}^\infty U_k$ belongs to $\mathbf{BV}(\mathbb{R}^2)$. Let

$$g(x) := \begin{cases} 0 & \text{if } |x| \ge 1, \\ |x|^{-1/2} - 1 & \text{if } 0 < |x| < 1, \\ \infty & \text{if } x = 0, \end{cases}$$

and calculate $\gamma = \int_{\mathbb{R}^2} |Dg(x)| \, dx < \infty$. The functions $g_i = \min\{g, i\}$ are Lipschitz for $i = 1, 2, \ldots$, and by (5.5.2),

$$\mathbf{V}(g_i) = \int_{\mathbb{R}^2} |Dg_i(x)| \, dx \le \gamma.$$

As $\lim \|g_i - g\|_{L^1(\mathbb{R}^2 S)} = 0$, Proposition 5.1.7 shows $g \in BV(\mathbb{R}^2)$, and a fortiori $g \upharpoonright U \in BV(U)$. Find $\varphi \in C^1_c(\mathbb{R}; [0, 1])$ so that $\varphi(t_{2k}) = 1$, and $\varphi(t) = 0$ for $t \le t_{2k+1}$. Since $v : x \mapsto \varphi(|x|) |x|^{-1} x$ belongs to $C^1_c(\mathbb{R}^2; \mathbb{R}^2)$ and $\|v\|_{L^\infty(\mathbb{R}^2; \mathbb{R}^2)} \le 1$, Proposition 2.1.4 yields

$$\mathbf{V}(g\chi_{U_k}) \ge \int_{\mathbb{R}^2} \chi_{U_k} g \operatorname{div} v = \int_{U_k} \operatorname{div}(gv) - \int_{U_k} v \cdot Dg$$

$$= \int_{\partial U_k} gv \cdot \nu_{U_k} \, d\mathcal{H}^1 - \int_{U_k} v \cdot Dg$$

$$= g(t_{2k}) \int_{\partial U(0, t_{2k})} d\mathcal{H}^1 - \int_{U_k} v \cdot Dg \ge 2\pi (2k)^{-3/4} - \gamma.$$

Thus $\sum_{k=1}^\infty \mathbf{V}(g\chi_{U_k}) = \infty$. As the sets U_k have disjoint closures, it is easy to infer that $\mathbf{V}(g\chi_U) = \infty$.

Proposition 5.5.9. *Let* $f \in BV_{\mathrm{loc}}(\Omega)$ *be such that* $\|Df\|(\Omega) = 0$. *If* Ω *is connected, then* f *is constant almost everywhere in* Ω.

PROOF. Observe that Ω is the union of a countable family

$$\mathcal{U} = \{U(x,r) \subset \Omega : x \in \Omega \cap \mathbb{Q}^n \text{ and } r \in \mathbb{Q}\}.$$

Given $U \in \mathcal{U}$, find $\varepsilon > 0$ with $B(U,\varepsilon) \subset \Omega$. If η is a mollifier, then $\lim_{k\to\infty} \|f - \eta_{\varepsilon/k} * f\|_{L^1_{\mathrm{loc}}(U)} = 0$. By Lemma 5.5.6,

$$\int_U |D(\eta_{\varepsilon/k} * f)| = \|D(\eta_{\varepsilon/k} * f)\|(U) \le \|Df\|(\Omega) = 0$$

for $k = 1, 2, \ldots$. As $\eta_{\varepsilon/k} * f$ are functions in $C^\infty(U)$, they are constant. It follows that there are $c_U \in \mathbb{R}$ and a negligible set $N_U \subset U$ such that $f(x) = c_U$ for all $x \in U - N_U$. If sets $U, V \in \mathcal{U}$ meet, then $c_U = c_V$ since $|U \cap V| > 0$. Thus $g : x \mapsto c_U$ for all $x \in U$ and all $U \in \mathcal{U}$ defines a locally constant function in Ω. As Ω is connected, g is constant. Hence f is constant outside the negligible set $\bigcup_{U \in \mathcal{U}} N_U$. $\qquad\square$

Lemma 5.5.10. *If* $f \in L^1_{\mathrm{loc}}(\mathbb{R}^n)$ *and* $y \in \mathbb{R}^n$, *then*

$$\int_{\mathbb{R}^n} |f(x - y) - f(x)| \, dx \le |y| \mathbf{V}(f).$$

PROOF. If $f \in C^1(\mathbb{R}^n)$, then

$$f(x - y) - f(x) = \int_0^1 \frac{d}{dt}[f(x - ty)] \, dt = - \int_0^1 y \cdot Df(x - ty) \, dt$$

for each $x \in \mathbb{R}^n$. By Fubini's theorem and (5.5.2),

$$\int_{\mathbb{R}^n} |f(x - y) - f(x)| \, dx \le |y| \int_0^1 \left(\int_{\mathbb{R}^n} |Df(x - ty)| dx \right) dt$$

$$= |y| \cdot \|Df\|_{L^1(\mathbb{R}^n;\mathbb{R}^n)} = |y| \cdot \|Df\|(\mathbb{R}^n).$$

If $f \in L^1_{\mathrm{loc}}(\mathbb{R}^n)$, select a mollifier η and let $f_k := \eta_{1/k} * f$. Then

$$\int_{B(0,r)} |f(x - y) - f(x)| \, dx \le \int_{B(0,r)} |f(x - y) - f_k(x - y)| \, dx$$

$$+ \int_{\mathbb{R}^n} |f_k(x - y) - f_k(x)| \, dx$$

$$+ \int_{B(0,r)} |f_k(x) - f(x)| \, dx$$

$$\le 2\|f - f_k\|_{L^1[B(0,r)]} + |y| \cdot \|Df_k\|(\mathbb{R}^n)$$

$$\le 2\|f - f_k\|_{L^1[B(0,r)]} + |y|\mathbf{V}(f)$$

by the first part of the proof and Lemma 5.5.6. The lemma follows by letting $k \to \infty$, and then $r \to \infty$. $\qquad\square$

Corollary 5.5.11. *If $f \in BV(\mathbb{R}^n)$ has compact support, then*

$$\|f\|_{L^1(\mathbb{R}^n)} \leq \tfrac{1}{2} d(\operatorname{spt} f)\|Df\|(\mathbb{R}^n).$$

PROOF. Choose $y \in \mathbb{R}^n$ so that $|y| > d(\operatorname{spt} f)$, and observe that

$$2\|f\|_{L^1(\mathbb{R}^n)} = \int_{\mathbb{R}^n} |f(x - y) - f(x)|\, dx \leq |y| \cdot \|Df\|$$

by Lemma 5.5.10. Letting $|y| \to d(\operatorname{spt} f)$ completes the proof. $\qquad\square$

Theorem 5.5.12. *Let $K \subset \mathbb{R}^n$ be a compact set and $\gamma \in \mathbb{R}_+$. Suppose $\{f_k\}$ is a sequence in $BV(\mathbb{R}^n)$ such that $\|Df_k\|(\mathbb{R}^n) \leq \gamma$ and $\operatorname{spt} f_k \subset K$ for $k = 1, 2, \ldots$. There are $f \in BV(\mathbb{R}^n)$ and subsequence $\{f_{k_j}\}$ satisfying*

$$\lim \|f_{k_j} - f\|_{L^1(\mathbb{R}^n)} = 0.$$

PROOF. Let $\beta = \gamma[d(K) + 1]$. Choose a mollifier η and $0 < \varepsilon < 1$. Define $f_{k,\varepsilon} := \eta_\varepsilon * f_k$ for $k = 1, 2, \ldots$. By Lemma 5.5.10,

$$\|f_{k,\varepsilon} - f_k\|_{L^1(\mathbb{R}^n)} = \int_{\mathbb{R}^n} \left| \int_{\mathbb{R}^n} \left(\eta_\varepsilon(y) f_k(x - y) - \eta_\varepsilon(y) f_k(x) \right) dy \right| dx$$

$$\leq \int_{\mathbb{R}^n} \eta_\varepsilon(y) \left(\int_{\mathbb{R}^n} |f_k(x - y) - f_k(x)|\, dx \right) dy$$

$$\leq \|Df_k\|(\mathbb{R}^n) \int_{B(0,\varepsilon)} \eta_\varepsilon(y)|y|\, dy \leq \varepsilon\beta.$$

Claim. For each $\delta > 0$, there is a subsequence $\{f_{k_i}\}$ such that

$$\|f_{k_i} - f_{k_j}\|_{L^1(\mathbb{R}^n)} \leq \delta, \quad i, j = 1, 2, \ldots.$$

Proof. Letting $\varepsilon := \delta/(3\beta)$, the previous inequality yields

$$\sup \|f_{k,\varepsilon} - f_k\|_{L^1(\mathbb{R}^n)} \leq \tfrac{\delta}{3} \tag{$*$}$$

for $k = 1, 2, \ldots$. By Lemma 5.5.6 and Corollary 5.5.11,

$$|f_{k,\varepsilon}(x)| \leq \varepsilon^{-n} \left| \int_{\mathbb{R}^n} \left| \eta\left(\tfrac{x-y}{\varepsilon}\right) f_k(y)\, dy \right| \right.$$

$$\leq \varepsilon^{-n} \|\eta\|_{L^\infty(\mathbb{R}^n)} \|f_k\|_{L^1(\mathbb{R}^n)} \leq \varepsilon^{-n} \|\eta\|_{L^\infty(\mathbb{R}^n)}\beta,$$

$$|Df_{k,\varepsilon}(x)| = |(f_k * D\eta_\varepsilon)(x)| \leq \varepsilon^{-n} \left| \int_{\mathbb{R}^n} D\left[\eta\left(\tfrac{x-y}{\varepsilon}\right)\right] f_k(y)\, dy \right|$$

$$\leq \varepsilon^{-n-1} \int_{\mathbb{R}^n} \left| D\eta\left(\tfrac{x-y}{\varepsilon}\right) f_k(y) \right| dy$$

$$\leq \varepsilon^{-n-1} \|D\eta\|_{L^\infty(\mathbb{R}^n;\mathbb{R}^n)} \|f_k\|_{L^1(\mathbb{R}^n)}$$

$$\leq \varepsilon^{-n-1} \|D\eta\|_{L^\infty(\mathbb{R}^n;\mathbb{R}^n)}\beta$$

for $k = 1, 2, \ldots$. The last two inequalities show that $\{f_{k,\varepsilon}\}$ is uniformly bounded and equicontinuous. Ascoli's theorem [64, Appendix A5] implies that $\{f_{k,\varepsilon}\}$ has a uniformly Cauchy subsequence. In particular, $\{f_{k,\varepsilon}\}$ has a subsequence $\{f_{k_i,\varepsilon}\}$ such that for $i,j = 1,2,\ldots,$

$$\|f_{k_i,\varepsilon} - f_{k_j,\varepsilon}\|_{L^1(\mathbb{R}^n)} \leq \|f_{k_i,\varepsilon} - f_{k_j,\varepsilon}\|_{L^\infty(\mathbb{R}^n)} |B(K,1)| < \tfrac{\delta}{3}.$$

Combining this with $(*)$ establishes the claim.

Using the claim construct recursively subsequences $s_i = \{f_{i,k}\}$ of $\{f_k\}$ so that s_{i+1} is a subsequence of s_i and

$$\|f_{i,j} - f_{i,k}\|_{L^1(\mathbb{R}^n)} \leq 1/i, \quad i,j,k = 1,2,\ldots.$$

Clearly $\{f_{k,k}\}$ is a subsequence of $\{f_k\}$ that is Cauchy in $L^1(\mathbb{R}^n)$. As $L^1(\mathbb{R}^n)$ is complete, $\{f_{k,k}\}$ converges in $L^1(\mathbb{R}^n)$ to $f \in L^1(\mathbb{R}^n)$. Since $\mathbf{V}(f) \leq \gamma$ by Proposition 5.1.7, we see that $f \in BV(\mathbb{R}^n)$. $\qquad\square$

5.6. Approximation theorem

Let $\mathcal{E}$ be a family of subsets of Ω and $A \subset \Omega$. We define

$$\mathbf{St}(A,\mathcal{E}) := \{E \in \mathcal{E} : A \cap E \neq \emptyset\}.$$

Recall from Section 2.2 that $\mathbf{St}(\{x\},\mathcal{E}) = \mathbf{St}(x,\mathcal{E})$ for every $x \in \Omega$. The family $\mathcal{E}$ is called *point-finite* if $\mathbf{St}(x,\mathcal{E})$ is finite for each $x \in \Omega$, and *locally finite* if every $x \in \Omega$ has a neighborhood U such that $\mathbf{St}(U,\mathcal{E})$ is finite. When there is $p \in \mathbb{N}$ such that $\mathbf{St}(x,\mathcal{E})$ contains at most p sets for every $x \in \Omega$, the family $\mathcal{E}$ is called *point-p-finite*.

Observation 5.6.1. *Let μ be a measure in Ω, and let $\mathcal{E}$ be a point-p-finite family of μ measurable subsets of Ω. If $A = \bigcup \mathcal{E}$, then*

$$\sum_{E \in \mathcal{E}} \mu(E) \leq p\mu(A).$$

PROOF. If $\mathcal{A} \subset \mathcal{E}$ is a finite family, then $\sum_{E \in \mathcal{A}} \chi_E \leq p\chi_A$. Thus

$$\sum_{E \in \mathcal{A}} \mu(E) = \sum_{E \in \mathcal{A}} \int_A \chi_E \, d\mu \leq p \int_A \chi_A \, d\mu = p\mu(A),$$

and the observation follows from the arbitrariness of $\mathcal{A}$. $\qquad\square$

If $C \subset \mathbb{R}^n$ is a cube with center x and $\varepsilon > 0$, let

$$C^\varepsilon := (1+\varepsilon)[\operatorname{int} C - x] + x.$$

Note that C^ε is an open cube of diameter $(1+\varepsilon)d(C)$ concentric with C, and hence $C \Subset C^\varepsilon$. If $\mathcal{C}$ is a family of cubes, we let $\mathcal{C}^\varepsilon := \{C^\varepsilon : C \in \mathcal{C}\}$.

Lemma 5.6.2 (Whitney). *Assume* $\partial\Omega \neq \emptyset$. *There is a family* $\mathcal{C}$ *of nonoverlapping dyadic cubes such that*

(i) $\Omega = \bigcup \mathcal{C}$;

(ii) $d(C) < \mathrm{dist}\,(C, \partial\Omega) \leq 4d(C)$ *for each* $C \in \mathcal{C}$.

Any family $\mathcal{C}$ *of nonoverlapping dyadic cubes which satisfies conditions* (i) *and* (ii) *has the following properties:*

(1) *If* $C, D \in \mathcal{C}$ *and* $C \cap D \neq \emptyset$, *then*

$$\tfrac{1}{4} d(D) \leq d(C) \leq 4d(D).$$

(2) *For each* $C \in \mathcal{C}$, *the number of cubes belonging to* $\mathrm{St}(C, \mathcal{C})$ *does not exceed* $N = (12)^n$. *In particular, the family* $\mathcal{C}$ *is point-N-finite.*

(3) *If* $0 < \varepsilon \leq 1/4$ *and* $C \in \mathcal{C}$, *then* $C^\varepsilon \Subset \Omega$ *and the number of cubes belonging to* $\mathrm{St}(C^\varepsilon, \mathcal{C}^\varepsilon)$ *does not exceed* N^2. *In particular, the family* $\mathcal{C}^\varepsilon$ *is locally finite.*

PROOF. Recall that $\mathfrak{D}_k$ denotes the family of all k-cubes, and let

$$\mathcal{C}_k := \left\{ C \in \mathfrak{D}_k : C \subset \Omega \text{ and } 2^{-k}\sqrt{n} < \mathrm{dist}\,(C, \partial\Omega) \leq 2^{-k+2}\sqrt{n} \right\}$$

for $k \in \mathbb{Z}$. There is a nonoverlapping family $\mathcal{C} \subset \bigcup_{k\in\mathbb{Z}} \mathcal{C}_k$ whose union is the same as that of $\bigcup_{k\in\mathbb{Z}} \mathcal{C}_k$. By the definition of $\mathcal{C}_k$, each $C \in \mathcal{C}$ satisfies condition (ii). Choose $x \in \Omega$. There are a unique $k \in \mathbb{Z}$ with

$$2^{-k+1}\sqrt{n} < \mathrm{dist}\,(x, \partial\Omega) \leq 2^{-k+2}\sqrt{n},$$

and a k-cube C containing x. The previous inequality implies

$$2^{-k}\sqrt{n} < \mathrm{dist}\,(C, \partial\Omega) \leq 2^{-k+2}\sqrt{n},$$

and hence $C \cap \partial\Omega = \emptyset$. Since C is connected and $x \in C \cap \Omega$, we see that $C \subset \Omega$. This establishes condition (i).

(1) Let $C, D \in \mathcal{C}$ and $C \cap D \neq \emptyset$. By condition (ii),

$$d(D) < \mathrm{dist}\,(D, \partial\Omega) \leq \mathrm{dist}\,(C, \partial\Omega) + d(C) \leq 5d(C).$$

As D and C are dyadic cubes, $d(D)/d(C) = 2^k$ for $k \in \mathbb{Z}$. From $2^k < 5$, we infer $k \leq 2$. Thus $d(D) \leq 4d(C)$ and by symmetry, $d(C) \leq 4d(D)$.

(2) By property (1), the smallest cubes in $\mathcal{C}$ that meet $C \in \mathcal{C}$ are of diameter $d(C)/4$. Hence the number of cubes in $\mathrm{St}(C, \mathcal{C})$ does not exceed the number of dyadic cubes of diameter $d(C)/4$ contained in the union of all dyadic cubes adjacent to C. Since there are 3^n dyadic cubes adjacent to C, and since each of them contains 4^n dyadic cubes of diameter $d(C)/4$, property (2) follows.

(3) Choose $C \in \mathcal{C}$, and observe that $C^\varepsilon \subset \mathrm{int} \bigcup \mathrm{St}(C, \mathcal{C}) \Subset \Omega$ according to properties (1) and (2). We infer that for each pair $D, Q \in \mathcal{C}$, the open cube D^ε meets Q if and only if $D \cap Q \neq \emptyset$. In particular, if $D \in \mathcal{C}$ meets

no $Q \in St(C, \mathcal{C})$, then $D^\varepsilon \cap Q = \emptyset$ for each $Q \in St(C, \mathcal{C})$, and consequently $D^\varepsilon \cap C^\varepsilon = \emptyset$. In other words, if $D^\varepsilon \in St(C^\varepsilon, \mathcal{C}^\varepsilon)$, then $D \in St(Q, \mathcal{C})$ for some $Q \in St(C, \mathcal{C})$. An application of property (2) completes the argument. $\square$

A collection $\mathcal{C}$ of dyadic cubes which satisfies conditions (i) and (ii) of Lemma 5.6.2 is called a *Whitney division* of Ω. We apply it to approximating functions in $BV_{\text{loc}}(\Omega)$ by C^∞ functions (Theorem 5.6.3 below). For other applications consult [70, Chapter 4].

The number $N = (12)^n$ in Lemma 5.6.2 is unnecessarily large. A better estimate can be obtained by counting more judiciously. However, what matters is that N is a finite number depending only on the dimension n.

Theorem 5.6.3. *Let $f \in L^1_{\text{loc}}(\Omega)$. There is $\{f_i\}$ in $C^\infty(\Omega)$ such that*

$$\lim \|f_i - f\|_{L^1(\Omega)} = 0 \quad \text{and} \quad \lim \mathbf{V}(f_i, \Omega) = \mathbf{V}(f, \Omega).$$

PROOF. Let $N = (12)^n$, and select a nonempty compact set $C_0 \subset \Omega$ that will be specified below. Let $U_0 := \Omega$, and let $V_0 \Subset \Omega$ be an open set containing C_0. Choose a Whitney division $\{C_k : k \in \mathbb{N}\}$ of $\Omega - C_0$, and define

$$U_k := C_k^{1/4} \quad \text{and} \quad V_k = C_k^{1/8}, \quad k = 1, 2, \ldots.$$

Mollifying the indicators χ_{V_k} of V_k, we obtain functions $\psi_k \in C_c^\infty(\mathbb{R}^n)$ such that $\chi_{C_k} \leq \psi_k \leq 1$ and $\operatorname{spt} \psi_k \subset U_k$ for $k = 0, 1, \ldots$. By Lemma 5.6.2, (3), the function $\psi = \sum_{k=0}^\infty \psi_k$ belongs to $C^\infty(\mathbb{R}^n)$ and $1 \leq \psi(x) \leq N^2 + 1$ for all $x \in \Omega$. Thus each function $\varphi_k := \psi_k / \psi$ belongs to $C_c^\infty(\Omega)$ and $\sum_{k=0}^\infty \varphi_k = 1$. Since $\{U_k : k = 0, 1, \ldots\}$ is a locally finite family, $\sum_{k=0}^\infty D\varphi_k = 0$.

Now select a mollifier η, and let $\phi_\theta := \eta_\theta * \phi$ for each $\phi \in L^1(\Omega; \mathbb{R}^m)$ and every $\theta > 0$. Choose $\varepsilon > 0$, and find $\varepsilon_k > 0$ so that the following conditions are satisfied for $k = 0, 1, \ldots$:

(a) $\operatorname{spt}(f\varphi_k)_{\varepsilon_k} \subset U_k$;

(b) $\left\|(f\varphi_k)_{\varepsilon_k} - f\varphi_k\right\|_{L^1(\Omega)} \leq \varepsilon 2^{-k-1}$;

(c) $\left\|(fD\varphi_k)_{\varepsilon_k} - fD\varphi_k\right\|_{L^1(\Omega)} \leq \varepsilon 2^{-k-1}$.

By Lemma 5.6.2, (3), the family $\{U_k : k = 0, 1, \ldots\}$ is locally finite. Hence $f^\varepsilon := \sum_{k=0}^\infty (f\varphi_k)_{\varepsilon_k}$ belongs to $C^\infty(\Omega)$ by (a). Moreover,

$$\|f^\varepsilon - f\|_{L^1(\Omega)} \leq \sum_{k=0}^\infty \left\|(f\varphi_k)_{\varepsilon_k} - f\varphi_k\right\|_{L^1(\Omega)} < \varepsilon$$

by (b). Thus $\lim_{\varepsilon \to 0} \|f^\varepsilon - f\|_{L^1(\Omega)} = 0$, and Proposition 5.1.7 shows

$$\mathbf{V}(f, \Omega) \leq \liminf_{\varepsilon \to 0} \mathbf{V}(f^\varepsilon, \Omega). \tag{$*$}$$

As there is nothing more to prove otherwise, assume $\mathbf{V}(f, \Omega) < \infty$. This implies $f \in BV_{\text{loc}}(\mathbb{R}^n)$, and we choose C_0 so that $\|Df\|(\Omega - C_0) < \varepsilon/N^2$.

Given $v \in C_c^1(\Omega; \mathbb{R}^n)$ with $\|v\|_{L^\infty(\Omega;\mathbb{R}^n)} \leq 1$, we calculate

$$\int_\Omega f^\varepsilon \operatorname{div} v = \sum_{k=0}^\infty \int_\Omega (f\varphi_k)_{\varepsilon_k} \operatorname{div} v = \sum_{k=0}^\infty \int_\Omega f\varphi_k \operatorname{div} v_{\varepsilon_k}$$

$$= \sum_{k=0}^\infty \int_\Omega f \operatorname{div}(\varphi_k v_{\varepsilon_k}) - \sum_{k=0}^\infty \int_\Omega f D\varphi_k \cdot v_{\varepsilon_k} = \Sigma_1 + \Sigma_2.$$

Since $\|\varphi_k v_{\varepsilon_k}\|_{L^\infty(\Omega;\mathbb{R}^n)} \leq 1$ and $\operatorname{spt}(\varphi_k v_{\varepsilon_k}) \subset U_k$,

$$\Sigma_1 = \int_\Omega f \operatorname{div}(\varphi_0 v_{\varepsilon_0}) + \sum_{k=1}^\infty \int_\Omega f \operatorname{div}(\varphi_k v_{\varepsilon_k})$$

$$\leq \|Df\|(\Omega) + \sum_{k=1}^\infty \|Df\|(U_k)$$

$$\leq \|Df\|(\Omega) + N^2 \|Df\|(\Omega - C_0) \leq \|Df\|(\Omega) + \varepsilon$$

according to Lemma 5.6.2, (3) and Observation 5.6.1. Condition (c) and the equality $\sum_{k=0}^\infty D\varphi_k = 0$ imply

$$\Sigma_2 = -\sum_{k=0}^\infty \int_\Omega f D\varphi_k \cdot v_{\varepsilon_k} + \sum_{k=0}^\infty \int_\Omega f D\varphi_k \cdot v$$

$$= \sum_{k=0}^\infty \int_\Omega \left[-(fD\varphi_k)_{\varepsilon_k} + fD\varphi_k\right] \cdot v$$

$$\leq \sum_{k=0}^\infty \|(fD\varphi_k)_{\varepsilon_k} - fD\varphi_k\|_{L^1(\Omega;\mathbb{R}^n)} < \varepsilon.$$

Consequently $\int_\Omega f^\varepsilon \operatorname{div} v \leq \|Df\|(\Omega) + 2\varepsilon$, and as v is arbitrary,

$$\mathbf{V}(f^\varepsilon, \Omega) \leq \|Df\|(\Omega) + 2\varepsilon.$$

Thus $f^\varepsilon \in BV(\Omega)$ and $\lim_{\varepsilon \to 0} \|Df^\varepsilon\|(\Omega) = \|Df\|(\Omega)$ by $(*)$. $\qquad\square$

5.7. Coarea theorem

W.H. Fleming and R. Rishel [34] have shown that the variation of a function $f \in L_{\mathrm{loc}}^1(\Omega)$ can be calculated from the variations of its level sets $\{f > t\}$.

Lemma 5.7.1. *There are $v_k \in C_c^1(\Omega; \mathbb{R}^n)$ such that $\|v_k\|_{L^\infty(\Omega;\mathbb{R}^n)} \leq 1$ for $k = 1, 2, \ldots,$ and for each $f \in L_{\mathrm{loc}}^1(\Omega)$,*

$$\mathbf{V}(f, \Omega) := \sup \int_\Omega f(x) \operatorname{div} v_k(x) \, dx.$$

PROOF. As $C_c(\Omega)$ equipped with the L^∞ norm is separable, so is the space

$$B := \{\operatorname{div} v : v \in C_c^1(\Omega; \mathbb{R}^n) \text{ and } \|v\|_{L^\infty(\Omega;\mathbb{R}^n)} \le 1\}.$$

Thus there are $v_k \in C_c^1(\Omega; \mathbb{R}^n)$ such that $\|v_k\|_{L^\infty(\Omega;\mathbb{R}^n)} \le 1$ and the set $\{\operatorname{div} v_k : k \in \mathbb{N}\}$ is dense in B. Since for each $f \in L_{\mathrm{loc}}^1(\Omega)$,

$$\phi : \varphi \mapsto \int_\Omega f(x)\varphi(x)\, dx : C_c(\Omega) \to \mathbb{R}$$

is a continuous linear functional with respect to the L^∞ norm in $C_c(\Omega)$, the set $\{\phi(\operatorname{div} v_k) : k \in \mathbb{N}\}$ is dense in $\phi(B)$, and the lemma follows. $\qquad\square$

Corollary 5.7.2. *If f is a measurable function defined on Ω, then*

$$t \mapsto \mathbf{V}(\{f > t\}, \Omega) : \mathbb{R} \to [0, \infty]$$

is an $\mathcal{L}^1$ measurable function.

PROOF. The set $E := \{(x, t) \in \Omega \times R : f(x) - t > 0\}$ is an $\mathcal{L}^{n+1}$ measurable subset of $\mathbb{R}^{n+1}$. Since

$$\{(x,t) \in \Omega \times R : \chi_{\{f>t\}}(x) > s\} = \begin{cases} \emptyset & \text{if } s \ge 1, \\ E & \text{if } 0 \le s < 1, \\ \Omega \times R & \text{if } s < 0. \end{cases}$$

$(x, t) \mapsto \chi_{\{f>t\}}(x) : \Omega \times R \to R$ is an $\mathcal{L}^{n+1}$ measurable function. Given $v \in C_c^1(\Omega; \mathbb{R}^n)$ and $k \in \mathbb{N}$, the function $(x, t) \mapsto \chi_{\{f>t\}}(x) \operatorname{div} v(x)$ belongs to $L^1(\Omega \times [-k, k], \mathcal{L}^{n+1})$. By Fubini's theorem, the function

$$F : t \mapsto \int_\Omega \chi_{\{f>t\}}(x)(\operatorname{div} v)^\pm(x)\, dx,$$

defined $\mathcal{L}^1$ almost everywhere in $[-k, k]$, is $\mathcal{L}^1$ measurable. As k is arbitrary, F is an $\mathcal{L}^1$ measurable function defined for almost all $t \in \mathbb{R}$, and the corollary follows from Lemma 5.7.1. $\qquad\square$

Theorem 5.7.3 (Coarea theorem). *If $f \in L_{\mathrm{loc}}^1(\Omega)$, then*

$$\mathbf{V}(f, \Omega) = \int_{\mathbb{R}} \mathbf{V}(\{f > t\}, \Omega)\, dt.$$

PROOF. *Case 1.* Let $f \in C^1(\Omega)$ and $\|Df\|(\Omega) < \infty$. Then

$$g : t \mapsto \int_{\{f<t\}} |Df(x)|\, dx$$

is an increasing real-valued function, $\lim_{t \to -\infty} g(t) = 0$, and

$$\int_{\mathbb{R}} g'(t)\, dt \le \lim_{t \to \infty} g(t) = \int_\Omega |Df(x)|\, dx = \|Df\|(\Omega). \qquad (*)$$

There is an $\mathcal{L}^1$ negligible set $N \subset \mathbb{R}$ such that $|\{f = t\}| = 0$ and $g'(t)$ exists for each $t \in \mathbb{R} - N$. Fix $t \in \mathbb{R} - N$, and for $h > 0$ define

$$\varphi_h(s) := \begin{cases} 0 & \text{if } s \leq t, \\ 1 & \text{if } s \geq t + h, \\ (s - t)/h & \text{if } t < s < t + h. \end{cases}$$

Choose $v \in C_c^1(\Omega; \mathbb{R}^n)$ with $\|v\|_{L^\infty(\Omega;\mathbb{R}^n)} \leq 1$. Since $\varphi_h \circ f \in Lip(\Omega)$ and $t \in \mathbb{R} - N$, integrating by parts gives

$$\int_\Omega (\varphi_h \circ f) \operatorname{div} v = -\int_\Omega (\varphi_h' \circ f) Df \cdot v = -\frac{1}{h} \int_{\{t < f < t+h\}} Df \cdot v$$

$$\leq \frac{1}{h} \int_{\{t \leq f < t+h\}} |Df| = \frac{g(t+h) - g(t)}{h}$$

for all $h > 0$. Letting $h \to 0$, we obtain

$$g'(t) \geq \lim_{h \to 0} \int_\Omega (\varphi_h \circ f) \operatorname{div} v$$

$$= \lim_{h \to 0} \int_{\{t < f < t+h\}} \frac{f - t}{h} \operatorname{div} v + \lim_{h \to 0} \int_{\{f \geq t+h\}} \operatorname{div} v$$

$$\geq -\lim_{h \to 0} \int_{\{t < f < t+h\}} |\operatorname{div} v| + \int_{\{f \geq t\}} \operatorname{div} v = \int_{\{f > t\}} \operatorname{div} v.$$

As v is arbitrary $g'(t) \geq \mathbf{V}(\{f > t\}, \Omega)$ for each $t \in \mathbb{R} - N$. By this and $(*)$,

$$\int_\mathbb{R} \mathbf{V}(\{f > t\}, \Omega) \, dt \leq \int_\mathbb{R} g'(t) \, dt \leq \mathbf{V}(f, \Omega).$$

Case 2. Let $f \in L_{\text{loc}}^1(\Omega)$. By Theorem 5.6.3, there is a sequence $\{f_k\}$ in $C^\infty(\Omega)$ such that

$$\lim \|f_k - f\|_{L^1(\Omega)} = 0 \quad \text{and} \quad \lim \|Df_k\|(\Omega) = \mathbf{V}(f, \Omega). \qquad (**)$$

If t lies between the values $f_k(x)$ and $f(x)$, then

$$\left| \chi_{\{f_k > t\}}(x) - \chi_{\{f > t\}}(x) \right| = 1;$$

otherwise $\left| \chi_{\{f_k > t\}}(x) - \chi_{\{f > t\}}(x) \right| = 0$. Hence

$$\int_\mathbb{R} \left| \chi_{\{f_k > t\}}(x) - \chi_{\{f > t\}}(x) \right| dt = \left| f_k(x) - f(x) \right|,$$

and for $k = 1, 2, \ldots$, Fubini's theorem yields

$$\int_\Omega \left| f_k(x) - f(x) \right| dx = \int_\Omega \left(\int_\mathbb{R} \left| \chi_{\{f_k > t\}}(x) - \chi_{\{f > t\}}(x) \right| dt \right) dx$$

$$= \int_\mathbb{R} \left(\int_\Omega \left| \chi_{\{f_k > t\}}(x) - \chi_{\{f > t\}}(x) \right| dx \right) dt.$$

By (∗∗), there is a subsequence of $\{f_k\}$, still denoted by $\{f_k\}$, such that

$$\lim \|\chi_{\{f_k>t\}} - \chi_{\{f>t\}}\|_{L^1(\Omega)} = 0$$

for $\mathcal{L}^1$ almost all $t \in \mathbb{R}$. In view of Proposition 5.1.7,

$$\mathbf{V}(\{f > t\}, \Omega) \leq \liminf \mathbf{V}(\{f_k > t\}, \Omega)$$

for $\mathcal{L}^1$ almost all $t \in \mathbb{R}$. Fatou's lemma and Case 1 yield

$$\int_{\mathbb{R}} \mathbf{V}(\{f > t\}, \Omega)\, dt \leq \liminf \int_{\mathbb{R}} \mathbf{V}(\{f_k > t\}, \Omega)\, dt$$

$$\leq \lim \|Df_k\|(\Omega) = \mathbf{V}(f, \Omega). \tag{†}$$

To complete the argument, observe that for each $x \in \Omega$,

$$f^+(x) = \int_0^\infty \chi_{[0, f^+(x))}(t)\, dt = \int_0^\infty \chi_{\{f^+>t\}}(x)\, dt.$$

Given $v \in C_c^1(\Omega; \mathbb{R}^n)$ with $\|v\|_{L^\infty(\Omega; \mathbb{R}^n)} \leq 1$, Fubini's theorem implies

$$\int_\Omega f^+(x) \operatorname{div} v(x)\, dx = \int_\Omega \left(\int_0^\infty \chi_{\{f^+>t\}}(x)\, dt \right) \operatorname{div} v(x)\, dx$$

$$= \int_0^\infty \left(\int_\Omega \chi_{\{f^+>t\}}(x) \operatorname{div} v(x)\, dx \right) dt$$

$$\leq \int_0^\infty \mathbf{V}(\{f^+ > t\}, \Omega)\, dt.$$

As v is arbitrary, $\mathbf{V}(f^+, \Omega) \leq \int_0^\infty \mathbf{V}(\{f^+ > t\}, \Omega)\, dt$. Similarly

$$\mathbf{V}(f^-, \Omega) \leq \int_0^\infty \mathbf{V}(\{f^- > t\}, \Omega)\, dt$$

$$= \int_{-\infty}^0 \mathbf{V}(\{f < t\}, \Omega)\, dt = \int_{-\infty}^0 \mathbf{V}(\{f > t\}, \Omega)\, dt,$$

since $\{f < t\} \cup \{f = t\} = \Omega - \{f > t\}$, and $|\{f = t\}| = 0$ for all but countably many $t \in \mathbb{R}$; see Proposition 5.1.2 and Lemma 4.5.5. In view of (†), the theorem follows:

$$\int_{\mathbb{R}} \mathbf{V}(\{f > t\}, \Omega)\, dt \leq \mathbf{V}(f, \Omega) \leq \mathbf{V}(f^+, \Omega) + \mathbf{V}(f^-, \Omega)$$

$$= \int_{\mathbb{R}} \mathbf{V}(\{f > t\}, \Omega)\, dt. \qquad \square$$

Remark 5.7.4. The last inequality of the previous proof shows that

$$\mathbf{V}(|f|, \Omega) \leq \mathbf{V}(f^+, \Omega) + \mathbf{V}(f^-, \Omega) = \mathbf{V}(f, \Omega)$$

holds for each $f \in L^1_{\mathrm{loc}}(\Omega)$. In particular, if f belongs to $BV(\Omega)$, then so do $f^\pm$ and $|f|$. A similar assertion holds for $BV_{\mathrm{loc}}(\Omega)$.

Proposition 5.7.5. *Let* $f \in BV_{\text{loc}}(\Omega)$. *The set* $E_t = \{f > t\}$ *belongs to* $\mathcal{BV}_{\text{loc}}(\Omega)$ *for* $\mathcal{L}^1$ *almost all* $t \in \mathbb{R}$, *and*

$$\|Df\|(B) = \int_{\mathbb{R}} \|D\chi_{E_t}\|(B)\, dt \tag{5.7.1}$$

for each $\|Df\|$ *measurable set* $B \subset \Omega$.

PROOF. For an open set $U \subset \Omega$, the coarea theorem and (5.1.1) imply

$$\int_{\mathbb{R}} \mathbf{V}(E_t, U)\, dt = \|D(f \upharpoonright U)\|(U) = \|Df\|(U), \tag{$*$}$$

since $E_t \cap U = \{f \upharpoonright U > t\}$. Find open sets $\Omega_1 \Subset \Omega_2 \Subset \cdots \Subset \Omega$ so that $\Omega = \bigcup_{k=1}^{\infty} \Omega_k$. By the previous equality

$$\int_{\mathbb{R}} \mathbf{V}(E_t, \Omega_k)\, dt = \|Df\|(\Omega_k) < \infty$$

for $k = 1, 2, \ldots$. There is an $\mathcal{L}^1$ negligible set $N \subset \mathbb{R}$ such that $\mathbf{V}(E_t, \Omega_k) < \infty$ for each $t \in \mathbb{R} - N$ and $k = 1, 2, \ldots$. If $U \Subset \Omega$ is an open set, then U is a subset of some Ω_k, and we see that

$$\mathbf{V}(E_t, U) \le \mathbf{V}(E_t, \Omega_k) < \infty \tag{$**$}$$

for every $t \in \mathbb{R} - N$. In other words, $E_t \in \mathcal{BV}_{\text{loc}}(\Omega)$ for all $t \in \mathbb{R} - N$, and $(*)$ shows that equality (5.7.1) holds for each open set $U \subset \Omega$.

If $K \subset \Omega$ is a compact set, there is a decreasing sequence of open sets $U_j \Subset \Omega$ such that $K = \bigcap_{j=1}^{\infty} U_j$. Then $\|D\chi_{E_t}\|(U_j)$ is a decreasing sequence converging to $\|D\chi_{E_t}\|(K)$ for each $t \in \mathbb{R} - N$, and by $(**)$,

$$\int_{\mathbb{R}} \|D\chi_{E_t}\|(K)\, dt = \lim \int_{\mathbb{R}} \|D\chi_{E_t}\|(U_j)\, dt$$
$$= \lim \|Df\|(U_j) = \|Df\|(K).$$

If $S \subset \Omega$ is a $\|Df\|$ negligible set, use Theorem 1.3.1, (1) to find a decreasing sequence $\{V_k\}$ of open subsets of Ω so that S is contained in $V = \bigcap_{k=1}^{\infty} V_k$ and $\|Df\|(V_k) < 1/k$ for $k = 1, 2, \ldots$. Since

$$\int_{\mathbb{R}} \|D\chi_{E_t}\|(V_1) = \|Df\|(V_1) < \infty,$$

$\|D\chi_{E_t}\|(V_1) < \infty$ for $\mathcal{L}^1$ almost all $t \in \mathbb{R}$, and hence

$$\int_{\mathbb{R}} \|D\chi_{E_t}\|(V)\, dt = \lim \int_{\mathbb{R}} \|D\chi_{E_t}\|(V_k) = \lim \|Df\|(V_k) = 0.$$

Thus $\|D\chi_{E_t}\|(S) = 0$ for $\mathcal{L}^1$ almost all $t \in \mathbb{R}$.

Let $B \Subset \Omega$ be $\|Df\|$ measurable. According to Theorem 1.3.1, (2), there is an increasing sequence $\{K_j\}$ of compact subsets of B such that the set

$S = B - \bigcup_{j=1}^{\infty} K_j$ is $\|Df\|$ is negligible. By the previous paragraphs,

$$\int_{\mathbb{R}} \|D\chi_{E_t}\|(B)\, dt = \lim \int_{\mathbb{R}} \|D\chi_{E_t}\|(K_j)$$
$$= \lim \|Df\|(K_j) = \|Df\|(B).$$

Finally, if $B \subset \Omega$ is an arbitrary $\|Df\|$ measurable set, we apply the previous paragraph to the sets $B \cap \Omega_k$, $k = 1, 2, \ldots$. $\square$

5.8. Bounded convex domains

If Ω is bounded and convex, the space $BV(\Omega)$ exhibits additional useful properties. Such properties remain valid for larger families of open sets [1, 75], e.g., for Lipschitz domains with compact boundary (Section 7.5 below). However, specializing to open sets that are bounded and convex is sufficient for our present tasks and simplifies the proofs.

Lemma 5.8.1. *Let Ω be bounded and convex, and let $g \in BV(\Omega)$. There is a sequence $\{g_k\}$ in $C^{\infty}(\Omega) \cap Lip(\Omega)$ such that*

$$\lim \|g_k - g\|_{L^1(\Omega)} = 0 \quad and \quad \lim \|Dg_k\|(\Omega) = \|Dg\|(\Omega).$$

PROOF. Translating Ω, we may assume $0 \in \Omega$. It follows from Theorem 5.6.3 that there is a sequence $\{f_k\}$ in $C^{\infty}(\Omega)$ such that

$$\lim \|f_k - g\|_{L^1(\Omega)} = 0 \quad and \quad \lim \|Df_k\|(\Omega) = \|Dg\|(\Omega). \tag{$*$}$$

For $k = 1, 2, \ldots$ and $x \in \mathbb{R}^n$, let $\phi_k(x) := kx/(k+1)$ and $\Omega_k := \phi_k^{-1}(\Omega)$. Since Ω is convex and bounded, $\Omega \Subset \Omega_k$ and the function

$$h_k : x \mapsto f_k[\phi_k(x)] : \Omega_k \to \mathbb{R}$$

belongs to $C^{\infty}(\Omega_k)$. Thus $g_k := h_k \upharpoonright \Omega$ belongs to $C^{\infty}(\Omega) \cap Lip(\Omega)$. In particular $g_k \in L^1(\Omega)$, since Ω is bounded. For $k = 1, 2, \ldots$,

$$\|f_k - g_k\|_{L^1(\Omega)} = \int_{\Omega} \left| f_k(x) - f_k\left(\tfrac{k}{k+1}x\right) \right| dx$$
$$= \int_{\Omega} \left| \int_0^1 \frac{d}{dt}\left[f_k\left(\tfrac{k+t}{k+1}x\right) \right] dt \right| dx$$
$$\leq \tfrac{1}{k+1} \int_0^1 \left(\int_{\Omega} |x| \cdot \left| Df_k\left(\tfrac{k+t}{k+1}x\right) \right| dx \right) dt$$
$$\leq \tfrac{1}{k+1} \left[\int_0^1 \left(\tfrac{k+1}{k+t}\right)^{n+1} dt \right] \left(\int_{\Omega} |y| \cdot \left| Df_k(y) \right| dy \right)$$
$$\leq \tfrac{1}{k+1} \cdot \left(\tfrac{k+1}{k}\right)^{n+1} d(\Omega) \|Df_k\|(\Omega),$$

$$\|Dg_k\|(\Omega) = \int_\Omega \left| D\Big[f_k\Big(\tfrac{kx}{k+1}\Big)\Big]\right| dx$$

$$= \big(\tfrac{k+1}{k}\big)^{n-1} \int_{\phi_k(\Omega)} |Df_k(x)|\, dx$$

$$\leq \big(\tfrac{k+1}{k}\big)^{n-1} \|Df_k\|(\Omega)$$

by the Fubini and change of variables theorems. These inequalities, $(*)$, and Proposition 5.1.7 imply $\lim \|g_k - g\|_{L^1(\Omega)} = 0$ and

$$\limsup \|Dg_k\|(\Omega) \leq \lim \|Df_k\|(\Omega) = \|Dg\|(\Omega) \leq \liminf \|Dg_k\|(\Omega). \qquad \square$$

Recall from (5.1.2) that a norm in $BV(\Omega)$ is defined by

$$\|f\|_{BV(\Omega)} = \|f\|_{L^1(\Omega)} + \|Df\|(\Omega)$$

for each $f \in BV(\Omega)$.

Lemma 5.8.2. *Let Ω be convex, and let $U \subset \mathbb{R}^n$ be an open set for which $\Omega \Subset U$. There is a constant $\kappa > 0$ such that each $g \in Lip(\Omega)$ has an extension $h \in Lip(\mathbb{R}^n)$ satisfying the following conditions:*

(i) $\operatorname{spt} h \subset U$,

(ii) $\|h\|_{L^\infty(\mathbb{R}^n)} \leq \|g\|_{L^\infty(\Omega)}$,

(iii) $\|h\|_{BV(\mathbb{R}^n)} \leq \kappa \|g\|_{BV(\Omega)}$.

PROOF. Select $z \in \Omega$, and let $\phi : \operatorname{cl}\Omega - \{z\} \to \mathbb{R}^n - \Omega$ be the radial reflection from z across $\partial\Omega$. More precisely: for $x \in \Omega - \{z\}$, the ray

$$\ell_x := \big\{z + t(x - z) : t \in \mathbb{R}_+\big\}$$

emanating from z and passing through x meets $\partial\Omega$ at a unique point x', and we let $\phi(x) := 2x' - x$. There is a compact set $K \subset \Omega$ such that $z \in \operatorname{int} K$ and the open set $V := \operatorname{cl}\Omega \cup \phi(\Omega - K)$ satisfies $\Omega \Subset V \subset U$. The restriction $\psi = \phi \restriction (\Omega - K)$ is a lipeomorphism from $\Omega - K$ onto $V - \operatorname{cl}\Omega$. Since g has a Lipschitz extension to $\operatorname{cl}\Omega$, still denoted by g,

$$g_1(x) := \begin{cases} g(x) & \text{if } x \in \operatorname{cl}\Omega, \\ g[\psi^{-1}(x)] & \text{if } x \in V - \operatorname{cl}\Omega \end{cases}$$

defines a Lipschitz function g_1 on V which extends g and satisfies

$$\|g_1\|_{L^\infty(V)} \leq \|g\|_{L^\infty(\Omega)},$$

$$\|g_1\|_{L^1(V)} \leq \beta \|g\|_{L^1(\Omega)}, \quad \|Dg_1\|_{L^1(V)} \leq \beta \|Dg\|_{L^1(\Omega)} \qquad (*)$$

where $\beta > 0$ is a constant depending only on $\operatorname{Lip}(\psi)$. Select open sets V_1, V_2 so that $\Omega \Subset V_1 \Subset V_2 \Subset V$, and find $\varphi \in C^1(\mathbb{R}^n)$ with $\chi_{V_1} \leq \varphi \leq \chi_{V_2}$. We see at once that $h : \mathbb{R}^n \to \mathbb{R}$ defined by the formula

$$h(x) := \begin{cases} g_1(x)\varphi(x) & \text{if } x \in V, \\ 0 & \text{if } x \in \mathbb{R}^n - V \end{cases}$$

is a Lipschitz extension of f satisfying (i) and (ii). By inequalities $(*)$,

$$\|h\|_{BV(\mathbb{R}^n)} \leq \|g_1\|_{L^1(V)} + \|D\varphi\|_{L^\infty(V)} \|g_1\|_{L^1(V)} + \|Dg_1\|_{L^1(V)}$$
$$\leq \beta[1 + \|D\varphi\|_{L^\infty(V)}] \|g\|_{L^1(\Omega)} + \beta\|Dg\|_{L^1(\Omega)} \leq \kappa\|g\|_{BV(\Omega)}$$

where $\kappa := \beta[1 + \|D\varphi\|_{L^\infty(\mathbb{R}^n)}]$. $\qquad\square$

Theorem 5.8.3. *Let $U \subset \mathbb{R}^n$ be an open set, and let $\Omega \Subset U$ be convex. There is a constant $\kappa > 0$ such that each $g \in BV(\Omega)$ has an extension $h \in BV(\mathbb{R}^n)$ that satisfies* $\operatorname{spt} h \subset U$ *and* $\|h\|_{BV(\mathbb{R}^n)} \leq \kappa\|g\|_{BV(\Omega)}$.

PROOF. Select an open set V so that $\Omega \Subset V \Subset U$. Combining Lemmas 5.8.1 and 5.8.2, find a sequence $\{h_k\}$ in $Lip(\mathbb{R}^n)$ such that

(i) $\lim \|h_k - g\|_{L^1(\Omega)} = 0$ and $\lim \|Dh_k\|(\Omega) = \|Dg\|(\Omega)$,

(ii) $\operatorname{spt} h_k \subset V$ and $\|h_k\|_{BV(\mathbb{R}^n)} \leq \kappa\|h_k\|_{BV(\Omega)}$ for $k = 1, 2, \ldots$.

It follows that $\sup \|Dh_k\|(\mathbb{R}^n) < \infty$. By Theorem 5.5.12, the sequence $\{h_k\}$ has a subsequence, still denoted by $\{h_k\}$, which converges in $L^1(\mathbb{R}^n)$ to a function $h \in BV(\mathbb{R}^n)$. This and (i) imply that a subsequence of $\{h_k\}$ converges to h almost everywhere in $\mathbb{R}^n$, and to g almost everywhere in Ω. By redefining h on a negligible set, we obtain $\operatorname{spt} h \subset U$ and $h \upharpoonright \Omega = g$. Moreover,

$$\|h\|_{BV(\mathbb{R}^n)} = \|h\|_{L^1(\mathbb{R}^n)} + \|Dh\|(\mathbb{R}^n)$$
$$\leq \lim \|h_k\|_{L^1(\mathbb{R}^n)} + \liminf \|Dh_k\|(\mathbb{R}^n)$$
$$\leq \liminf \|h_k\|_{BV(\mathbb{R}^n)}$$
$$\leq \kappa \lim \|h_k\|_{BV(\Omega)} = \kappa\|g\|_{BV(\Omega)}$$

by Proposition 5.1.7. $\qquad\square$

Corollary 5.8.4. *Let Ω be bounded and convex, and let $\{f_k\}$ be a sequence in $BV(\Omega)$ such that* $\sup \|f_k\|_{BV(\Omega)} < \infty$. *There is a subsequence of $\{f_k\}$ converging in $L^1(\Omega)$ to a function $f \in BV(\Omega)$.*

PROOF. Choose an open set U with $\Omega \Subset U \Subset \mathbb{R}^n$. Using Theorem 5.8.3, find h_k in $BV(\mathbb{R}^n)$ so that $h_k \upharpoonright \Omega = f_k$, $\operatorname{spt} h_k \subset U$, and $\sup \|Dh_k\|(\mathbb{R}^n) < \infty$. By Theorem 5.5.12, there is a subsequence $\{h_{k_j}\}$ and $h \in BV(\mathbb{R}^n)$ such that $\lim \|h - h_{k_j}\|_{L^1(\mathbb{R}^n)} = 0$. Now it is clear that $\{f_{k_j}\}$ and $f := h \upharpoonright \Omega$ satisfy the corollary's assertion. $\qquad\square$

Corollary 5.8.5. *Let Ω be bounded. If $f \in BV(\Omega)$ has an extension to a BV function in $\mathbb{R}^n$, then it has an extension to a BV function in $\mathbb{R}^n$ with compact support.*

PROOF. Assume $g \in BV(\mathbb{R}^n)$ extends f, and select $U(0, r)$ containing $\operatorname{cl}\Omega$. By Theorem 5.8.3, there is $h \in BV(\mathbb{R}^n)$ such that $h \upharpoonright U = g \upharpoonright U$ and $\operatorname{spt} h \subset U(0, r+1)$. Clearly h is the desired extension of f. $\qquad\square$

The following examples show that various assumptions of Corollary 5.8.4 cannot be omitted.

Example 5.8.6. Given a nonempty Ω that is bounded and convex, the functions $f_k := k\chi_\Omega \upharpoonright \Omega$ belong to $BV(\Omega)$ and $\|Df_k\|(\Omega) = 0$ for $k = 1, 2, \ldots$. However, no subsequence of $\{f_k\}$ converges in $L^1(\Omega)$ to a function $f \in BV(\Omega)$, since $\lim \|f_k\|_{L^1(\Omega)} = \infty$.

Example 5.8.7. Let $n = 1$. The function $f_k(x) = \sin kx$ defined on an interval $\Omega = (0, \pi)$ belongs to $BV(\Omega)$, and a direct calculation shows that $\|f_k\|_{L^1(\Omega)} = 2$ and $\|Df_k\|(\Omega) = 2k$ for $k = 1, 2, \ldots$. In addition, $\lim \int_a^b f_k = 0$ for each interval $(a, b) \subset \Omega$. If a subsequence of $\{f_k\}$ converges in $L^1(\Omega)$ to $f \in L^1(\Omega)$, then $\|f\|_{L^1(\Omega)} = 2$ and $\int_a^b f = 0$ for each $(a, b) \subset \Omega$, which is a contradiction.

Example 5.8.8. Assume $n \geq 2$, and choose a closed ball $B \subset \mathbb{R}^n$. Let Ω be the union of a family $\{U_k \subset B : k \in \mathbb{N}\}$ of open balls whose closures are disjoint (cf. Example 5.2.6). Functions $f_k := |U_k|^{-1}\chi_{U_k} \upharpoonright \Omega$ belong to $BV(\Omega)$ and $\|f_k\|_{BV(\Omega)} = 1$ for $k = 1, 2, \ldots$. If some $\{f_{k_j}\}$ converges in $L^1(\Omega)$ to $f \in L^1(\Omega)$, then $\|f\|_{L^1(\Omega)} = 1$ and a subsequence of $\{f_{k_j}\}$ converges to f almost everywhere in Ω. Since $\lim f_k(x) = 0$ for each $x \in \Omega$, we have a contradiction.

Proposition 5.8.9. *Let Ω be arbitrary, and let $\{f_k\}$ be a sequence in $L^1_{\mathrm{loc}}(\Omega)$ such that $\limsup \|f_k\|_{BV(U)} < \infty$ for each open ball $U \Subset \Omega$. There is a subsequence of $\{f_k\}$ that converges in $L^1_{\mathrm{loc}}(\Omega)$, as well as almost everywhere, to a function $f \in BV_{\mathrm{loc}}(\Omega)$.*

Proof. There are open balls $U_i \Subset \Omega$ such that $\Omega = \bigcup_{i=1}^\infty U_i$. Using Corollary 5.8.4, construct recursively subsequences $s_i = \{g_{i,k}\}$ of $\{f_k\}$ so that s_{i+1} is a subsequence of $\{s_i\}$ and $\{g_{i,k} \upharpoonright U_i\}$ converges in $L^1(U_i)$, as well as almost everywhere, to a function $g_i \in BV(U_i)$. Then $\{g_{k,k}\}$ is a subsequence of $\{f_k\}$ which converges in $L^1(U_i)$, as well as almost everywhere, to g_i for $i = 1, 2, \ldots$. There is a negligible set $N \subset \Omega$ and a function f defined on Ω such that $f(x) = g_i(x)$ for each $x \in U_i - N$ and $i = 1, 2, \ldots$. In particular, $\lim g_{k,k}(x) = f(x)$ for all $x \in \Omega - N$. As every $U \Subset \Omega$ is covered by finitely many balls U_i, we see that $f \in BV_{\mathrm{loc}}(\Omega)$ and $\lim g_{k,k} = f$ in $L^1_{\mathrm{loc}}(\Omega)$. $\square$

Example 5.8.10. Let $\{B_k\}$ be a sequence of disjoint closed balls, each of radius 1. Then $\{\chi_{B_k}\}$ is a sequence in $BV(\mathbb{R}^n)$ which converges to 0 in $L^1_{\mathrm{loc}}(\mathbb{R}^n)$, but does not converge in $L^1(\mathbb{R}^n)$, since $i \neq j$ implies

$$\|\chi_{B_i} - \chi_{B_j}\|_{L^1(\mathbb{R}^n)} = |B_i| + |B_j| = 2\alpha(n) > 0.$$

Thus L^1_{loc} convergence is substantially weaker than L^1 convergence.

5.9. Inequalities

Definition 5.9.1. Let E be a measurable set such that $0 < |E| < \infty$. The *mean value* of $f \in L^1(E)$ is the number

$$(f)_E := \frac{1}{|E|} \int_E f(x)\, dx.$$

Throughout this book, when using the symbol $(f)_E$, we tacitly assume that E is a measurable set with $0 < |E| < \infty$ and $f \in L^1(E)$.

Lemma 5.9.2. *Let Ω be bounded and convex. There is $\beta > 0$ such that*

$$\left\| f - (f)_\Omega \right\|_{L^1(\Omega)} \le \beta \|Df\|(\Omega)$$

for each $f \in BV(\Omega)$.

PROOF. Seeking a contradiction, suppose there are $f_k \in BV(\Omega)$ with

$$\left\| f_k - (f_k)_\Omega \right\|_{L^1(\Omega)} > k \|Df_k\|(\Omega) \tag{*}$$

for $k = 1, 2, \dots$. Define $g_k \in BV(\Omega)$ by letting

$$g_k(x) := \frac{f_k(x) - (f_k)_\Omega}{\left\| f_k - (f_k)_\Omega \right\|_{L^1(\Omega)}}$$

for each $x \in \Omega$, and observe that $\|g_k\|_{L^1(\Omega)} = 1$, $\int_\Omega g_k = 0$, and by $(*)$,

$$\|Dg_k\|(\Omega) = \frac{\|Df_k\|(\Omega)}{\left\| f_k - (f_k)_\Omega \right\|_{L^1(\Omega)}} < \frac{1}{k}. \tag{**}$$

Thus $\sup \|g_k\|_{BV(\Omega)} \le 2$. By Corollary 5.8.4, there are $g \in BV(\Omega)$ and a subsequence of $\{g_{k_j}\}$ such that $\lim \|g_{k_j} - g\|_{L^1(\Omega)} = 0$. We infer $\|g\|_{L^1(\Omega)} = 1$ and $\int_\Omega g(x)\, dx = 0$. By Proposition 5.1.7 and $(**)$,

$$\|Dg\|(\Omega) \le \liminf \|Dg_k\|(\Omega) = 0$$

and a contradiction follows from Proposition 5.5.9. $\qquad\qquad\square$

A convex set $C \subset \mathbb{R}^n$ is *centrally symmetric* if there is $z \in C$, called the *center* of C, such that $C = 2z - C$. For such a set and $r > 0$, we let

$$C_r := z + r(C - z).$$

Corollary 5.9.3. *Let Ω be bounded, convex, and centrally symmetric. There is $\beta > 0$ such that for each $r > 0$ and each $f \in BV(\Omega_r)$,*

$$\left\| f - (f)_{\Omega_r} \right\|_{L^1(\Omega_r)} \le \beta r \|Df\|(\Omega_r).$$

PROOF. As this is our first scaling argument, we present it in detail. Via translation, we may assume that 0 is the center of Ω. Given $r > 0$ and $f \in BV(\Omega_r)$, let $\phi(x) := rx$ and $g := f \circ \phi$. Clearly $g \in L^1(\Omega)$, and we show that $g \in BV(\Omega)$. To this end, choose $v \in C_c^1(\Omega; \mathbb{R}^n)$ with $\|v\|_{L^\infty(\Omega;\mathbb{R}^n)} \leq 1$, and observe that $w := v \circ \phi^{-1}$ belongs to $C_c^1(\Omega_r; \mathbb{R}^n)$ and $\|w\|_{L^\infty(\Omega_r;\mathbb{R}^n)} \leq 1$. Using the area theorem,

$$\int_\Omega g \operatorname{div} v = r^{-n} \int_{\Omega_r} (g \circ \phi^{-1}) \cdot \left[(\operatorname{div} v) \circ \phi^{-1} \right]$$

$$= r^{-n} \int_{\Omega_r} f \cdot [r \operatorname{div} w] \leq r^{-n+1} \|Df\|(\Omega_r).$$

The arbitrariness of v implies $\mathbf{V}(g, \Omega) \leq r^{-n+1} \|Df\|(\Omega_r) < \infty$, which means $g \in BV(\Omega)$. By the area theorem and Lemma 5.9.2,

$$\|f - (f)_{\Omega_r}\|_{L^1(\Omega_r)} = \|f - (g)_\Omega\|_{L^1(\Omega_r)}$$

$$= r^n \|g - (g)_\Omega\|_{L^1(\Omega)}$$

$$\leq r^n \beta \|Dg\|(\Omega) \leq r\beta \|Df\|(\Omega_r)$$

where $\beta > 0$ is a constant independent of f. □

Lemma 5.9.4. *Let $E \subset \mathbb{R}^n$. If U and $U \cap E$ are measurable sets, then*

$$\min\{|U \cap E|, |U - E|\}^{\frac{1}{p}} \leq 2 \|\chi_E - (\chi_E)_U\|_{L^p(U)}$$

for each real number $p \geq 1$.

PROOF. Let $I := \int_U |\chi_E(x) - (\chi_E)_U|^p \, dx$ and calculate

$$I = \int_{U \cap E} \left| \chi_E(x) - \frac{|U \cap E|}{|U|} \right|^p dx + \int_{U - E} \left| \chi_E(x) - \frac{|U \cap E|}{|U|} \right|^p dx$$

$$= \left(1 - \frac{|U \cap E|}{|U|} \right)^p |U \cap E| + \left(\frac{|U \cap E|}{|U|} \right)^p |U - E|$$

$$= \left(\frac{|U - E|}{|U|} \right)^p |U \cap E| + \left(\frac{|U \cap E|}{|U|} \right)^p |U - E|.$$

If $|U - E| \geq |U \cap E|$ then $2|U - E| \geq |U \cap E| + |U - E| = |U|$, and hence

$$I \geq \left(\frac{|U - E|}{|U|} \right)^p |U \cap E| \geq (\tfrac{1}{2})^p |U \cap E| = (\tfrac{1}{2})^p \min\{|U \cap E|, |U - E|\}.$$

Similarly, if $|U \cap E| \geq |U - E|$ then $2|U \cap E| \geq |U|$, and hence

$$I \geq \left(\frac{|U \cap E|}{|U|} \right)^p |U - E| \geq (\tfrac{1}{2})^p \min\{|U \cap E|, |U - E|\}. \quad □$$

Lemma 5.9.5. *Let Ω be a bounded centrally symmetric convex set. There is $\beta > 0$ such that*

$$\min\{(\chi_E)_{\Omega_r}, 1 - (\chi_E)_{\Omega_r}\} \leq \beta \frac{2r}{|\Omega_r|} \mathbf{V}(E, \Omega_r)$$

for each $r > 0$ and every $E \subset \mathbb{R}^n$ for which $\Omega_r \cap E$ is measurable.

PROOF. Choose $r > 0$ and $E \subset \mathbb{R}^n$ so that $\Omega_r \cap E$ is measurable. As Ω_r is bounded, $\chi_E \in BV(\Omega_r)$. By Lemma 5.9.4 with $p = 1$ and Corollary 5.9.3, there is $\beta > 0$ such that

$$\min\{(\chi_E)_{\Omega_r}, 1 - (\chi_E)_{\Omega_r}\} = \frac{1}{|\Omega_r|} \min\{|\Omega_r \cap E|, |\Omega_r - E|\}$$

$$\leq \frac{2}{|\Omega_r|} \|\chi_E - (\chi_E)_{\Omega_r}\|_{L^1(\Omega_r)}$$

$$\leq \beta \frac{2r}{|\Omega_r|} \|D\chi_E\|(\Omega_r) = \beta \frac{2r}{|\Omega_r|} \mathbf{V}(E, \Omega_r). \qquad \square$$

Theorem 5.9.6 (Isoperimetric inequality). *Let $n \geq 2$. There is a constant $\gamma = \gamma(n) > 0$ such that for each measurable set E,*

$$\min\{|E|, |\mathbb{R}^n - E|\}^{\frac{n-1}{n}} \leq \gamma \mathbf{V}(E).$$

If $\mathbf{V}(E) < \infty$, then either E or $\mathbb{R}^n - E$ is a BV set.

PROOF. As there is nothing to prove otherwise, assume $\mathbf{V}(E) < \infty$. Hence $E \in \mathbf{BV}_{\mathrm{loc}}(\mathbb{R}^n)$, and we can use the variational measure $\|D\chi_E\|$. For $x = (x_1, \ldots, x_n)$ in $\mathbb{R}^n$ and $r > 0$, a cube

$$Q_{x,r} := \prod_{i=1}^{n} (x_i - r, x_i + r)$$

is an open centrally symmetric convex set, and $Q_{x,r} = x + (Q_{0,1})_r$. A constant $\beta > 0$ associated with $Q_{0,1}$ according to Lemma 5.9.5 depends only on the dimension n. By translation we obtain

$$\min\{(\chi_E)_{Q_{x,r}}, 1 - (\chi_E)_{Q_{x,r}}\} \leq \frac{\beta}{(2r)^{n-1}} \|D\chi_E\|(Q_{x,r}) \qquad (*)$$

for each open cube $Q_{x,r}$. Let $Q_x := Q_{x,R}$ where $R := \frac{1}{2}[3\beta \mathbf{V}(E)]^{\frac{1}{n-1}}$. For this choice of R, inequality $(*)$ yields

$$\min\{(\chi_E)_{Q_x}, 1 - (\chi_E)_{Q_x}\} \leq \frac{1}{3\mathbf{V}(E)} \|D\chi_E\|(Q_x) \leq \frac{1}{3}.$$

Thus for each $x \in \mathbb{R}^n$, we have two mutually exclusive cases

$$(\chi_E)_{Q_x} < 1/2 \quad \text{or} \quad (\chi_E)_{Q_x} > 1/2,$$

and the case that holds for one $x \in \mathbb{R}^n$, holds for all $x \in \mathbb{R}^n$, since

$$x \mapsto (\chi_E)_{Q_x} : \mathbb{R}^n \to [0, 1/2) \cup (1/2, 1]$$

is a continuous function. If $(\chi_E)_{Q_x} < 1/2$, inequality $(*)$ implies

$$\frac{|Q_x \cap E|}{(2R)^n} = (\chi_E)_{Q_x} \leq \frac{\beta}{(2R)^{n-1}} \|D\chi_E\|(Q_x),$$

or equivalently, $|Q_x \cap E| \leq 2R\beta \|D\chi_E\|(Q_x)$. There is a sequence $\{x_i\}$ such that $\{Q_{x_i}\}$ is a disjoint family which covers almost all of $\mathbb{R}^n$. Thus

$$|E| = \sum_{i=1}^{\infty} |Q_{x_i} \cap E| \leq 2R\beta \sum_{i=1}^{\infty} \|D\chi_E\|(Q_{x_i}) \leq 2R\beta\, \mathbf{V}(E),$$

and the choice of R yields $|E| \leq \gamma \mathbf{V}(E)^{\frac{n}{n-1}}$ where $\gamma := 3^{\frac{1}{n-1}}\beta^{\frac{n}{n-1}}$. When $(\chi_E)_{Q_x} > 1/2$, inequality $(*)$ implies

$$\frac{|Q_x - E|}{(2R)^n} = 1 - (\chi_E)_{Q_x} \leq \frac{\beta}{(2R)^{n-1}} \|D\chi_E\|(Q_x),$$

or equivalently, $|Q_x - E| \leq 2R\beta\|D\chi_E\|(Q_x)$. Proceeding as in the previous case, we obtain $|\mathbb{R}^n - E| \leq \gamma \mathbf{V}(E)^{\frac{n}{n-1}}$. $\qquad\square$

Corollary 5.9.7. *Let $n \geq 2$. If $E \subset \mathbb{R}^n$ and $P(E) < \infty$, then either E or $\mathbb{R}^n - E$ has finite measure.*

The corollary follows from Proposition 4.5.3 and Remark 4.6.8. The set $\mathbb{R}_+ \subset \mathbb{R}^1$ shows that it is false in dimension one.

Theorem 5.9.8 (Sobolev's inequality). *Let $n \geq 2$, and let γ be the constant from the isoperimetric inequality. If $f \in BV(\mathbb{R}^n)$, then*

$$\|f\|_{L^{\frac{n}{n-1}}(\mathbb{R}^n)} \leq \gamma\|Df\|(\mathbb{R}^n).$$

PROOF. Suppose first that $f \geq 0$. For $t \in \mathbb{R}_+$, the function $f_t := \min\{f, t\}$ belongs to $L^{\frac{n}{n-1}}(\mathbb{R}^n)$. Indeed,

$$\int_{\mathbb{R}^n} [f_t(x)]^{\frac{n}{n-1}} dx = \int_{\{f<1\}} [f_t(x)]^{\frac{n}{n-1}} dx + \int_{\{f\geq 1\}} [f_t(x)]^{\frac{n}{n-1}} dx$$

$$\leq \int_{\{f<1\}} f(x)\, dx + t^{\frac{n}{n-1}} |\{f \geq 1\}|$$

$$\leq \left(1 + t^{\frac{n}{n-1}}\right) \int_{\mathbb{R}^n} f(x)\, dx < \infty.$$

Thus $g : t \mapsto \|f_t\|_{L^{\frac{n}{n-1}}(\mathbb{R}^n)}$ is an increasing real-valued function defined on $\mathbb{R}_+$, $g(0) = 0$, and $\lim_{t\to\infty} g(t) = \|f\|_{L^{\frac{n}{n-1}}(\mathbb{R}^n)}$. If $h > 0$, then

$$g(t + h) - g(t) = \|f_{t+h}\|_{L^{\frac{n}{n-1}}(\mathbb{R}^n)} - \|f_t\|_{L^{\frac{n}{n-1}}(\mathbb{R}^n)}$$

$$\leq \|f_{t+h} - f_t\|_{L^{\frac{n}{n-1}}(\mathbb{R}^n)} \leq \|h\chi_{\{f>t\}}\|_{L^{\frac{n}{n-1}}(\mathbb{R}^n)}$$

$$= h|\{f > t\}|^{\frac{n-1}{n}}.$$

This and the isoperimetric inequality yield

$$0 \leq g'(t) \leq |\{f > t\}|^{\frac{n-1}{n}} \leq \gamma \mathbf{V}(\{f > t\})$$

for almost all $t \geq 0$. By the coarea theorem (Theorem 5.7.3),

$$\|f\|_{L^{\frac{n}{n-1}}(\mathbb{R}^n)} = \int_0^\infty g'(t)\,dt \leq \gamma \int_0^\infty \mathbf{V}(\{f > t\})\,dt = \gamma \mathbf{V}(f).$$

If f is arbitrary, then Remark 5.7.4 implies

$$\|f\|_{L^{\frac{n}{n-1}}(\mathbb{R}^n)} \leq \|f^+\|_{L^{\frac{n}{n-1}}(\mathbb{R}^n)} + \|f^-\|_{L^{\frac{n}{n-1}}(\mathbb{R}^n)}$$
$$= \gamma \mathbf{V}(f^+) + \gamma \mathbf{V}(f^-) = \gamma \mathbf{V}(f) = \|Df\|(\mathbb{R}^n). \qquad \square$$

The following corollary improves on Corollary 5.5.11.

Corollary 5.9.9. *If $f \in BV(\mathbb{R}^n)$ has compact support, then*

$$\|f\|_{L^1(\mathbb{R}^n)} \leq \gamma |\operatorname{spt} f|^{\frac{1}{n}} \|Df\|(\mathbb{R}^n)$$

where $\gamma = \gamma(n) > 0$.

PROOF. Choose an open set $U \subset \mathbb{R}^n$ containing $\operatorname{spt} f$. If $n \geq 2$, then the Hölder and Sobolev inequalities yield

$$\|f\|_{L^1(\mathbb{R}^n)} = \|f\|_{L^1(U)} \leq \|1\|_{L^n(U)} \|f\|_{L^{\frac{n}{n-1}}(U)}$$
$$= |U|^{\frac{1}{n}} \|f\|_{L^{\frac{n}{n-1}}(\mathbb{R}^n)} \leq \gamma(n) |U|^{\frac{1}{n}} \|Df\|(\mathbb{R}^n)$$

where $\gamma(n)$ is the constant from the isoperimetric inequality. Let $n = 1$. If $g \in C_c^1(\mathbb{R})$, then $g(t) = \int_{-\infty}^t g'$. Hence $|g(t)| \leq \int_{\mathbb{R}} |g'|$, and

$$\|g\|_{L^1(U)} \leq |U| \int_U |g'| = |U| \cdot \|Dg\|(U)$$

whenever $\operatorname{spt} g \subset U$. In view of this, $\|f\|_{L^1(U)} \leq |U| \cdot \|Df\|(U)$ by Theorem 5.6.3, since $f \upharpoonright U$ belongs to $BV(U)$. Thus with $\gamma(1) = 1$,

$$\|f\|_{L^1(U)} \leq \gamma(n) |U|^{\frac{1}{n}} \|Df\|(\mathbb{R}^n)$$

holds in each dimension $n \geq 1$. The corollary follows from the arbitrariness of U and Theorem 1.3.1, (1). $\qquad \square$

Example 5.9.10. Assume $n \geq 2$, and let Ω be a bounded set that is the union of disjoint open balls $U_k = U(x_i, k^{-2})$, $k = 1, 2, \ldots$. Define a function $f \in BV(\Omega)$ by letting $f(x) = k^{2n-2}$ for each $x \in U_k$. Since

$$\int_{U_k} f^{\frac{n}{n-1}}(x)\,dx = \alpha(n) > 0,$$

we see $\|f\|_{L^{\frac{n}{n-1}}(U)} = \infty$. Suppose f has an extension $g \in BV(\mathbb{R}^n)$. In view of Corollary 5.8.5, we may assume that g has compact support, and a contradiction follows from Sobolev's inequality:

$$\infty = \|f\|_{L^{\frac{n}{n-1}}(U)} \leq \|g\|_{L^{\frac{n}{n-1}}(\mathbb{R}^n)} \leq \gamma \|Dg\|(\mathbb{R}^n) < \infty.$$

Theorem 5.9.11 (Poincaré's inequality). *Let $n \geq 2$, and let Ω be convex, bounded, and centrally symmetric. There is $\theta > 0$ such that*

$$\left\| f - (f)_U \right\|_{L^{\frac{n}{n-1}}(U)} \leq \theta \left\| Df \right\|(U)$$

for each $U = x + \Omega_r$ and each $f \in BV(U)$.

PROOF. If g is a BV function in Ω, then so is $g - (g)_\Omega$. By Theorem 5.8.3, $g - (g)_\Omega$ has an extension $h \in BV(\mathbb{R}^n)$ with compact support such that

$$\left\| h \right\|_{BV(\mathbb{R}^n)} \leq \kappa \left\| g - (g)_\Omega \right\|_{BV(\Omega)}$$

where $\kappa > 0$ is a constant independent of g. From Sobolev's inequality and Lemma 5.9.2, we obtain

$$\left\| g - (g)_\Omega \right\|_{L^{\frac{n}{n-1}}(\Omega)} = \left\| h \right\|_{L^{\frac{n}{n-1}}(\mathbb{R}^n)} \leq \gamma \left\| Dh \right\|(\mathbb{R}^n)$$

$$\leq \gamma \left\| h \right\|_{BV(\mathbb{R}^n)} \leq \gamma \kappa \left\| g - (g)_\Omega \right\|_{BV(\Omega)}$$

$$= \gamma \kappa \left[\left\| g - (g)_\Omega \right\|_{L^1(\Omega)} + \left\| Dg \right\|(\Omega) \right]$$

$$\leq \gamma \kappa (\beta + 1) \left\| Dg \right\|(\Omega).$$

Here $\gamma = \gamma(n) > 0$ and $\beta > 0$ are constants independent of g, and so is the constant $\theta := \gamma \kappa (\beta + 1)$.

Let $U = x + \Omega_r$ and $f \in BV(U)$. In view of translation invariance, we may assume that $x = 0$. For a map $\phi : x \mapsto rx : \Omega \to \Omega_r$, the scaling argument, similar to that used in the proof of Corollary 5.9.3, shows that $g = f \circ \phi$ belongs to $BV(\Omega)$ and $\left\| Dg \right\|(\Omega) \leq r^{-n+1} \left\| Df \right\|(\Omega_r)$. Thus

$$\left\| g - (g)_\Omega \right\|_{L^{\frac{n}{n-1}}(\Omega)} \leq \theta \left\| Dg \right\|(\Omega) \leq r^{-n+1} \left\| Df \right\|(\Omega_r)$$

by the previous paragraph. Since the area theorem yields

$$\left\| g - (g)_\Omega \right\|_{L^{\frac{n}{n-1}}(\Omega)} = \left\| g - (f)_{\Omega_r} \right\|_{L^{\frac{n}{n-1}}(\Omega)} = r^{-n+1} \left\| f - (f)_{\Omega_r} \right\|_{L^{\frac{n}{n-1}}(\Omega_r)},$$

Poincaré's inequality follows. $\qquad \square$

Theorem 5.9.12 (Relative isoperimetric inequality). *Let $n \geq 2$, and let Ω be bounded, convex, and centrally symmetric. There is $\eta > 0$ such that*

$$\min \left\{ |U \cap E|, |U - E| \right\}^{\frac{n-1}{n}} \leq \eta \left\| D\chi_{E \cap U} \right\|(U)$$

for each $U = x + \Omega_r$ and each set $E \subset \mathbb{R}^n$ with $E \cap U \in \mathcal{BV}(U)$.

PROOF. By Lemma 5.9.4 with $p = \frac{n}{n-1}$, and Poincaré's inequality,

$$\min \left\{ |U \cap E|, |U - E| \right\}^{\frac{n-1}{n}} \leq 2 \left\| \chi_E - (\chi_E)_U \right\|_{L^{\frac{n}{n-1}}(U)} \leq 2\theta \left\| D\chi_{E \cap U} \right\|(U)$$

where $\theta > 0$ is the constant occurring in Poincaré's inequality. $\qquad \square$

Corollary 5.9.13. *Let* $n \geq 2$, *and let* $U \subset \mathbb{R}^n$ *be either an open ball or the interior of a cube. There is* $\kappa = \kappa(n) > 0$ *such that*

$$\left\|f - (f)_U\right\|_{L^{\frac{n}{n-1}}(U)} \leq \kappa \|Df\|(U),$$

$$\min\{|U \cap E|, |U - E|\}^{\frac{n-1}{n}} \leq \kappa \|D\chi_{E \cap U}\|(U)$$

for each $f \in BV(U)$ *and each* $E \subset \mathbb{R}^n$ *with* $E \cap U \in \mathcal{BV}(U)$.

PROOF. Let κ be the largest of the constants θ_U, θ_Q and η_U, η_Q associated with the ball $U = U(0,1)$ and cube $Q = (0,1)^n$ in the Poincaré and relative isoperimetric inequalities, respectively. Since each open ball in $\mathbb{R}^n$ has the form $x + U_r$ and each open cube in $\mathbb{R}^n$ has the form $x + Q_r$, the corollary follows from Theorems 5.9.11 and 5.9.12.   □

Considering $U_s := (-s, s)$ and $E_{s,t} := (0, s+t)$ for $s, t \in \mathbb{R}_+$ shows that the relative isoperimetric inequality provides no immediate information when $n = 1$. For a more detailed treatment of isoperimetric inequalities, we refer to [75, Section 5.11] and [1, Section 3.4].

Lemma 5.9.14. *If* $E \in \mathcal{BV}_{\mathrm{loc}}(\Omega)$, *then for each* $x \in \partial_* E \cap \Omega$,

$$\limsup_{r \to 0} \frac{\|D\chi_E\|[B(x,r)]}{r^{n-1}} > 0.$$

PROOF. Choose $x \in \partial_* E \cap \Omega$, and note that the numbers

$$t := \limsup_{r \to 0} \frac{|B(x,r) \cap E|}{|B(x,r)|} \quad \text{and} \quad s := \limsup_{r \to 0} \frac{|B(x,r) - E|}{|B(x,r)|}$$

are positive. By Corollary 5.9.13,

$$\frac{\|D\chi_E\|[U(x,r)]}{r^{n-1}} \geq \frac{\alpha(n)^{\frac{n-1}{n}}}{\kappa} \min\left\{\frac{|B(x,r) \cap E|}{|B(x,r)|}, \frac{|B(x,r) - E|}{|B(x,r)|}\right\}^{\frac{n-1}{n}}$$

whenever $U(x,r) \subset \Omega$. Hence

$$\limsup_{r \to 0} \frac{\|D\chi_E\|[B(x,r)]}{r^{n-1}} \geq \limsup_{r \to 0} \frac{\|D\chi_E\|[U(x,r)]}{r^{n-1}}$$

$$= \frac{\alpha(n)^{\frac{n-1}{n}}}{\kappa} \min\{t, s\}^{\frac{n-1}{n}} > 0. \qquad \square$$

Definition 5.9.15. If μ and ν are measures in an open set $\Omega \subset \mathbb{R}^n$, we say that ν is *absolutely continuous with respect to* μ, and write $\nu \ll \mu$, if $\mu(E) = 0$ implies $\nu(E) = 0$ for each $E \subset \Omega$. If ν is an $\mathbb{R}^m$-valued measure in Ω, we say that ν is absolutely continuous with respect to μ, and write again $\nu \ll \mu$, whenever $\|\nu\| \ll \mu$. A measure, or an $\mathbb{R}^m$-valued measure, in Ω that is absolutely continuous with respect to Lebesgue measure $\mathcal{L}^n$ in Ω is called *absolutely continuous*.

Proposition 5.9.16. *For every* $E \in \mathcal{BV}_{\mathrm{loc}}(\Omega)$,

$$\mathcal{H}^{n-1} \llcorner (\partial_* E \cap \Omega) \ll \|D\chi_E\| \llcorner (\partial_* E \cap \Omega).$$

PROOF. Select $A \subset \partial_* E \cap \Omega$ with $\|D\chi_E\|(A) = 0$. Letting

$$A_k := \left\{ x \in A : \limsup_{r \to 0} \frac{\|D\chi_E\|\big[B(x,r)\big]}{r^{n-1}} > \frac{1}{k} \right\},$$

Lemma 5.9.14 implies $A = \bigcup_{k=1}^{\infty} A_k$. Fix $k \in \mathbb{N}$, and choose $\varepsilon > 0$. By Theorem 1.3.1, (1), there is an open set $U \subset \Omega$ such that $A_k \subset U$ and $\|D\chi_E\|(U) < \varepsilon$. Select $\delta > 0$, and denote by $\mathcal{B}$ the family of all closed balls $B(x,r) \subset U$ satisfying

$$x \in A_k, \quad 0 < r < \delta/10, \quad \text{and} \quad \|D\chi_E\|\big[B(x,r)\big] \ge r^{n-1}/k.$$

By Vitali's theorem (Theorem 4.3.2), there are disjoint balls $B(x_i, r_i)$ in $\mathcal{B}$ such that $A_k \subset \bigcup_i B(x_i, 5r_i)$. Thus

$$\mathcal{H}_\delta^{n-1}(A_k) \le \sum_i \alpha(n-1)(5r_i)^{n-1}$$

$$\le 5^{n-1} k\alpha(n-1) \sum_i \|D\chi_E\|\big[B(x_i, r_i)\big]$$

$$\le 5^{n-1} k\alpha(n-1)\|D\chi_E\|(U) < 5^{n-1} k\alpha(n-1)\varepsilon.$$

As δ and ε are arbitrary, $\mathcal{H}^{n-1}(A_k) = 0$ and the proposition follows. $\square$

5.10. Lipschitz maps

Throughout this section, $\phi = (f_1, \ldots, f_n)$ is a Lipschitz map from $\mathbb{R}^n$ to $\mathbb{R}^n$. Let $g \in L^1(\mathbb{R}^n)$. Since g sign det $D\phi$ belongs to $L^1(\mathbb{R}^n)$, it follows from the area theorem (Theorem 1.5.6) that the function

$$\phi_\# g : y \mapsto \sum \left\{ g(x) \big[\, \text{sign det } D\phi(x) \big] : x \in \phi^{-1}(y) \right\} \tag{5.10.1}$$

belongs to $L^1(\mathbb{R}^n)$. Indeed, inequality (1.5.2) implies

$$\|\phi_\# g\|_{L^1(\mathbb{R}^n)} \le \int_{\mathbb{R}^n} \sum_{x \in \phi^{-1}(y)} |g(x)| \, dy$$

$$= \int_{\mathbb{R}^n} |g(x)| J_\phi(x) \, dx \le (\text{Lip } \phi)^n \|g\|_{L^1(\mathbb{R}^n)}. \tag{5.10.2}$$

We show that if $g \in BV(\mathbb{R}^n)$ has compact support, then $\phi_\# g \in BV(\mathbb{R}^n)$. To this end, denote by adj $D\phi$ the *adjoint matrix* of $D\phi$, and recall that

$$(\text{adj } D\phi) \cdot D\phi = D\phi \cdot (\text{adj } D\phi) = (\det D\phi)I \tag{5.10.3}$$

where I is the identity $n \times n$ matrix [40, Section 5.4].

Lemma 5.10.1. *Let* $u \in C^1(\mathbb{R}^n; \mathbb{R}^n)$ *and* $v = (\operatorname{adj} D\phi)(u \circ \phi)$. *Then*

$$-\int_{\mathbb{R}^n} v \cdot d(Dg) = \int_{\mathbb{R}^n} g(x) \left[\det D\phi(x)\right] (\operatorname{div} u) \left[\phi(x)\right] dx.$$

for each $g \in BV(\mathbb{R}^n)$ *with compact support.*

PROOF. Assume first that $\phi \in C^2(\mathbb{R}^n; \mathbb{R}^n)$. For $i, j = 1, \ldots, n$, denote by $a_{i,j}$ the elements of $\operatorname{adj} D\phi$ so that the vector fields

$$R_i = (a_{i,1}, \ldots, a_{i,n}) \quad \text{and} \quad C_j = (a_{1,j} \ldots, a_{n,j})$$

are the i-th row and j-th column of $\operatorname{adj} D\phi$, respectively. A straightforward but tedious calculation establishes that $\operatorname{div} C_j = 0$ for $j = 1, \ldots, n$. Letting $u = (u_1, \ldots, u_n)$ and $v = (v_1, \ldots, v_n)$, we calculate

$$\operatorname{div} v = \sum_i \left[D_i R_i \cdot (u \circ \phi) + R_i \cdot D_i(u \circ \phi)\right]$$

$$= \sum_i \sum_j D_i a_{i,j} u_j + \sum_i R_i \cdot D_i(u \circ \phi)$$

$$= \sum_j u_j \operatorname{div} C_j + \sum_i R_i \cdot D_i(u \circ \phi)$$

$$= \sum_i R_i \cdot \left(\sum_k \left[(D_k u) \circ \phi\right] D_i f_k\right)$$

$$= \sum_i \sum_j a_{i,j} \sum_k \left[(D_k u_j) \circ \phi\right] D_i f_k$$

$$= \sum_j \sum_k \left[(D_k u_j) \circ \phi\right] \sum_i a_{i,j} D_i f_k$$

$$= \sum_j \sum_k \left[(D_k v_j) \circ \phi\right] C_j \cdot D f_k$$

$$= \sum_j \left[(D_j u_j) \circ \phi\right] (\det D\phi) = (\det D\phi) \left[(\operatorname{div} u) \circ \phi\right];$$

the equality before last follows from (5.10.3). Since g has compact support, Theorem 5.5.1 implies

$$-\int_{\mathbb{R}^n} v \cdot d(Dg) = \int_{\mathbb{R}^n} g(x) \operatorname{div} v(x) \, dx$$

$$= \int_{\mathbb{R}^n} g(x) \left[\det D\phi(x)\right] (\operatorname{div} u) \left[\phi(x)\right] dx.$$

If ϕ is merely Lipschitz, select a mollifier η and define $\phi_k := \eta_{1/k} * \phi$ and $w_k := (\operatorname{adj} D\phi_k)(u \circ \phi_k)$, $k = 1, 2, \ldots$. By the first part of the proof,

$$-\int_{\mathbb{R}^n} w_k \cdot d(Dg) = \int_{\mathbb{R}^n} g(x) \left[\det D\phi_k(x)\right] (\operatorname{div} u) \left[\phi_k(x)\right] dx$$

for $k = 1, 2, \ldots$. As the convergence $\lim \phi_k = \phi$ and $\lim D\phi_k = D\phi$ is uniform on a compact neighborhood of spt g, the desired equality follows. $\qquad \square$

Theorem 5.10.2. *If $g \in BV(\mathbb{R}^n)$ has compact support, then*

$$\mathbf{V}(\phi_\# g) \leq (\operatorname{Lip} \phi)^{n-1} \|Dg\|(\mathbb{R}^n);$$

in particular $\phi_\# g \in BV(\mathbb{R}^n)$.

PROOF. Choose $u \in C^1(\mathbb{R}^n; \mathbb{R}^n)$ with $\|u\|_{L^\infty(\mathbb{R}^n;\mathbb{R}^n)} \leq 1$. If

$$v = (\operatorname{adj} D\phi)(u \circ \phi),$$

then $\|v\|_{L^\infty(\mathbb{R}^n;\mathbb{R}^n)} \leq (\operatorname{Lip} \phi)^{n-1}$ by the polar decomposition theorem for linear maps [29, Section 3.2, Theorem 2]. Since

$$\phi_\# g(y) \operatorname{div} u(y) = \sum \left\{ g(x) \left[\operatorname{sign} \det D\phi(x) \right] (\operatorname{div} u)[\phi(x)] : x \in \phi^{-1}(y) \right\}$$

for each $y \in \mathbb{R}^n$, the area theorem and Lemma 5.10.1 yield

$$\int_{\mathbb{R}^n} \phi_\# g(y) \operatorname{div} u(y) \, dy = \int_{\mathbb{R}^n} g(x) \left[\operatorname{sign} \det D\phi(x) \right] (\operatorname{div} u)[\phi(x)] J_\phi(x) \, dx$$

$$= \int_{\mathbb{R}^n} g(x) \det D\phi(x) (\operatorname{div} u)[\phi(x)] \, dx$$

$$= - \int_{\mathbb{R}^n} v \cdot d(Dg) \leq (\operatorname{Lip} \phi)^{n-1} \|Dg\|(\mathbb{R}^n).$$

Thus $\mathbf{V}(\phi_\# g) \leq (\operatorname{Lip} \phi)^{n-1} \|Dg\|(\mathbb{R}^n)$ by the arbitrariness of u. $\qquad \square$

Corollary 5.10.3. *Let $E \in \mathbf{BV}(\mathbb{R}^n)$ be bounded. If $\phi \restriction E$ is a lipeomorphism, then $\phi(E) \in \mathbf{BV}(\mathbb{R}^n)$.*

PROOF. As χ_E belongs to $BV(\mathbb{R}^n)$ and has compact support, Theorem 5.10.2 shows $\phi_\# \chi_E \in BV(\mathbb{R}^n)$. Thus $|\phi_\# \chi_E|$ belongs to $BV(\mathbb{R}^n)$ by Remark 5.7.4. Since $\phi \restriction E$ is a lipeomorphism, $\det D\phi(x) \neq 0$ for almost all $x \in E$. It follows that $|\phi_\# \chi_E| = \chi_{\phi(E)}$ almost everywhere. $\qquad \square$

Remark 5.10.4. Observe that $\operatorname{sign} \det D\phi(x)$ equals 1 or -1 according to whether the vectors $D\phi(x)e_1, \ldots, D\phi(x)e_n$ form a base in $\mathbb{R}^n$ whose orientation is, respectively, the same as or opposite to the orientation of the standard base $e_1, \ldots, e_n$. We abbreviate this by saying that the point x has, respectively, a *positive* or *negative multiplicity* with respect to ϕ. Thus given $E \subset \mathbb{R}^n$ and $y \in \mathbb{R}^n$, the function $\phi_\# \chi_E$ sums the multiplicities of points $x \in E$ that are mapped by ϕ onto y. If E is measurable and bounded, then inequality (5.10.2) shows that $\phi_\# \chi_E(y)$ is an integer for almost all $y \in \mathbb{R}^n$.

In Section 7.5 we show that Theorem 5.10.2 and Corollary 5.10.3 hold, respectively, for *any* function $g \in BV(\mathbb{R}^n)$ and *any* set $E \in \mathbf{BV}(\mathbb{R}^n)$.

Chapter 6

Locally BV sets

This chapter is devoted to deeper results about locally BV sets obtained by E. De Giorgi [18, 19]. We prove that the families of all locally BV sets and all sets with locally finite perimeter coincide. In addition, we show that the variational measure $D\chi_E$ of a locally BV set E lives in $\partial_* E$ and has the polar decomposition $(\mathcal{H}^{n-1} \mathbin{\llcorner} \partial_* E) \mathbin{\llcorner} (-\nu_E)$ where $\nu_E \upharpoonright \partial_* E$ is the measure-theoretic exterior normal of E. Throughout this chapter Ω denotes an open subset of the ambient space $\mathbb{R}^n$.

6.1. Dimension one

Although the approach developed in Sections 6.2–6.5 below is valid for all dimensions, it is beneficial to present separately a substantially simpler proof of the one-dimensional case.

Proposition 6.1.1. *Let $E \in \mathcal{BV}_{\mathrm{loc}}(\Omega)$. There is a locally finite family $\mathfrak{J}$ of open intervals $J \subset \Omega$ with disjoint closures such that*

$$\mathrm{int}_* E = \bigcup \mathfrak{J} \quad and \quad \partial_* E \cap \Omega = \bigcup \{\partial_* J \cap \Omega : J \in \mathfrak{J}\}.$$

Let $\nu_E(x) = 0$ if $x \in \Omega - \partial_* E$, and let $\nu_E(x)$ equal 1 or -1 according to whether $x \in \partial_* E \cap \Omega$ is the right or left endpoint of an interval $J \in \mathfrak{J}$, respectively. Then $D\chi_E = \mathcal{H}^0 \mathbin{\llcorner} (-\nu_E)$ and $\|D\chi_E\| = \mathcal{H}^0 \mathbin{\llcorner} \partial_* E$.

PROOF. As it suffices to prove the proposition for each connected component of Ω, we may assume from the onset that $\Omega = (a, b)$ where $-\infty \le a < b \le \infty$. In view of translation invariance, we may also assume that $a < 0 < b$. Let $D\chi_E = \mu_1 - \mu_2$ be the Jordan decomposition of the signed measure $D\chi_E$; see (5.3.2). For $i = 1, 2$, the formulae

$$f_i(x) := \begin{cases} \mu_i([0, x)) & \text{if } 0 \le x < b, \\ -\mu_i([x, 0)) & \text{if } a < x < 0 \end{cases}$$

define increasing left-continuous functions $f_i : \Omega \to \mathbb{R}$. For $v \in C_c^1(\Omega)$, Fubini's theorem yields

$$
\int_0^b f_i(x)v'(x)\,dx = \int_0^b v'(x)\left(\int_{[0,x)} d\mu_i(y)\right) dx
$$

$$
= \int_{[0,b)}\left(\int_y^b v'(x)\,dx\right) d\mu_i(y) = -\int_{[0,b)} v(y)\,d\mu_i(y),
$$

and similarly,

$$
\int_a^0 f_i(x)v'(x)\,dx = -\int_{(a,0)} v(y)\,d\mu_i(y).
$$

Summing the previous equalities and letting $f := f_1 - f_2$, we obtain

$$
\int_\Omega f(x)v'(x)\,dx = -\int_\Omega v\,d(D\chi_E).
$$

Proposition 5.5.3 shows that $f \in BV_{\mathrm{loc}}(\Omega)$ and $D\chi_E = Df$. By Proposition 5.5.9, there are a real number c and a negligible set $N \subset \Omega$ such that $\chi_E(x) = f(x)+c$ for every $x \in \Omega - N$. Each $x \in \Omega$ is a two-sided cluster point of $\Omega - N$, and $f(x) + c$ equals 0 or 1 for every $x \in \Omega - N$. Since the function $f + c$ is left-continuous, $f(x)+c$ equals 0 or 1 for every $x \in \Omega$. In other words, $f + c = \chi_A$ where $A \subset \Omega$ is equivalent to E. From the left-continuity of χ_A it is easy to deduce that A is the union of nonempty intervals $(x,y] \cap (a,b)$ whose closures are disjoint. The interiors of these intervals form a countable family $\mathcal{J} = \{J_k : k \in \mathbb{N}\}$ whose union is equivalent to E.

Let $J_k = (a_k, b_k)$, and choose an open interval $U = (\alpha,\beta)$ so that $U \Subset \Omega$. Replacing α by $\alpha' < \alpha$, we may assume $\inf\{a_k : a_k \in U\} > \alpha$. Since χ_A is left-continuous at $a_k \in U$, there is $B_k = B(a_k, r_k)$ such that $B_k \subset U$ and $B_k \cap A \subset J_k$. Define

$$
v_k(x) := \begin{cases} \dfrac{|x-a_k|}{r_k} - 1 & \text{if } x \in B_k, \\ 0 & \text{if } x \in U - B_k, \end{cases}
$$

and let $v := \sum_{a_k \in U} v_k$. As $\{B_k : a_k \in U\}$ is a disjoint family, $v \in Lip(U)$ and $\|v\|_{L^\infty(U)} = 1$. The choice of α shows that $\mathrm{spt}\,v \subset U$. Thus

$$
\|D\chi_E\|(U) \geq \int_E v'(x)\,dx = \sum_{a_k \in U}\int_{a_k}^{a_k+r_k} v_k'(x)\,dx = -\sum_{a_k \in U} v_k(a_k).
$$

As $\|D\chi_E\|(U) < \infty$ and $v_k(a_k) = -1$ for each $a_k \in U$, we see that U meets only finitely many intervals J_k. This implies that $\mathrm{int}_* E = \bigcup_{k \in \mathbb{N}}(a_k, b_k)$ and $\partial_* E \cap \Omega = \{a_k, b_k : k \in \mathbb{N}\} \cap \Omega$.

Define ν_E as in the proposition, and select $\varphi \in C_c^1(\Omega)$. There is $p \in \mathbb{N}$ such that $J_k \cap \mathrm{spt}\,\varphi = \emptyset$ for all integers $k > p$, and an open interval $U \Subset \Omega$

containing spt φ. By the previous paragraph,

$$\int_\Omega \chi_E(x)\varphi'(x)\,dx = \sum_{j=1}^{p}\big[\varphi(b_j) - \varphi(a_j)\big]$$

$$= \sum_{x\in\partial_*E\cap\Omega} \varphi(x)\nu_E(x) = \int_\Omega \varphi\,d\big(\mathcal{H}^0 \mathop{\llcorner} \nu_E\big).$$

Thus $D\chi_E = \mathcal{H}^0 \mathop{\llcorner} (-\nu_E)$ by Proposition 5.5.3, and $\|D\chi_E\| = \mathcal{H}^0 \mathop{\llcorner} \partial_* E$ follows from the definition of ν_E. $\qquad\square$

Remark 6.1.2. An immediate consequence of Proposition 6.1.1 is that each $E \in \boldsymbol{BV}(\mathbb{R})$ is equivalent to a one-dimensional figure. In this case, ν_E defined in Proposition 6.1.1 is the zero extension of the exterior normal ν_E of E defined in Section 2.1.

6.2. Besicovitch's covering theorem

An extension of Section 6.1 to higher dimensions is not easy. To begin with, we need a covering theorem, similar to Vitali's theorem (Theorem 4.3.2), that holds for any Borel measure in $\mathbb{R}^n$. This result, obtained by A.S. Besicovitch [3, 4], is based on a combinatorial lemma, quoted below without proof. The lemma is proved in many standard texts, e.g., in [29, Section 1.5.2, Theorem 2] or [75, Theorem 1.3.5]. A very detailed and transparent presentation is given in [46, Chapter 2]. The most general version can be found in [33, Theorem 2.8.14].

Lemma 6.2.1. *Let $E \subset \mathbb{R}^n$, and let $\boldsymbol{B}$ be a family of closed balls whose centers cover E. If $\sup\{d(B) : B \in \boldsymbol{B}\} < \infty$, then there are a positive integer $N = N(n)$ and subfamilies $\mathcal{E}_1,\ldots,\mathcal{E}_N$ of $\boldsymbol{B}$ such that each $\mathcal{E}_i$ is a disjoint family and $\bigcup_{i=1}^{N} \mathcal{E}_i$ covers E.*

To see that the assumption $\sup\{d(B) : B \in \boldsymbol{B}\} < \infty$ is essential, consider the set $\mathbb{N} \subset \mathbb{R}$ covered by the family $\{B(k,k) : k \in \mathbb{N}\}$.

Theorem 6.2.2 (Besicovitch). *Let μ be a Borel measure in Ω, and let $E \subset \Omega$. Suppose $\boldsymbol{B}$ is a family of closed balls contained in Ω such that for each $x \in E$ and each $\eta > 0$ there is $B(x,r) \in \boldsymbol{B}$ with $r < \eta$. If $\mu(E) < \infty$, there is a disjoint family $\mathcal{C} \subset \boldsymbol{B}$ satisfying*

$$\mu\Big(E - \bigcup\mathcal{C}\Big) = 0.$$

PROOF. As there is nothing to prove otherwise, assume $\mu(E) > 0$. Eliminating larger balls, we may assume that $d(B) \leq 1$ for every $B \in \boldsymbol{B}$. By

Lemma 6.2.1, there are subfamilies $\mathcal{E}_1, \dots, \mathcal{E}_N$ of $\mathcal{B}$, each consisting of disjoint sets, whose union covers E. Thus

$$\mu(E) \le \sum_{i=1}^{N} \mu\left(E \cap \bigcup \mathcal{E}_i\right),$$

and there is an integer k with $1 \le k \le N$ and $\mu(E)/N \le \mu(E \cap \bigcup \mathcal{E}_k)$. The family $\mathcal{E}_k$ contains a finite subfamily $\mathcal{C}_1$ such that

$$\frac{1}{N+1}\mu(E) < \mu\left(E \cap \bigcup \mathcal{C}_1\right).$$

Since $C_1 := \bigcup \mathcal{C}_1$ is a closed set, and since μ is a Borel measure,

$$\mu(E) = \mu(E \cap C_1) + \mu(E - C_1) > \frac{1}{N+1}\mu(E) + \mu(E - C_1).$$

Hence $\mu(E - C_1) < c\mu(E)$ where $c = N/(N+1)$.

If $\mu(E - C_1) = 0$, then letting $\mathcal{C} := \mathcal{C}_1$ completes the argument. If $\mu(E - C_1) > 0$, let $\mathcal{B}_1 := \{B \in \mathcal{B} : B \cap C_1 = \emptyset\}$. As C_1 is a closed set, $\mathcal{B}_1$ satisfies the hypothesis of the theorem with respect to $E_1 = E - C_1$. Applying the previous result, we obtain a finite disjoint family $\mathcal{C}_2 \subset \mathcal{B}_1$ such that for $C_2 = \bigcup \mathcal{C}_2$,

$$\mu\left[E - (C_1 \cup C_2)\right] = \mu(E_1 - C_2) < c\mu(E_1) < c^2\mu(E).$$

The set $C_1 \cup C_2$ is closed, and $\mathcal{C}_1 \cup \mathcal{C}_2$ is a disjoint subfamily of $\mathcal{B}$. Proceeding recursively, we construct finite subfamilies $\mathcal{C}_1, \mathcal{C}_2, \dots$ of $\mathcal{B}$ so that $\mathcal{C} := \bigcup_i \mathcal{C}_i$ is a disjoint family, and

$$\mu\left(E - \bigcup \mathcal{C}\right) \le \mu\left[E - \bigcup_{i=1}^{p}\left(\bigcup \mathcal{C}_i\right)\right] < c^p\mu(E)$$

for $p = 1, 2, \dots$. As $0 < c < 1$, we conclude $\mu(E - \bigcup \mathcal{C}) = 0$. $\square$

With some extra work, Besicovitch's theorem for a Radon measure μ can be proved without the assumption $\mu(E) < \infty$.

Theorem 6.2.3. *Let μ be a Radon measure in Ω. If $f \in L^1_{\mathrm{loc}}(\Omega, \mu)$, then*

$$\lim_{r \to 0} \frac{1}{\mu[B(x,r)]} \int_{B(x,r)} f \, d\mu = f(x) \tag{6.2.1}$$

for μ almost all $x \in \Omega$.

PROOF. The proof is similar to that of Theorem 4.3.4. We may assume that f is a real-valued function. Let $F(B) := \int_B f \, d\mu$ for each ball $B \subset \Omega$, and denote by E the set of all $x \in \Omega$ at which the limit

$$\lim_{r \to 0} \frac{F[B(x,r)]}{\mu[B(x,r)]}$$

either does not exist, or differs from $f(x)$. Given $x \in E$, find $\gamma_x > 0$ so that for each $\eta > 0$, there is $B = B(x, r)$ with $r < \eta$ and

$$\left| F(B) - f(x)\mu(B) \right| \geq \gamma_x \mu(B). \tag{$*$}$$

Select open sets $\Omega_k \Subset \Omega$ so that $\Omega = \bigcup_{k=1}^{\infty} \Omega_k$. It suffices to show that each $E_k := \{x \in E \cap \Omega_k : \gamma_x > 1/k\}$ is a μ negligible set. To this end, fix $k \in \mathbb{N}$ and choose $\varepsilon > 0$. By Henstock's lemma there is $\delta : \Omega_k \to \mathbb{R}_+$ such that

$$\sum_{i=1}^{p} \left| f(x_i)\mu(B_i) - F(B_i) \right| < \varepsilon \tag{$**$}$$

whenever $B_i \subset \Omega_k$ are disjoint balls with $x_i \in B_i$ and $d(B_i) < \delta(x_i)$ for $i = 1, \ldots, p$. Let $\mathcal{B}$ be the family of all closed balls $B(x, r) \subset \Omega_k$ such that $x \in E_k$, $r < \delta(x)$, and inequality $(*)$ holds for $B = B(x, r)$. As $\mathcal{B}$ and E_k satisfy the hypotheses of Besicovitch's theorem, μ almost all of E_k is covered by disjoint balls $B_1, B_2, \ldots$ from $\mathcal{B}$. In view of $(*)$ and $(**)$,

$$\mu(E_k) \leq \sum_i \mu(B_i) \leq k \sum_i \left| f(x_i)\mu(B_i) - F(B_i) \right| < k\varepsilon,$$

and $\mu(E_k) = 0$ by the arbitrariness of ε. $\square$

Corollary 6.2.4. *Let μ be a Radon measure in Ω, and let $1 \leq p < \infty$. If $f \in L^p_{\mathrm{loc}}(\Omega, \mu)$, then for μ almost all $x \in \Omega$,*

$$\lim_{r \to 0} \frac{1}{\mu[B(x, r)]} \int_{B(x,r)} \left| f(y) - f(x) \right|^p d\mu(y) = 0. \tag{6.2.2}$$

PROOF. For all nonnegative real numbers a and b,

$$(a + b)^p \leq \left(2 \max\{a, b\} \right)^p = 2^p \max\{a^p, b^p\} \leq 2^p (a^p + b^p). \tag{$*$}$$

Thus the function $|f - t|^p$ belongs to $L^1_{\mathrm{loc}}(\Omega, \mu)$ for each $t \in \mathbb{R}$. Since $\mathbb{Q}$ is a countable set, it follows from Theorem 6.2.3 that there is a μ negligible set $E \subset \Omega$ such that for each $t \in \mathbb{Q}$ and each $x \in \Omega - E$,

$$\lim_{r \to 0} \frac{1}{\mu[B(x, r)]} \int_{B(x,r)} |f - t|^p d\mu = \left| f(x) - t \right|^p. \tag{$**$}$$

Fix $x \in \Omega - E$, and for $r > 0$, let $B_r := B(x, r)$. Given $\varepsilon > 0$, find $t \in \mathbb{Q}$ with $\left| f(x) - t \right|^p < \varepsilon$. By inequality $(*)$,

$$\int_{B_r} \left| f(y) - f(x) \right|^p d\mu(y)$$

$$\leq 2^p \int_{B_r} \left| f(y) - t \right|^p d\mu(y) + 2^p \int_{B_r} \left| t - f(x) \right|^p d\mu(y)$$

$$\leq 2^p \int_{B_r} \left| f(y) - t \right|^p d\mu(y) + 2^p \varepsilon \mu(B_r)$$

for each $r > 0$. From this and inequality $(**)$, we obtain

$$\limsup_{r \to 0} \frac{1}{\mu(B_r)} \int_{B_r} |f(y) - f(x)|^p \, d\mu(y) \le 2^p |f(x) - t|^p + 2^p \varepsilon \le 2^{p+1} \varepsilon$$

and the corollary follows from the arbitrariness of ε. $\qquad\square$

Each point $x \in \Omega$ at which equality (6.2.2) holds is called the *Lebesgue point* of $f \in L^p(\Omega, \mu)$.

Remark 6.2.5. Since

$$\left| \frac{1}{\mu[B(x,r)]} \int_{B(x,r)} f \, d\mu - f(x) \right| \le \frac{1}{\mu[B(x,r)]} \int_{B(x,r)} |f(y) - f(x)| \, d\mu(y),$$

it is clear that equality (6.2.1) holds at every Lebesgue point of $f \in L^1_{\mathrm{loc}}(\Omega, \mu)$. However, the converse is false: it suffices to define $f(0) = 0$ and $f(x) = x/|x|$ for $x \in \mathbb{R} - \{0\}$, and consider the point $x = 0$ with respect to $\mathcal{L}^1$.

Proposition 6.2.6. *Let σ and μ be Radon measures in Ω. If $E \subset \mathrm{spt}\,\mu$ is a Borel set and*

$$\lim_{r \to 0} \frac{\sigma[B(x,r)]}{\mu[B(x,r)]} = 1$$

for each $x \in E$, then $\sigma \llcorner E = \mu \llcorner E$.

PROOF. By Theorem 1.3.1, it suffices to show $\sigma(K) = \mu(K)$ for each compact set $K \subset E$. Select a compact set $K \subset E$, choose $\varepsilon > 0$, and find an open set U so that $K \subset U \Subset \Omega$. Denote by $\mathcal{B}$ the family of all $B(x,r) \subset U$ such that

$$\sigma[B(x,r)] \le (1 + \varepsilon)\mu[B(x,r)]$$

for each $x \in K$. Besicovitch's theorem shows that there are disjoint balls $B_1, B_2, \ldots$ in $\mathcal{B}$ such that $\sigma(K - \bigcup_i B_i) = 0$. Consequently,

$$\sigma(K) \le \sum_i \sigma(B_i) \le (1 + \varepsilon) \sum_i \mu(B_i) \le (1 + \varepsilon)\mu(U)$$

and $\sigma(K) \le \mu(K)$ by the arbitrariness of U and ε. As our assumptions imply $E \subset \mathrm{spt}\,\sigma$, the reverse inequality follows by symmetry. $\qquad\square$

6.3. The reduced boundary

While the results of Sections 6.3–6.7 are correct in any dimension, the proofs we present assume tacitly that the dimension is greater than one. The proofs in dimension $n = 1$ are left to the reader: they are either trivial or follow from Proposition 6.1.1.

Throughout this section we fix a set $E \in \mathcal{BV}_{\mathrm{loc}}(\Omega)$ of positive measure. We let $\mu_E := \|D\chi_E\|$, and define a μ_E almost everywhere unique Borel map

$\nu_E : \mathbb{R}^n \to \mathbb{R}^n$ so that $D\chi_E = \mu_E \llcorner (-\nu_E)$ is the polar decomposition of $D\chi_E$; see Proposition 5.3.9. Theorem 5.5.1 shows that for each $v \in Lip_c(\Omega; \mathbb{R}^n)$,

$$\int_E \operatorname{div} v(x)\, dx = \int_\Omega v \cdot \nu_E \, d\mu_E. \tag{6.3.1}$$

Aside from simplifying notation, the introduction of symbols μ_E and ν_E enhances the geometric essence of the measure $D\chi_E$, whose complete description is given in Theorem 6.5.2 below. Verifying equalities

$$\mu_{\Omega - E} = \mu_E \quad \text{and} \quad \nu_{\Omega - E} = -\nu_E \tag{6.3.2}$$

is easy: the first is a consequence of Proposition 5.1.2; the second follows from (6.3.1), since $\int_{\Omega - E} \operatorname{div} v = - \int_E \operatorname{div} v$ for each $v \in Lip_c(\Omega)$ by Theorem 2.3.7.

Remark 6.3.1. In Section 2.1, we denoted by ν_A and ν_B, respectively, the unit exterior normal of a figure A and a ball B in $\Omega = \mathbb{R}^n$. This is consistent with the above notation, provided ν_A and ν_B are extended arbitrarily to $\mathbb{R}^n$. Indeed, let C be a figure or a ball, and let ν_C be the unit exterior normal of C whose extension to $\mathbb{R}^n$ is denoted by $\tilde{\nu}_C$. By Propositions 2.1.2 and 2.1.4,

$$\int_C \operatorname{div} v(x)\, dx = \int_{\partial C} v \cdot \nu_C \, d\mathcal{H}^{n-1} = \int_{\mathbb{R}^n} v \cdot \tilde{\nu}_C \, d(\mathcal{H}^{n-1} \llcorner \partial C)$$

for each $v \in C_c^1(\mathbb{R}^n; \mathbb{R}^n)$. Hence $\mu_C = \mathcal{H}^{n-1} \llcorner \partial C$ and $D\chi_C = \mu_C \llcorner (-\tilde{\nu}_C)$ is the polar decomposition of $D\chi_C$; see Propositions 5.5.3 and 5.3.7.

Definition 6.3.2 (De Giorgi). The *reduced boundary* of E is the set $\partial^* E$ of all $x \in \operatorname{spt}\mu_E$ such that $|\nu_E(x)| = 1$ for $\nu_E(x)$ defined by

$$\nu_E(x) := \lim_{r \to 0} \frac{1}{\mu_E[B(x,r)]} \int_{B(x,r)} \nu_E \, d\mu_E. \tag{6.3.3}$$

Remark 6.3.3. Note that $|\nu_E(x)| = 1$ for μ_E almost all $x \in \Omega$ by Proposition 5.3.9, and that equality (6.3.3) holds for μ_E almost all $x \in \Omega$ by Theorem 6.2.3. Consequently

$$\mu_E(\Omega - \partial^* E) = 0. \tag{6.3.4}$$

In particular, $\partial^* E$ is μ_E measurable, and so is $\nu_E \upharpoonright \partial^* E$. In addition,

$$\partial^*(\Omega - E) = \partial^* E \tag{6.3.5}$$

by the identities (6.3.2).

In the line of motivation, we describe the geometric meaning of ν_E, established in Corollary 6.4.4 below. For each $x \in \partial^* E$, denote by

$$H_\pm(E, x) := \{ y \in \mathbb{R}^n : \pm \nu_E(x) \cdot (y - x) \geq 0 \} \tag{6.3.6}$$

the closed half-spaces determined by the hyperplane

$$H(E, x) := \{ y \in \mathbb{R}^n : \nu_E(x) \cdot (y - x) = 0 \} \tag{6.3.7}$$

perpendicular to $\nu_E(x)$ and passing through x. We show that $\nu_E(x)$ is the *unit exterior normal* of E at the point $x \in \partial^* E$ by proving that $H(x)$ is tangent to E in the following measure-theoretic sense:

$$\Theta(H_+(E,x) \cap E, x) = \Theta(H_-(E,x) - E, x) = 0.$$

Recall that the density $\Theta(A,x)$ was defined for any set $A \subset \mathbb{R}^n$ in Section 4.4.

Observation 6.3.4. *If* $A \Subset \Omega$ *is a measurable set, then*

$$\mathbf{V}(A, \Omega) = \mathbf{V}(A).$$

PROOF. Find an open set U so that $A \Subset U \Subset \Omega$, and find $\varphi \in C^1(\mathbb{R}^n)$ so that $\chi_A \le \varphi \le \chi_U$. If $v \in C_c^1(\mathbb{R}^n; \mathbb{R}^n)$ and $\|v\|_{L^\infty(\mathbb{R}^n;\mathbb{R}^n)} \le 1$, then

$$\int_A \operatorname{div} v(x)\, dx = \int_A \operatorname{div}(\varphi v)(x)\, dx \le \mathbf{V}(A, \Omega).$$

The arbitrariness of v implies $\mathbf{V}(A) \le \mathbf{V}(A, \Omega)$, and the reverse inequality is obvious. $\qquad\square$

Lemma 6.3.5. *If* $x \in \Omega$, *then*

$$\mathbf{V}[E \cap B(x,r)] \le \mu_E[B(x,r)] + \mathcal{H}^{n-1}[\partial B(x,r)] \tag{1}$$

for all $r > 0$ *with* $B(x,r) \subset \Omega$, *and*

$$\mathbf{V}[E \cap B(x,r)] \le \mu_E[B(x,r)] + \mathcal{H}^{n-1}[E \cap \partial B(x,r)] \tag{2}$$

for $\mathcal{L}^1$ *almost all* $r > 0$ *with* $B(x,r) \subset \Omega$. *If* $x \in \partial^* E$, *then*

$$\mu_E[B(x,r)] \le 2\mathcal{H}^{n-1}[E \cap \partial B(x,r)] \tag{3}$$

for $\mathcal{L}^1$ *almost all sufficiently small* $r > 0$.

PROOF. If $\varphi \in Lip_c(\Omega)$ and $v \in C_c^1(\mathbb{R}^n; \mathbb{R}^n)$, then by (6.3.1),

$$\int_\Omega (\varphi v) \cdot \nu_E\, d\mu_E = \int_E \operatorname{div}(\varphi v)\, d\mathcal{L}^n$$
$$= \int_E \varphi \operatorname{div} v\, d\mathcal{L}^n + \int_E v \cdot D\varphi\, d\mathcal{L}^n. \tag{$*$}$$

In view of translation invariance, we may assume that $x = 0$. Given a ball $B_r := B(0,r)$ contained in Ω, choose $\varepsilon > 0$ so that $B_{r+\varepsilon} \subset \Omega$. Define

$$\varphi^\varepsilon(y) := \begin{cases} 1 & \text{if } |y| \le r, \\ \frac{1}{\varepsilon}(r + \varepsilon - |y|) & \text{if } r < |y| < r + \varepsilon, \\ 0 & \text{if } |y| \ge r + \varepsilon, \end{cases}$$

and observe that $\varphi^\varepsilon \in Lip_c(\Omega)$, and that for each $y \in \Omega$,

$$D\varphi^\varepsilon(y) = \begin{cases} -\frac{1}{\varepsilon} \frac{y}{|y|} & \text{if } r < |y| < r + \varepsilon, \\ 0 & \text{if } |y| < r \text{ or } |y| > r + \varepsilon. \end{cases}$$

Consider $(*)$ with $\varphi = \varphi^\varepsilon$ and $\|v\|_{L^\infty(\mathbb{R}^n;\mathbb{R}^n)} \leq 1$. Employing the spherical coordinates [29, Section 3.4.4, A], calculate

$$\int_\Omega (\varphi^\varepsilon v) \cdot \nu_E \, d\mu_E$$

$$= \int_E \varphi^\varepsilon \operatorname{div} v \, d\mathcal{L}^n - \frac{1}{\varepsilon} \int_{E \cap (B_{r+\varepsilon} - B_r)} v(y) \cdot \frac{y}{|y|} \, dy$$

$$= \int_E \varphi^\varepsilon \operatorname{div} v \, d\mathcal{L}^n - \frac{1}{\varepsilon} \int_r^{r+\varepsilon} \left(\int_{E \cap \partial B_s} v \cdot \nu_{B_s} \, d\mathcal{H}^{n-1} \right) ds \qquad (**)$$

$$\geq \int_E \varphi^\varepsilon \operatorname{div} v \, d\mathcal{L}^n - \frac{1}{\varepsilon} \int_r^{r+\varepsilon} \left(\int_{\partial B_s} d\mathcal{H}^{n-1} \right) ds$$

$$= \int_E \varphi^\varepsilon \operatorname{div} v \, d\mathcal{L}^n - \frac{1}{\varepsilon} |B_{r+\varepsilon} - B_r|.$$

Note that $\|\varphi^\varepsilon\|_{L^\infty(\mathbb{R}^n)} \leq 1$, $\lim_{\varepsilon \to 0} \varphi^\varepsilon(y) = \chi_{B_r}(y)$ for all $y \in \mathbb{R}^n$, and

$$\lim_{\varepsilon \to 0} \frac{1}{\varepsilon} |B_{r+\varepsilon} - B_r| = \frac{d}{dr} |B_r| = \mathcal{H}^{n-1}(\partial B_r).$$

Thus letting $\varepsilon \to 0$ in inequality $(**)$, we obtain

$$\int_{B_r} v \cdot \nu_E \, d\mu_E \geq \int_{E \cap B_r} \operatorname{div} v \, d\mathcal{L}^n - \mathcal{H}^{n-1}(\partial B_r),$$

$$\int_{B_r} v \cdot \nu_E \, d\mu_E = \int_{E \cap B_r} \operatorname{div} v \, d\mathcal{L}^n - \int_{E \cap \partial B_r} v \cdot \nu_{B_r} \, d\mathcal{H}^{n-1} \qquad (\dagger)$$

where the inequality holds for all $r > 0$ with $B_r \subset \Omega$, and the equality $(\dagger)$ holds for $\mathcal{L}^1$ almost all $r > 0$ with $B_r \subset \Omega$. The qualifier "almost all" appears because we applied Theorem 4.3.4 to the second term in the line opposite to the marker $(**)$. Hence

$$\int_{E \cap B_r} \operatorname{div} v \, d\mathcal{L}^n \leq \mu_E(B_r) + \mathcal{H}^{n-1}(\partial B_r),$$

$$\int_{E \cap B_r} \operatorname{div} v \, d\mathcal{L}^n \leq \mu_E(B_r) + \mathcal{H}^{n-1}(E \cap \partial B_r),$$

and inequalities (1) and (2) follow from the arbitrariness of v.

Now assume $x = 0$ lies in $\partial^* E$. According to identity (6.3.3),

$$\nu_E(0) \cdot \lim_{r \to 0} \frac{1}{\mu_E(B_r)} \int_{B_r} \nu_E \, d\mu_E = |\nu_E(0)|^2 = 1.$$

Select $\delta > 0$ so that $0 < r < \delta$ implies $B_r \subset \Omega$ and

$$\nu_E(0) \cdot \int_{B_r} \nu_E \, d\mu_E \geq \frac{1}{2} \mu_E(B_r).$$

There is $w \in C_c^1(\Omega; \mathbb{R}^n)$ such that $\|w\|_{L^\infty(\Omega;\mathbb{R}^n)} \leq 1$ and $w(y) = \nu_E(0)$ for each $y \in B_\delta$. Equality (†) with $v = w$ implies

$$\frac{1}{2}\mu_E(B_r) \leq \nu_E(0) \cdot \int_{B_r} \nu_E \, d\mu_E = \int_{B_r} w \cdot \nu_E$$

$$= -\int_{E \cap \partial B_r} w \cdot \nu_{B_r} \, d\mathcal{H}^{n-1} \leq \mathcal{H}^{n-1}(E \cap \partial B_r)$$

for $\mathcal{L}^1$ almost all positive $r < \delta$. This establishes inequality (3). $\qquad\square$

Lemma 6.3.6. *There are positive constants* $\beta_i = \beta_i(n)$, $i = 1, 2, 3$, *such that given* $x \in \partial^* E$, *the inequalities*

$$\beta_1 \leq \frac{|E \cap B(x,r)|}{|B(x,r)|} \leq 1 - \beta_1 \tag{1}$$

$$\beta_2 \leq \frac{\mu_E[B(x,r)]}{r^{n-1}} \leq \beta_3 \tag{2}$$

$$\frac{\mathbf{V}[E \cap B(x,r)]}{r^{n-1}} \leq \beta_3 \tag{3}$$

hold for all sufficiently small $r > 0$.

PROOF. Choose $x \in \partial^* E$, and let $B_r := B(x,r)$. By Lemma 6.3.5, (3), there is $\delta > 0$ such that $B_\delta \subset \Omega$ and

$$\mu_E(B_r) \leq 2\mathcal{H}^{n-1}(E \cap \partial B_r) \leq 2\mathcal{H}^{n-1}(\partial B_r) = 2n\alpha(n)r^{n-1}$$

for $\mathcal{L}^1$ almost all positive $r < \delta$. As $r \mapsto \mu_E(B_r)$ is a right-continuous function defined in $(0, \delta)$, the inequality $\mu_E(B_r) \leq 2n\alpha(n)r^{n-1}$ holds for *all* sufficiently small $r > 0$. In view of Lemma 6.3.5, (1), so does

$$\max\{\mu_E(B_r), \mathbf{V}(E \cap B_r)\} \leq 3n\alpha(n)r^{n-1}.$$

With $\beta_3 = 3n\alpha(n)$, this establishes inequality (3) and the second inequality in (2). Inequalities (2) and (3) of Lemma 6.3.5 yield

$$\mathbf{V}(E \cap B_r) \leq 3\mathcal{H}^{n-1}(E \cap \partial B_r) \tag{$*$}$$

for $\mathcal{L}^1$ almost all sufficiently small $r > 0$. Since

$$|E \cap B_r| = \int_0^r \mathcal{H}^{n-1}(E \cap \partial B_s) \, ds,$$

inequality $(*)$ and the isoperimetric inequality (Theorem 5.9.6) imply

$$\frac{d}{dr}\left(|E \cap B_r|^{\frac{1}{n}}\right) = |E \cap B_r|^{\frac{1}{n}-1}\mathcal{H}^{n-1}(E \cap \partial B_r)$$

$$\geq \frac{1}{3}|E \cap B_r|^{-\frac{n-1}{n}}\mathbf{V}(E \cap B_r) \geq \frac{1}{3\gamma}$$

for $\mathcal{L}^1$ almost all sufficiently small $r > 0$; here $\gamma = \gamma(n) > 0$ is the constant appearing in the isoperimetric inequality. By integration,

$$|E \cap B_r| \geq \frac{1}{(3\gamma)^n} r^n = \frac{1}{(3\gamma)^n \alpha(n)} |B_r| \qquad (**)$$

for all sufficiently small $r > 0$. Letting $\beta_1 = \left[(3\gamma)^n \alpha(n)\right]^{-1}$ and applying inequality $(**)$ to $\Omega - E$ yield

$$\beta_1 |B_r| \leq \left|(\Omega - E) \cap B_r\right| = |B_r| - |E \cap B_r| \qquad (\dagger)$$

for all sufficiently small $r > 0$. Inequality (1) follows from $(**)$ and $(\dagger)$. Combining $(**)$ and $(\dagger)$ with the relative isoperimetric inequality of Corollary 5.9.13, we obtain that for all sufficiently small $r > 0$,

$$\frac{\mu_E(B_r)}{r^{n-1}} \geq \frac{\mu_E\left[U(x,r)\right]}{r^{n-1}} \geq \frac{1}{\kappa} \left[\frac{1}{r^n} \min\{|B_r \cap E|, |B_r - E|\}\right]^{\frac{n-1}{n}}$$

$$= \frac{1}{\kappa} \min\left[\alpha(n)\left\{\frac{|B_r \cap E|}{|B_r|}, 1 - \frac{|B_r \cap E|}{|B_r|}\right\}\right]^{\frac{n-1}{n}}$$

$$\geq \frac{1}{\kappa} \left[\alpha(n)\beta_1\right]^{\frac{n-1}{n}} = \frac{1}{\kappa}(3\gamma)^{1-n} = \beta_2;$$

here $\kappa = \kappa(n) > 0$ is the constant from Corollary 5.9.13. This establishes the first inequality in (2). $\qquad \square$

Proposition 6.3.7. $\partial^* E \subset \partial_* E \cap \Omega$ and $\mathcal{H}^{n-1}(\partial_* E \cap \Omega - \partial^* E) = 0$.

PROOF. The inclusion $\partial^* E \subset \partial_* E \cap \Omega$ follows from Lemma 6.3.6, (1). Since $\mu_E = \mu_E \llcorner \partial^* E$ by (6.3.4), we have $\mu_E(\partial_* E \cap \Omega - \partial^* E) = 0$. An application of Proposition 5.9.16 completes the argument. $\qquad \square$

6.4. Blow-up

In this section, we again fix a set $E \in \mathcal{BV}_{\mathrm{loc}}(\Omega)$. For $z \in \Omega$ and $r > 0$, we define a *blow-up* of E about z by r as the set

$$E_{z,r} := z + r^{-1}(E - z).$$

We say blow-*up* because we are interested only in *small* $r > 0$. Clearly $E_{z,r} \subset \Omega_{z,r}$, and it follows by translation and scaling that $E_{z,r}$ belongs to $\mathcal{BV}_{\mathrm{loc}}(\Omega_{z,r})$. Observe $B(z,s) = \left[B(z,rs)\right]_{z,r}$ for each $s > 0$.

Lemma 6.4.1. Let $z \in \partial^* E$. Then equalities

$$\left|B(z,s) \cap E_{z,r}\right| = \frac{1}{r^n}\left|B(z,rs) \cap E\right|, \qquad (1)$$

$$\mu_{E_{z,r}}\left[B(z,s)\right] = \frac{1}{r^{n-1}}\mu_E\left[B(z,rs)\right], \qquad (2)$$

$$\int_{B(z,s)} \nu_{E_{z,r}} \, d\mu_{E_{z,r}} = \frac{1}{r^{n-1}} \int_{B(z,rs)} \nu_E \, d\mu_E \tag{3}$$

hold for each pair $r, s \in \mathbb{R}_+$ *for which* $B(z, rs) \subset \Omega$.

PROOF. We may assume $z = 0$. To simplify the notation, let

$$U_r = U(0, r), \quad B_r = B(0, r), \quad \Omega_r = \Omega_{0,r}, \quad E_r = E_{0,r}.$$

Observe that the linear map $\phi : x \mapsto r^{-1}x$ maps U_{rs} onto U_r, B_{rs} onto B_r, Ω onto Ω_r, and E onto E_r. Equality (1) is a direct consequence of the area theorem, and holds for all $r, s \in \mathbb{R}_+$. For the proof of remaining equalities, fix $r, s \in \mathbb{R}_+$ so that $B_{rs} \subset \Omega$ and consequently $B_s \subset \Omega_r$.

Choose $v \in C_c^1(U_s; \mathbb{R}^n)$ with $\|v\|_{L^\infty(U_s;\mathbb{R}^n)} \leq 1$. Since $\chi_E = \chi_{E_r} \circ \phi$,

$$\begin{aligned}
\int_{U_s} \chi_{E_r} \operatorname{div} v &= \frac{1}{r^n} \int_{U_{rs}} (\chi_{E_r} \operatorname{div} v) \circ \phi \\
&= \frac{1}{r^{n-1}} \int_{U_{rs}} \chi_E \operatorname{div}(v \circ \phi) \leq \mu_E(U_{rs}).
\end{aligned} \tag{6.4.1}$$

Consequently $\mu_{E_r}(U_s) \leq \mu_E(U_{rs})$, and $\mu_{E_r}(U_s) = \mu_E(U_{rs})$ by symmetry. Since $B_s = \bigcap_{k=1}^\infty U_{s(k+1)/k}$ and $B_{rs} = \bigcap_{k=1}^\infty U_{rs(k+1)/k}$, and since both μ_{E_r} and μ_E are Radon measures in Ω_r and Ω, respectively, identity (2) follows.

Denote by ν_1 and $\nu_{r,1}$ the first component of ν_E and ν_{E_r}, respectively. Find $\{\varphi_k\}$ in $Lip_c(\Omega_r; [0,1])$ that converges pointwise to χ_{B_s}. Then $\{\varphi_k \circ \phi\}$ is a sequence in $Lip_c(\Omega; [0,1])$ that converges pointwise to $\chi_{B_{rs}}$. Since the vector field $w_k := \varphi_k e_1$ belongs to $Lip_c(\Omega_r, \mathbb{R}^n)$, the dominated convergence theorem and the equality in (6.3.1) imply

$$\begin{aligned}
\int_{B_s} \nu_{r,1} \, d\mu_{E_r} &= \lim \int_{\Omega_r} \varphi_k \nu_{r,1} \, d\mu_{E_r} \\
&= \lim \int_{\Omega_r} w_k \cdot \nu_{E_r} \, d\mu_{E_r} = \lim \int_{E_r} \operatorname{div} w_k \\
&= \frac{1}{r^n} \lim \int_E (\operatorname{div} w_k) \circ \phi = \frac{1}{r^{n-1}} \lim \int_E \operatorname{div}(w_k \circ \phi) \\
&= \frac{1}{r^{n-1}} \lim \int_\Omega (w_k \circ \phi) \cdot \nu_E \, d\mu_E \\
&= \frac{1}{r^{n-1}} \lim \int_\Omega (\varphi_k \circ \phi) \nu_1 \, d\mu_E = \frac{1}{r^{n-1}} \int_{B_{rs}} \nu_1 \, d\mu_E.
\end{aligned}$$

Identity (3) follows by symmetry. $\qquad \square$

Lemma 6.4.2. *The half-space*

$$H_- = \{(\xi_1, \ldots, \xi_n) \in \mathbb{R}^n : \xi_n \leq 0\}$$

belongs to $BV_{\text{loc}}(\mathbb{R}^n)$ *and* $\mu_{H_-} = \mathcal{H}^{n-1} \llcorner \partial H_-$.

PROOF. The first assertion follows from Propositions 4.5.8 and 5.1.3. Select an open set $U \Subset \mathbb{R}^n$, $v \in Lip_c(U; \mathbb{R}^n)$ with $\|v\|_{L^\infty(U;\mathbb{R}^n)} \leq 1$, and a dyadic figure $B \subset U$ such that spt $v \subset$ int B. Since $B \cap H_-$ is still a dyadic figure, Theorem 2.3.7 implies

$$\int_U \chi_{H_-} \operatorname{div} v(x)\,dx = \int_{B \cap H_-} \operatorname{div} v(x)\,dx$$

$$= \int_{\partial(B \cap H_-)} v \cdot \nu_{B \cap H_-}\,d\mathcal{H}^{n-1}$$

$$= \int_{U \cap \partial H_-} v \cdot e_n\,d\mathcal{H}^{n-1} \leq \mathcal{H}^{n-1}(U \cap \partial H_-).$$

As v is arbitrary, $\mu_{H_-}(U) \leq \mathcal{H}^{n-1}(\partial H_- \cap U)$. For the reverse inequality, choose $\varepsilon > 0$ and find $w \in Lip_c(U; \mathbb{R}^n)$ so that $\|w\|_{L^\infty(U;\mathbb{R}^n)} \leq 1$ and

$$\mathcal{H}^{n-1}\Big(\{x \in \partial H_- \cap U : w(x) \neq e_n\}\Big) < \varepsilon;$$

cf. Example 5.1.4. If $B \subset U$ is a dyadic figure containing spt w, then calculating as above but backward, we obtain

$$\mathcal{H}^{n-1}(U \cap \partial H_-) \leq \int_{U \cap \partial H_-} w \cdot e_n\,d\mathcal{H}^{n-1} + 2\varepsilon$$

$$= \int_U \chi_{H_-} \operatorname{div} w(x)\,dx + 2\varepsilon \leq \mu_{H_-}(U) + 2\varepsilon.$$

The lemma follows from the arbitrariness of ε and Theorem 1.3.1. $\qquad\square$

Theorem 6.4.3. *Let $z \in \partial^* E$. Then*

$$\lim_{r \to 0} \chi_{E_{z,r}} = \chi_{H_-(E,z)}$$

in $L^1_{\mathrm{loc}}(\mathbb{R}^n)$, and for each $B(z,s) \subset \Omega_{z,r}$,

$$\lim_{r \to 0} \mu_{E_{z,r}}\big[B(z,s)\big] = \alpha(n-1)s^{n-1}.$$

PROOF. In view of the translation and rotation invariance, we may assume that $z = 0$ and $\nu_E(0) = e_n$. We simplify the notation by letting

$$U_r = U(0,r), \quad B_r = B(0,r), \quad H_- = H_-(E,0).$$

Recall from (6.3.6) that H_- is the same as that defined in Lemma 6.4.2. Select $R > 0$, and let $\phi_r(x) := r^{-1}x$ for $x \in \mathbb{R}^n$ and $r > 0$. Given $v \in C^1_c(U_R; \mathbb{R}^n)$ with $\|v\|_{L^\infty(U_R;\mathbb{R}^n)} \leq 1$, the equality in (6.4.1) and Lemma 6.3.6, (3) yield

$$\int_{U_R} \chi_{E_{0,r}} \operatorname{div} v = \frac{1}{r^{n-1}} \int_{B_{rR}} \chi_E \operatorname{div}(v \circ \phi_r)$$

$$\leq \frac{\mathbf{V}(E \cap B_{rR})}{r^{n-1}} \leq \beta_3 R^{n-1}$$

for all sufficiently small $r > 0$; here $\beta_3 = \beta_3(n) > 0$. As v is arbitrary,

$$\limsup_{r \to 0} \|\chi_{E_0,r}\|_{BV(U_R)} \le \alpha(n)R^n + \beta_3 R^{n-1} < \infty. \tag{*}$$

Choose a sequence $\{r_k\}$ in $\mathbb{R}_+$ converging to zero. Since R is arbitrary, Proposition 5.8.9 shows that $\{r_k\}$ has a subsequence, still denoted by $\{r_k\}$, such that $\{\chi_{E_0,r_k}\}$ converges in $L^1_{\text{loc}}(\mathbb{R}^n)$, as well as almost everywhere, to a function $f \in BV_{\text{loc}}(\mathbb{R}^n)$. The convergence almost everywhere shows that f is equivalent to χ_A for a set A in $\mathbf{BV}_{\text{loc}}(\mathbb{R}^n)$. As we have started with an arbitrary sequence $\{r_k\}$, the first assertion of the theorem will be established by showing that A is equivalent to H_-. We split the proof into three claims. To simplify the notation further, we let $\Omega_k = \Omega_{0,r_k}$ and $E_k = E_{0,r_k}$ for $k = 1, 2, \dots$.

Claim 1. $\lim \mu_{E_k}(B_s) = \mu_A(B_s)$ for all but countably many $s > 0$, and $\nu_A(x) = e_n$ for μ_A almost all $x \in \mathbb{R}^n$.

Proof. Choose $R > 0$. Eliminating finitely many members of $\{r_k\}$, we may assume that $U_R \Subset \Omega_k$ for $k = 1, 2, \dots$. As each E_k belongs to $\mathbf{BV}_{\text{loc}}(\Omega_k)$, our assumption implies that each χ_{E_k} belongs to $BV(U_R)$. Now we have $\lim \|\chi_{E_k} - \chi_A\|_{L^1(U_R)} = 0$, and

$$\limsup \mu_{E_k}(U_R) \le \limsup \|\chi_{E_k}\|_{BV(U_R)} < \infty$$

according to inequality $(*)$. It follows from Proposition 5.5.4 that

$$\text{w-lim}(\mu_{E_k} \, \llcorner \, \nu_{E_k}) = \mu_A \, \llcorner \, \nu_A.$$

Theorem 5.4.8 shows that $\{E_k\}$ has a subsequence, still denoted by $\{E_k\}$, such that $\{\mu_{E_k}\}$ converges weakly in U_R to a Radon measure μ. Select $0 < s < R$ so that $\mu(\partial B_s) = 0$. Then

$$\lim \mu_{E_k}(B_s) = \mu(B_s) \tag{**}$$

by Theorem 5.4.3. Proposition 5.4.9 yields $\mu_A \le \mu$ and

$$\lim(\mu_{E_k} \, \llcorner \, \nu_{E_k})(B_s) = (\mu_A \, \llcorner \, \nu_A)(B_s).$$

Writing out the n-th component of the previous equality gives

$$\lim \int_{B_s} e_n \cdot \nu_{E_k} \, d\mu_{E_k} = \int_{B_s} e_n \cdot \nu_A \, d\mu_A. \tag{†}$$

As each B_{sr_k} is contained in Ω, we can apply Lemma 6.4.1: dividing the n-th component of (3) by (2), we obtain

$$\frac{1}{\mu_{E_k}(B_s)} \int_{B_s} e_n \cdot \nu_{E_k} \, d\mu_{E_k} = \frac{1}{\mu_E(B_{sr_k})} \int_{B_{sr_k}} e_n \cdot \nu_E \, d\mu_E.$$

Since $0 \in \partial^* E$ and $\nu_E(0) = e_n$, Definition 6.3.2 implies

$$\lim \frac{1}{\mu_{E_k}(B_s)} \int_{B_s} e_n \cdot \nu_{E_k} \, d\mu_{E_k} = e_n \cdot \lim \frac{1}{\mu_E(B_{sr_k})} \int_{B_{sr_k}} \nu_E \, d\mu_E$$

$$= e_n \cdot \nu_E(0) = 1.$$

According to (∗∗), the previous equality, and (†),

$$\mu_A(B_s) \le \mu(B_s) = \lim \mu_{E_k}(B_s) = \lim \int_{B_s} e_n \cdot \nu_{E_k} \, d\mu_{E_k}$$

$$= \int_{B_s} e_n \cdot \nu_A \, d\mu_A \le \mu_A(B_s).$$

Summarizing, we have established that

$$\lim \mu_{E_k}(B_s) = \mu_A(B_s) \quad \text{and} \quad \int_{B_s} e_n \cdot \nu_A \, d\mu_A = \mu_A(B_s) \qquad (\dagger\dagger)$$

for each positive $s < R$ with $\mu(B_s) = 0$. However, $\mu(\partial B_s) = 0$ for all but countably many positive $s < R$; see Lemma 4.5.5. Thus (††) holds for all but countably many positive $s < R$. Since the second equality in (††) implies $\nu_A(x) = e_n$ for μ_A almost all $x \in B_s$, we see that $\nu_A(x) = e_n$ for almost all $x \in U_R$. Claim 1 follows from the arbitrariness of R.

Claim 2. $\mu_A(U_s) \le \lim \mu_{E_k}(B_s) \le \mu_A(B_s)$ for each $s > 0$.

Proof. Choose $s > 0$ and $\varepsilon > 0$. Denote by D the set of all $s \in \mathbb{R}_+$ such that $\mu_A(B_s) = \lim \mu_{E_k}(B_s)$. Since $\mathbb{R}_+ - D$ is a countable set by Claim 1, there are sequences $\{t_j\}$ and $\{u_j\}$ in D such that $\{t_j\}$ is strictly increasing, $\{u_j\}$ is decreasing, and $\lim t_j = \lim u_j = s$. As

$$U_s = \bigcup \{B_{t_j} : j \in \mathbb{N}\} \quad \text{and} \quad B_s = \bigcap \{B_{u_j} : j \in \mathbb{N}\},$$

there is $p \in \mathbb{N}$ satisfying

$$\mu_A(U_s) < \mu_A(B_{t_p}) + \varepsilon = \lim_{k \to \infty} \mu_{E_k}(B_{t_p}) + \varepsilon \le \lim_{k \to \infty} \mu_{E_k}(B_s) + \varepsilon,$$

$$\mu_A(B_s) > \mu_A(B_{u_p}) - \varepsilon = \lim_{k \to \infty} \mu_{E_k}(B_{u_p}) - \varepsilon \ge \lim_{k \to \infty} \mu_{E_k}(B_s) - \varepsilon.$$

The claim follows from the arbitrariness of ε.

Claim 3. There is $c \in \mathbb{R}$ such that A is equivalent to the set

$$\{(\xi_1, \ldots, \xi_n) \in \mathbb{R}^n : \xi_n < c\}.$$

Proof. Choose a mollifier η and let $f_k := \eta_{1/k} * \chi_A$ for $k = 1, 2, \ldots$. Then $\{f_k\}$ is a sequence in $C^1(\mathbb{R}^n)$ which converges in $L^1_{\text{loc}}(\mathbb{R}^n)$ to χ_A. The standard diagonal construction (see the proof of Proposition 5.8.9) produces a subsequence of $\{f_k\}$, still denoted by $\{f_k\}$, which converges to χ_A almost everywhere in $\mathbb{R}^n$. If $\varphi \in C^1_c(\mathbb{R}^n)$, then

$$\int_{\mathbb{R}^n} \varphi D_i f_k = \int_{\mathbb{R}^n} (\varphi e_i) \cdot D f_k = -\int_{\mathbb{R}^n} f_k \operatorname{div}(\varphi e_i)$$

$$= -\int_{\mathbb{R}^n} \chi_A \operatorname{div}\left[(\eta_{1/k} * \varphi)e_i\right] = \int_{\mathbb{R}^n} \left[(\eta_{1/k} * \varphi)e_i\right] \cdot d(D\chi_A)$$

$$= -\int_{\mathbb{R}^n} (\eta_{1/k} * \varphi)e_i \cdot \nu_A \, d\mu_A$$

and consequently,

$$\int_{\mathbb{R}^n} \varphi D_i f_k = \begin{cases} 0 & \text{if } i = 1, \ldots, n-1, \\ -\int_{\mathbb{R}^n} \eta_{1/k} * \varphi \, d\mu_A & \text{if } i = n. \end{cases}$$

As $D_i f_k$ is continuous and φ is arbitrary, we infer $D_n f_k \leq 0$ and $D_i f_k = 0$ for $i = 0, \ldots, n-1$. Consequently f_k is constant with respect to the first $n-1$ variables, and decreasing with respect to the n-th variable. By Fubini's theorem, there is an $\mathcal{L}^{n-1}$ negligible set $M \subset \mathbb{R}^{n-1}$ so that for each point $x \in \mathbb{R}^{n-1} - M$, the function $t \mapsto \chi_A(x, t)$ is decreasing $\mathcal{L}^1$ almost everywhere in $\mathbb{R}$. Thus for every $x \in \mathbb{R}^{n-1} - M$, there is a $t_x \in \overline{\mathbb{R}}$ such that $\chi_A(x, t) = 1$ for $\mathcal{L}^1$ almost all $t < t_x$ and $\chi_A(x, t) = 0$ for $\mathcal{L}^1$ almost all $t > t_x$. Using Fubini's theorem again, we find an $\mathcal{L}^1$ negligible set $N \subset \mathbb{R}$ so that for each $t \in \mathbb{R} - N$, the function $x \mapsto \chi_A(x, t)$ is constant $\mathcal{L}^{n-1}$ almost everywhere in $\mathbb{R}^{n-1}$. This implies that the function $x \mapsto t_x$ is constant $\mathcal{L}^{n-1}$ almost everywhere in $\mathbb{R}^{n-1} - M$, and Claim 2 follows.

If $c \neq 0$, select a positive $s < |c|$ and observe that either $|B_s \cap A| = 0$ or $|B_s \cap A| = |B_s|$. Observe that $\lim \chi_{E_k} = \chi_A$ in L^1_{loc} implies

$$\lim |B_s \cap E_k| = |B_s \cap A|.$$

Using Lemma 6.4.1, (1), we obtain

$$\lim \frac{|B_{r_k} \cap E|}{|B_{r_k}|} = \lim \frac{|B_{r_k s} \cap E|}{|B_{r_k s}|} = \frac{1}{|B_s|} \lim \frac{|B_{r_k s} \cap E|}{r^n}$$

$$= \frac{1}{|B_s|} \lim |B_s \cap E_k| = \frac{|B_s \cap A|}{|B_s|} = 0 \text{ or } 1.$$

As $0 \in \partial^* E$, a contradiction follows from Lemma 6.3.6, (1). Thus $c = 0$ and A is equivalent to H_-, which is the first assertion of the theorem.

An immediate consequence of Lemma 6.4.2 is that

$$\mu_{H_-}(B_s) = \mu_{H_-}(U_s) = \alpha(n-1)s^{n-1}$$

for every $s > 0$. An application of Claim 2 completes the proof. □

Corollary 6.4.4. *If $z \in \partial^* E$, then*

$$\Theta\big(H_+(E, z) \cap E, z\big) = \Theta\big(H_-(E, z) - E, z\big) = 0, \tag{1}$$

$$\Theta(E, z) = \tfrac{1}{2}, \tag{2}$$

$$\lim_{r \to 0} \frac{\mu_E\big[B(z, r)\big]}{\alpha(n-1)r^{n-1}} = 1. \tag{3}$$

PROOF. If $B_\pm(z, r) := B(z, r) \cap H_\pm(E, z)$, then Lemma 6.4.1, (1) with $s = 1$ and Theorem 6.4.3 imply

$$\Theta\big(H_+(E, z) \cap E, z\big) = \lim_{r \to 0} \frac{|B_+(z, r) \cap E|}{\alpha(n)r^n}$$

$$= \frac{1}{\alpha(n)} \lim_{r \to 0} \left| B_+(z,1) \cap E_{z,r} \right|$$

$$= \frac{1}{\alpha(n)} \left| B_+(z,1) \cap H_-(E,z) \right| = 0,$$

and similarly Θ $(H_-(z,E) - E, z) = 0$. Now (2) follows from (1) and the obvious equality:

$$\left| B(z,r) \cap E \right| = \left| B_-(z,r) \right| - \left| B_-(z,r) - E \right| + \left| B_+(z,r) \cap E \right|.$$

Finally, by Lemma 6.4.1, (2) and Theorem 6.4.3,

$$\lim_{r \to 0} \frac{1}{r^{n-1}} \mu_E \big[B(z,r) \big] = \lim_{r \to 0} \mu_{E_{x,r}} \big[B(z,1) \big] = \alpha(n-1). \qquad \square$$

6.5. Perimeter and variation

As in the previous section, a set $E \in \mathcal{BV}_{\mathrm{loc}}(\Omega)$ is fixed. We employ the notation introduced in Section 6.3. In particular, we use the polar decomposition $D\chi_E = \mu_E \llcorner \nu_E$, and the symbols defined in (6.3.6) and (6.3.7).

Lemma 6.5.1. *Let $n \geq 2$. For $k = 1, 2, \ldots,$ the functions*

$$f_{k,+} : z \mapsto \sup_{0 < r < 1/k} \frac{\left| B(z,r) \cap H_+(E,z) \cap E \right|}{\left| B(z,r) \right|} : \partial^* E \to \mathbb{R},$$

$$f_{k,-} : z \mapsto \sup_{0 < r < 1/k} \frac{\left| B(z,r) \cap H_-(E,z) - E \right|}{\left| B(z,r) \right|} : \partial^* E \to \mathbb{R}$$

are μ_E measurable.

PROOF. For $v : \partial^* E \to \mathbb{R}^n$ and $z \in \partial^* E$, let

$$H_+(E,z;v) := \left\{ y \in \mathbb{R}^n : v(z) \cdot (y-z) \geq 0 \right\}$$

and note $H_+(E,z;\nu_E) = H_+(E,z)$. If $v \in C(\partial^* E; \mathbb{R}^n)$ and $r > 0$, then

$$g_{v,r} : z \mapsto \frac{\left| B(z,r) \cap H_+(E,z;v) \cap E \right|}{\left| B(z,r) \right|} : \partial^* E \to \mathbb{R}$$

is a continuous function. As ν_E is μ_E measurable, it follows from Luzin's theorem (Theorem 1.3.4) that there is a sequence $\{v_k\}$ in $C(\partial^* B; \mathbb{R}^n)$ such that $\lim v_k(z) = \nu_E(z)$ for μ_E almost all $z \in \partial^* E$. Given $r > 0$, we see that $\lim g_{v_k,r}(z) = g_{\nu_E,r}(z)$ for μ_E almost all $z \in \partial^* E$. Thus $g_{\nu_E,r}$ is μ_E measurable for every $r > 0$, and so is

$$f_{k,+} = \sup \{ g_{\nu_E,r} : r \in \mathbb{Q} \cap (0,1/k) \}.$$

The μ_E measurability of $f_{k,-}$ is established similarly. $\qquad \square$

Theorem 6.5.2. *The variational measure* $\|D\chi_E\|$ *is equal to the measure* $\mathcal{H}^{n-1} \llcorner (\partial_* E \cap \Omega)$ *restricted to subsets of* Ω.

PROOF. Denote by $f_{k,\pm}$, $k = 1, 2, \ldots$, the μ_E measurable functions defined in Lemma 6.5.1. By Corollary 6.4.4, (1), the sequences $\{f_{k,\pm}\}$ converge pointwise to zero. Using Egoroff's theorem (Theorem 1.3.3), construct recursively disjoint μ_E measurable sets $A_i \subset \partial^* E$ so that the sequences $\{f_{k,\pm}\}$ converge uniformly to zero on each A_i, and

$$\mu_E \left(\partial^* E - \bigcup_{i=1}^{\infty} A_i \right) = 0.$$

It follows from Luzin's theorem (Theorem 1.3.4) that there are disjoint compact sets $K_{i,j} \subset A_i$ such that ν_E is continuous on each $K_{i,j}$, and

$$\mu_E \left(A_i - \bigcup_{j=1}^{\infty} K_{i,j} \right) = 0$$

for $i = 1, 2, \ldots$. Organizing $K_{i,j}$ into a sequence $\{K_i\}$, the following conditions are satisfied:

(i) $\partial^* E = N \cup \bigcup_{i=1}^{\infty} K_i$ where $\mu_E(N) = 0$,
(ii) $\lim f_{k,\pm} = 0$ uniformly in each K_i,
(iii) ν_E is continuous in each K_i.

Choose a positive $\varepsilon < 1$, and for a fixed $i \in \mathbb{N}$, let $K := K_i$. There is $k \in \mathbb{N}$ such that $f_{k,\pm}(z) < \varepsilon^n 2^{-n-2}$ for every $z \in K$. Thus for all positive $r < 1/k$, the definitions of $f_{k,\pm}$ imply

$$\left| B(z,r) \cap H_+(E,z) \cap E \right| < \frac{\varepsilon^n}{2^{n+2}} \left| B(z,r) \right|, \tag{1}$$

$$\left| B(z,r) \cap H_-(E,z) - E \right| < \frac{\varepsilon^n}{2^{n+2}} \left| B(z,r) \right|. \tag{2}$$

Observe that for each $r > 0$,

$$\frac{1}{2} \left| B(z,r) \right| = \left| B(z,r) \cap H_-(E,z) \right|$$

$$= \left| B(z,r) \cap H_-(E,z) \cap E \right| + \left| B(z,r) \cap H_+(E,z) - E \right|.$$

Thus if $0 < r < 1/k$, inequalities (1) and (2) imply

$$\left| B(z,r) \cap H_-(E,z) \cap E \right| > \left(\frac{1}{2} - \frac{\varepsilon^n}{2^{n+2}} \right) \left| B(z,r) \right|, \tag{3}$$

$$\left| B(z,r) \cap H_+(E,z) - E \right| > \left(\frac{1}{2} - \frac{\varepsilon^n}{2^{n+2}} \right) \left| B(z,r) \right|. \tag{4}$$

Claim 1. For each pair $x, y \in K$ with $|y - x| < 1/(2k)$,

$$\left| \nu_E(x) \cdot (y - x) \right| \le \varepsilon |y - x|.$$

Proof. Seeking a contradiction, assume there are $x, y \in K$ such that

$$0 < |y - x| < 1/(2k) \quad \text{and} \quad |\nu_E(x) \cdot (y - x)| > \varepsilon|y - x|.$$

Suppose first $\nu_E(x) \cdot (y - x) > \varepsilon|y - x|$. If $u \in B(y, \varepsilon|x - y|)$, then

$$\nu_E(x) \cdot (u - x) = \nu_E(x) \cdot (u - y) + \nu_E(x) \cdot (y - x)$$
$$> -|u - y| + \varepsilon|y - x| \geq 0$$

and consequently $u \in H_+(E, x)$. As $\varepsilon < 1$,

$$B(y, \varepsilon|x - y|) \subset B(x, 2|x - y|) \cap H_+(E, x)$$

and intersecting this inclusion with E, we obtain

$$B(y, \varepsilon|x - y|) \cap E \subset B(x, 2|x - y|) \cap H_+(E, x) \cap E. \tag{$*$}$$

For $z = x$ and $r = 2|x - y|$, inequality (1) yields

$$\left| B(x, 2|x - y|) \cap H_+(E, x) \cap E \right| < \frac{\varepsilon^n}{2^{n+2}} \alpha(n) (2|x - y|)^n$$
$$= \tfrac{1}{4} \alpha(n) (\varepsilon|x - y|)^n,$$

and for $z = y$ and $r = \varepsilon|x - y|$ inequality (3) yields

$$\left| B(y, \varepsilon|x - y|) \cap E \right| \geq \left| B(y, \varepsilon|x - y|) \cap H_-(E, y) \cap E \right|$$
$$> \alpha(n) (\varepsilon|x - y|)^n \left(\frac{1}{2} - \frac{\varepsilon^n}{2^{n+2}} \right)$$
$$> \tfrac{1}{4} \alpha(n) (\varepsilon|x - y|)^n,$$

since $\varepsilon < 1$. The previous two inequalities contradict inclusion ($*$).

If $\nu_E(x) \cdot (y - x) < -\varepsilon|y - x|$, then

$$B(y, \varepsilon|x - y|) \subset B(x, 2|x - y|) \cap H_-(E, x)$$

by the same calculation as above. Subtracting E gives

$$B(y, \varepsilon|x - y|) - E \subset B(x, 2|x - y|) \cap H_-(E, x) - E,$$

and a contradiction follows by letting $z = x$ and $r = 2|x - y|$ in inequality (2), and $z = y$ and $r = \varepsilon|x - y|$ in inequality (4). Claim 1 is thus established.

In view of Claim 1, Whitney's extension theorem (Theorem 1.5.5) applies to the function $f \equiv 0$ on K and $v = \nu_E \restriction K$. There is $g \in C^1(\mathbb{R}^n)$ such that $g(x) = 0$ and $Dg(x) = \nu_E(x)$ for each $x \in K$. By the implicit function theorem [62, Theorem 9.28], the set

$$S := \left\{ x \in \mathbb{R}^n : g(x) = 0 \text{ and } Dg(x) > 1/2 \right\}$$

is an $(n-1)$-dimensional C^1 submanifold of $\mathbb{R}^n$ without boundary (in the sense of differential topology). Note that $K \subset S$, and that $Dg(x)$ is the normal vector of S at each $x \in S$.

Claim 2. $\mu_E(B) = \mathcal{H}^{n-1}(B)$ for each Borel set $B \subset K$.

Proof. Fix $x \in K$. Via translation and rotation, we may assume that $x = 0$ and $Dg(x) = e_n$. Let $\Pi = \Pi_n$ and $\pi = \pi_n$ be as in Section 1.1. There are $R > 0$ and a function $\varphi \in C^1(U(0, R) \cap \Pi)$ such that the map

$$\phi : u \mapsto \big(u, \varphi(u)\big) : U(0, R) \cap \Pi \to S \cap \pi^{-1}\big[U(0, R) \cap \Pi\big]$$

is a bijective C^1 diffeomorphism. Letting $B_r := B(0, r)$, the inequality

$$\mathcal{H}^{n-1}\big[\pi(S \cap B_r)\big] \leq \mathcal{H}^{n-1}(S \cap B_r) \leq \mathcal{H}^{n-1}\Big(\phi\big[\pi(B_r)\big]\Big) \qquad (*)$$

holds for $0 < r < R$. By [29, Section 3.3.4, B],

$$\mathcal{H}^{n-1}\Big(\phi\big[\pi(B_r)\big]\Big) = \int_{B_r \cap \Pi} \sqrt{1 + \big|D\varphi(u)\big|^2}\, du.$$

Since $Dg(0) = e_n$ and $g\big[u, \varphi(u)\big] = 0$ for each u in $U(0, R) \cap \Pi$, we calculate $D\varphi(0) = 0$. The continuity of $D\varphi$ implies

$$\lim_{r \to 0} \frac{\mathcal{H}^{n-1}\Big(\phi\big[\pi(B_r)\big]\Big)}{\alpha(n-1)r^{n-1}} = 1. \qquad (**)$$

In addition, $\big|\varphi(y')\big| \leq \varepsilon(r)|y'|$ where $\lim_{r \to 0} \varepsilon(r) = 0$. It follows that for $s(r) = r/\sqrt{1 + \varepsilon(r)^2}$, we have $B_{s(r)} \cap \Pi \subset \pi(S \cap B_r)$ for $0 < r < R$. Thus

$$\lim_{r \to 0} \frac{\mathcal{H}^{n-1}\big[\pi(S \cap B_r)\big]}{\alpha(n-1)r^{n-1}} \geq \lim_{r \to 0} \frac{\mathcal{H}^{n-1}\big[\pi(B_{s(r)})\big]}{\alpha(n-1)r^{n-1}} = 1. \qquad (\dagger)$$

Summarizing our results,

$$\lim_{r \to 0} \frac{(\mathcal{H}^{n-1} \llcorner S)\big[B(x, r)\big]}{\alpha(n-1)r^{n-1}} = 1 \quad \text{and} \quad \lim_{r \to 0} \frac{\mu_E\big[B(x, r)\big]}{\alpha(n-1)r^{n-1}} = 1.$$

Indeed, the first equality follows from $(*)$, $(**)$, and $(\dagger)$, and the second is implied by Corollary 6.4.4, (3). Consequently

$$\lim_{r \to 0} \frac{(\mathcal{H}^{n-1} \llcorner S)\big[B(x, r)\big]}{\mu_E\big[B(x, r)\big]} = 1.$$

As $x \in K$ is arbitrary, Claim 2 follows from Proposition 6.2.6.

By condition (i), each Borel set $B \subset \partial_* E \cap \Omega$ is the union of disjoint Borel sets $B_i \subset K_i$, $i = 1, 2, \ldots$, and the set

$$B \cap \big[N \cup (\partial_* E \cap \Omega - \partial^* E)\big],$$

which is both μ_E and $\mathcal{H}^{n-1}$ negligible by (6.3.4) and Propositions 5.9.16 and 6.3.7. In view of identity (6.3.4) and Proposition 6.3.7, the theorem follows from Claim 2. $\qquad \square$

Remark 6.5.3. In proving Theorem 6.5.2, we have shown that up to an $\mathcal{H}^{n-1}$ negligible set, $\partial_* E \cap \Omega$ is the union of countably many pieces of $(n-1)$-dimensional C^1 manifolds (the compact sets $K_1, K_2, \ldots$). Sets of this type are called *rectifiable*. For the precise definition and terminology we refer to [33, Section 3.2.14].

An instant benefit afforded by Theorems 6.5.2 is that in $\partial_* E \cap \Omega$ we can switch freely between the measures μ_E and $\mathcal{H}^{n-1}$. We will do it at will throughout the remainder of this book, often tacitly without any references.

If $v \in L^\infty(\Omega, \mathcal{H}^{n-1}; \mathbb{R}^n)$ has compact support, we call a function

$$F : B \to \int_{\partial_* B \cap \Omega} v \cdot \nu_B \, d\mathcal{H}^{n-1} : \mathcal{BV}_{\mathrm{loc}}(\Omega) \to \mathbb{R} \qquad (6.5.1)$$

the *flux* of v. This extends the concept of flux defined in Section 2.1 for figures $A \subset \Omega$. The next divergence theorem is an immediate consequence of Theorems 5.5.1 and 6.5.2: it follows at once from identity (6.3.1) by substituting $\mathcal{H}^{n-1} \llcorner (\partial_* E \cap \Omega)$ for the measure μ_E.

Theorem 6.5.4. *If $E \in BV_{\mathrm{loc}}(\Omega)$ and $v \in Lip_c(\Omega; \mathbb{R}^n)$, then*

$$\int_E \mathrm{div}\, v(x)\, dx = \int_{\partial_* E \cap \Omega} v \cdot \nu_E \, d\mathcal{H}^{n-1}.$$

Theorem 6.5.4 is due to H. Federer [31, 32]. For Lipschitz vector fields, it achieves the maximal generality with respect to the integration domains. More general integration domains are available for continuously differentiable vector fields [39, 38], but we will not pursue this direction.

Theorem 6.5.5. $\mathcal{BV}(\Omega) = \mathcal{P}(\Omega)$ *and* $\mathcal{BV}_{\mathrm{loc}}(\Omega) = \mathcal{P}_{\mathrm{loc}}(\Omega)$. *Moreover,*

$$\mathbf{V}(B, \Omega) = \mathbf{P}(B, \Omega)$$

for each $B \subset \mathbb{R}^n$ such that $B \cap \Omega$ is measurable.

PROOF. The inclusion $\mathcal{BV}(\Omega) \subset \mathcal{P}(\Omega)$ is a direct consequence of Theorem 6.5.5. Given $B \in \mathcal{P}(\Omega)$, choose a vector field $v \in C_c^1(\Omega; \mathbb{R}^n)$ with $\|v\|_{L^\infty(\Omega; \mathbb{R}^n)} \leq 1$, and find a figure $A \subset \Omega$ so that $\mathrm{spt}\, v \subset \mathrm{int}\, A$. As $\partial_*(B \cap A) \subset (\partial_* B \cap \Omega) \cup \partial A$, we see that $B \cap A$ belongs to $\mathcal{P}(\mathbb{R}^n)$ and by Proposition 5.1.3, also to $\mathcal{BV}(\mathbb{R}^n)$. By Theorem 6.5.4,

$$\int_B \mathrm{div}\, v(x)\, dx = \int_{B \cap A} \mathrm{div}\, v(x)\, dx = \int_{\partial_*(B \cap A)} v \cdot \nu_{B \cap A} \, d\mathcal{H}^{n-1}$$

$$\leq \int_{\partial_*(B \cap A)} |v|\, d\mathcal{H}^{n-1} \leq \int_{(\partial_* B \cap \Omega) \cup \partial A} |v|\, d\mathcal{H}^{n-1}$$

$$= \int_{\partial_* B \cap \Omega} |v|\, d\mathcal{H}^{n-1} \leq \mathcal{H}^{n-1}(\partial_* B \cap \Omega).$$

Since v is arbitrary, $\mathbf{V}(B, \Omega) \leq \mathbf{P}(B, \Omega) < \infty$ and hence $B \in \mathcal{BV}(\Omega)$. Thus $\mathcal{BV}(\Omega) = \mathcal{P}(\Omega)$ and consequently $\mathcal{BV}_{\mathrm{loc}}(\Omega) = \mathcal{P}_{\mathrm{loc}}(\Omega)$. Now select $B \subset \mathbb{R}^n$

so that $B \cap \Omega$ is measurable, and observe that

$$\mathbf{V}(B \cap \Omega, \Omega) = \mathbf{V}(B, \Omega) \quad \text{and} \quad \mathbf{P}(B \cap \Omega, \Omega) = \mathbf{P}(B, \Omega)$$

according to (5.1.1) and (4.5.1). Thus if $\mathbf{V}(B, \Omega) < \infty$ or $\mathbf{P}(B, \Omega) < \infty$, then $B \cap \Omega \in \mathbf{BV}(\Omega)$. In this case, the equality $\mathbf{V}(B, \Omega) = \mathbf{P}(B, \Omega)$ follows from Theorem 6.5.2. $\qquad\square$

Proposition 6.5.6. *Let $E \in \mathbf{BV}(\Omega)$ and $\varepsilon > 0$. There is an open set $U \subset \Omega$ such that the following conditions are satisfied:*

(1) $\partial_* E \cap U = \partial_* E \cap \Omega$;
(2) $\mathbf{P}(U) \le \beta \big[\mathbf{P}(E, \Omega) + \varepsilon \big]$ *where* $\beta = \beta(n) > 0$;
(3) *the sets U and $E \cap U$ belong to $\mathbf{BV}(\mathbb{R}^n)$;*
(4) *the measures $\|D\chi_E\|$ and $\|D\chi_{E \cap U}\|$ coincide on subsets of $\partial_* E \cap \Omega$.*

PROOF. For each $x \in \partial_* E \cap \Omega$, select $0 < \delta_x \le 1$ with $B(x, \delta_x) \subset \Omega$. By Lemma 2.2.1, there is a family $\mathcal{C}$ of dyadic cubes such that

(a) $\mathcal{C}$ is a star cover of $\partial_* E \cap \Omega$;
(b) for each $C \in \mathcal{C}$ there is $x_C \in C \cap \partial_* E \cap \Omega$ with $d(C) < \delta_{x_C}$;
(c) $\sum_{C \in \mathcal{C}} d(C)^{n-1} \le \kappa \big[\mathcal{H}^{n-1}(\partial_* E \cap \Omega) + \varepsilon \big]$ where $\kappa = \kappa(n) > 0$.

Let $U := \mathrm{int}\,(\bigcup \mathcal{C})$. Conditions (a) and (b) imply, respectively, the inclusions $\partial_* E \cap \Omega \subset U$ and $U \subset \Omega$. Thus $\partial_* E \cap \Omega = \partial_* E \cap U$. Since

$$\bigcup \{\mathrm{int}\, C : C \in \mathcal{C}\} \subset U \subset \bigcup \{C : C \in \mathcal{C}\},$$

the sets U and $\bigcup \mathcal{C}$ are equivalent. By Theorem 6.5.5 and Propositions 5.1.7,

$$\mathbf{P}(U) = \mathbf{V}\Big(\bigcup \mathcal{C}\Big) \le \sum_{C \in \mathcal{C}} \mathbf{V}(C) = \sum_{C \in \mathcal{C}} \mathbf{P}(C) = 2n \sum_{C \in \mathcal{C}} d(C)^{n-1}$$

and (c) shows that (2) is satisfied with $\beta := 2n\kappa$. Thus $U \in \mathbf{BV}(\mathbb{R}^n)$, since (c) implies $|U| < \infty$. From the inclusion $\partial_*(E \cap U) \subset (\partial_* E \cap \Omega) \cup \partial_* U$ we obtain $E \cap U \in \mathbf{BV}(\mathbb{R}^n)$. As $\partial_* E \cap \Omega \subset \partial_*(E \cap U)$, condition (4) follows. $\quad\square$

Proposition 6.5.6 allows us to derive many properties of sets in $\mathbf{BV}(\Omega)$ from those in $\mathbf{BV}(\mathbb{R}^n)$. Accordingly, in the remainder of this book we concentrate on the family $\mathbf{BV}(\mathbb{R}^n)$.

6.6. Properties of BV sets

The additivity of flux, which is trivial for figures (Observation 2.1.1), holds also for BV sets, albeit nontrivially.

Theorem 6.6.1. *If $A, B \in \mathcal{BV}_{\text{loc}}(\mathbb{R}^n)$ and $|A \cap B| = 0$, then*

$$\int_{\partial_*(A \cup B)} v \cdot \nu_{A \cup B} \, d\mathcal{H}^{n-1} = \int_{\partial_* A} v \cdot \nu_A \, d\mathcal{H}^{n-1} + \int_{\partial_* B} v \cdot \nu_B \, d\mathcal{H}^{n-1}$$

for each $v \in L^1(\partial_* A \cup \partial_* B, \mathcal{H}^{n-1}; \mathbb{R}^n)$.

PROOF. From $|A \cap B| = 0$, we obtain $\chi_{A \cup B} = \chi_A + \chi_B$ almost everywhere, and hence $D\chi_{A \cup B} = D\chi_A + D\chi_B$. As Theorem 6.5.2 implies

$$D\chi_{A \cup B} = \left[\mathcal{H}^{n-1} \llcorner \partial_*(A \cup B)\right] \llcorner \nu_{A \cup B},$$

$$D\chi_A = (\mathcal{H}^{n-1} \llcorner \partial_* A) \llcorner \nu_A, \quad D\chi_B = (\mathcal{H}^{n-1} \llcorner \partial_* B) \llcorner \nu_B,$$

the theorem follows by integrating over $\mathbb{R}^n$ the zero extension $\bar{v}$ of v. $\quad\square$

Remark 6.6.2. Let $A, B \in \mathcal{BV}_{\text{loc}}(\mathbb{R}^n)$ and $|A \cap B| = 0$. Since

$$\Theta(A \cup B, x) = \Theta(A, x) + \Theta(B, x)$$

for each $x \in \mathbb{R}^n$, using Corollary 6.4.4, (2), it is easy to show that the sets $\partial_*(A \cup B)$ and $\partial_* A \triangle \partial_* B$ are $\mathcal{H}^{n-1}$ equivalent. By the uniqueness of polar decomposition, the equalities

$$\nu_{A \cup B} \upharpoonright (\partial_* A - \partial_* B) = \nu_A \upharpoonright (\partial_* A - \partial_* B),$$

$$\nu_{A \cup B} \upharpoonright (\partial_* B - \partial_* A) = \nu_B \upharpoonright (\partial_* B - \partial_* A),$$

$$\nu_A \upharpoonright (\partial_* A \cap \partial_* B) = -\nu_B \upharpoonright (\partial_* A \cap \partial_* B)$$

hold $\mathcal{H}^{n-1}$ almost everywhere. From this we see that the geometric reasons for the additivity of flux for BV sets are the same as those for figures.

The next proposition provides a useful and intuitive identity.

Proposition 6.6.3. *If $A, B \in \mathcal{BV}(\mathbb{R}^n)$ and $B \subset A$, then*

$$\mathbf{P}(A - B) = \mathbf{P}(A) + \mathbf{P}(B) - 2\mathcal{H}^{n-1}(\partial_* A \cap \partial_* B).$$

PROOF. If $E \in \mathcal{BV}_{\text{loc}}(\mathbb{R}^n)$, then Proposition 6.3.7 and Corollary 6.4.4, (2) imply that for $\mathcal{H}^{n-1}$ almost all $x \in \mathbb{R}^n$, the density $\Theta(E, x)$ attains only the values 0, 1/2, and 1. Since

$$\Theta(A - B, x) = \Theta(A, x) - \Theta(B, x)$$

for each $x \in \mathbb{R}^n$, a direct calculation shows that the sets

$$\partial_*(A - B) \quad \text{and} \quad (\partial_* A - \text{cl}_* B) \cup (\text{int}_* A \cap \partial_* B)$$

are $\mathcal{H}^{n-1}$ equivalent. As $\partial_* A \cap \text{int}_* B = \emptyset$ and $\partial_* B \subset \text{cl}_* A$, we obtain

$$(\partial_* A - \text{cl}_* B) \cup (\text{int}_* A \cap \partial_* B) = (\partial_* A - \partial_* B) \cup (\partial_* B - \partial_* A)$$

and the proposition follows. $\quad\square$

Proposition 6.6.4. *Let $E \in \mathcal{BV}(\mathbb{R}^n)$. If $C \subset \mathbb{R}^n$ is a convex set, then*

$$\mathbf{P}(E \cap C) \leq \mathbf{P}(E).$$

PROOF. With no loss of generality, we may assume that C is open. Enumerate $C \cap \mathbb{Q}^n$ as $\{x_1, x_2, \dots\}$, and let $p \geq n+1$ be the least integer such that the convex hull of $\{x_1, \dots, x_p\}$ is a polytope (see the paragraph following Lemma 4.6.9). Denote by C_k the convex hull of $\{x_1, \dots, x_{p+k}\}$, and observe that $\{C_k\}$ is an increasing sequence of polytopes whose union contains $C \cap \mathbb{Q}^n$. As C is an open set, each $x \in C$ has a neighborhood $U(x, r) \subset C$. Since $U(x, r) \cap \mathbb{Q}^n$ contains a set $\{x_{i_0}, \dots, x_{i_j}\}$ whose convex hull contains x, it is clear that $C = \bigcup_{k=1}^{\infty} C_k$. Thus

$$\mathbf{P}(E \cap C) \leq \liminf \mathbf{P}(E \cap C_k) \leq \mathbf{P}(E)$$

by Propositions 4.6.10 and 5.1.7 and Theorem 6.5.2.   □

A different proof of Proposition 6.6.4 is given in [51, Corollary 1.9.4]. It is based on calculating perimeters via the $(n-1)$-dimensional *integral-geometric measure*; see [46, Section 5.14] or [47, Section 2.4].

Proposition 6.6.5. *If $E \in \mathcal{BV}_{\mathrm{loc}}(\mathbb{R}^n)$, then*

$$\mathbf{P}(E) \leq \sum_{i=1}^{n} \int_{\Pi_i} \mathcal{H}^0\left[\partial_*^{L_u}(L_u \cap E)\right] d\mathcal{L}^{n-1}(u),$$

$$\mathbf{P}(E) \geq \int_{\Pi_i} \mathcal{H}^0\left[\partial_*^{L_u}(L_u \cap E)\right] d\mathcal{L}^{n-1}(u), \quad i = 1, \dots, n.$$

PROOF. Since Theorem 4.6.6 and Proposition 4.6.7 imply

$$\int_E \operatorname{div} v(x)\, dx \leq \sum_{i=1}^{n} \int_{\Pi_i} \mathcal{H}^0\left[\partial_*^{L_u}(L_u \cap E)\right] d\mathcal{L}^{n-1}(u)$$

for each $v \in Lip(\mathbb{R}^n; \mathbb{R}^n)$ with $\|v\|_{L^\infty(\mathbb{R}^n; \mathbb{R}^n)} \leq 1$, the first inequality follows directly from Definition 5.1.1 and Theorem 6.5.5. The second inequality is a mere reformulation of inequality (4.6.1).   □

Proposition 6.6.6. *Let $E \in \mathcal{BV}_{\mathrm{loc}}(\mathbb{R}^n)$, and assume $\mathbf{P}(E) < \infty$. If $\{E_k\}$ is a sequence in $\mathcal{BV}(\mathbb{R}^n)$ such that $\lim \mathbf{P}(E_k) = 0$, then*

$$\lim \mathbf{P}(E_k \cap E) = \lim \mathbf{P}(E_k - E) = 0.$$

PROOF. For $i = 1, \dots, n$ and $k = 1, 2, \dots,$

$$E_k^+ := \left\{u \in \Pi_i : \mathcal{H}^0\left[\partial_*^{L_u}(L_u \cap E_k)\right] > 0\right\}$$

$$= \left\{u \in \Pi_i : \mathcal{H}^0\left[\partial_*^{L_u}(L_u \cap E_k)\right] \geq 1\right\}.$$

By Theorem 4.6.6, for $\mathcal{L}^{n-1}$ almost all $u \in \Pi_i$, the intersection $L_u \cap E_k$ is $\mathcal{L}^1$ equivalent to the union of finitely many open segments. It follows that for

$\mathcal{L}^{n-1}$ almost all $u \in \Pi_i - E_k^+$, the set $L_u \cap E_k$ is $\mathcal{L}^1$ negligible, and so is the set $L_u \cap E_k \cap E$. Using this and Proposition 6.6.5,

$$\int_{\Pi_i} \mathcal{H}^0 \left[\partial_*^{L_u} (L_u \cap E_k \cap E) \right] d\mathcal{L}^{n-1}(u)$$

$$= \int_{E_k^+} \mathcal{H}^0 \left[\partial_*^{L_u} (L_u \cap E_k \cap E) \right] d\mathcal{L}^{n-1}(u) \qquad (*)$$

$$\leq \int_{E_k^+} \mathcal{H}^0 \left[\partial_*^{L_u} (L_u \cap E_k) \cup \partial_*^{L_u} (L_u \cap E) \right] d\mathcal{L}^{n-1}(u)$$

$$\leq P(E_k) + \int_{E_k^+} \mathcal{H}^0 \left[\partial_*^{L_u} (L_u \cap E) \right] d\mathcal{L}^{n-1}(u).$$

Applying Proposition 6.6.5 again,

$$\mathcal{L}^{n-1}(E_k^+) \leq \int_{\Pi_i} \mathcal{H}^0 \left[\partial_*^{L_u} (L_u \cap E_k) \right] d\mathcal{L}^{n-1}(u) \leq \mathbf{P}(E_k).$$

According to our assumption, $\lim \mathcal{L}^{n-1}(E_k^+) = 0$. As $\mathbf{P}(E) < \infty$,

$$\lim \int_{E_k^+} \mathcal{H}^0 \left[\partial_*^{L_u} (L_u \cap E) \right] d\mathcal{L}^{n-1}(u) = 0.$$

From this and $(*)$ we obtain

$$\lim \sum_{i=1}^{n} \int_{\Pi_i} \mathcal{H}^0 \left[\partial_*^{L_u} (L_u \cap (E_k \cap E)) \right] d\mathcal{L}^{n-1}(u) = 0,$$

and Proposition 6.6.5 implies $\lim \mathbf{P}(E_k \cap E) = 0$. The remaining equality $\lim \mathbf{P}(E_k - E) = 0$ follows from the identity $E_k - E = E_k - (E_k \cap E)$ by Observation 4.2.1. $\qquad \square$

Considering sets $E = B(0,1)$ and $E_k = \mathbb{R}^n - B(0, 1/k)$, we see that Proposition 6.6.6 is false for $E_k \in \mathcal{P}_{\text{loc}}(\mathbb{R}^n)$. The following example shows that the assumption $\mathbf{P}(E) < \infty$ cannot be omitted either.

Example 6.6.7. For $j, k \in \mathbb{N}$, let $x_k = (k, 0, \ldots, 0)$ and

$$B_{k,j} := \left\{ x \in \mathbb{R}^n : \frac{1}{(2j+1)k} < |x - x_k|^{n-1} < \frac{1}{2jk} \right\}.$$

Since each sum $\sum_{j=1}^{\infty} \mathbf{P}(B_{k,j})$ diverges, there are integers $p_k \geq 1$ such that $\sum_{j=1}^{p_k} \mathbf{P}(B_{k,j}) \geq k$. It follows that $B := \bigcup_{k=1}^{\infty} \bigcup_{j=1}^{p_k} B_{k,j}$ is a locally BV set with $\mathbf{P}(B) = \infty$. Now $\lim \mathbf{P} \left[U(x_k, 1/k) \right] = 0$ and

$$\lim_{k \to \infty} \mathbf{P} \left[U(x_k, 1/k) \cap B \right] = \lim_{k \to \infty} \sum_{j=1}^{p_k} \mathbf{P}(B_{k,j}) = \infty.$$

6.7. Approximating by figures

In Section 2.4 we extended the divergence theorem from dyadic figures to the family $\overline{\mathcal{DF}}$, which was the completion of $\mathcal{DF}$ with respect to the convergence of dyadic figures introduced in Section 2.4. We denote by $\mathcal{BV}_c(\mathbb{R}^n)$ the family of all *bounded* BV sets in $\mathbb{R}^n$, and show that $\overline{\mathcal{DF}} = \mathcal{BV}_c(\mathbb{R}^n)$.

The convergence of dyadic figures defined in Section 2.4 extends verbatim to a convergence of bounded BV sets. We say that a sequence $\{E_k\}$ in $\mathcal{BV}_c(\mathbb{R}^n)$ *converges* to a set $E \subset \mathbb{R}^n$ if the following conditions are satisfied:

(i) each E_k is contained in a fixed compact set $K \subset \mathbb{R}^n$;

(ii) $\lim |E_k \bigtriangleup E| = 0$ and $\sup \mathbf{P}(E_k) < \infty$.

Proposition 6.7.1. *Let $\{E_k\}$ be a sequence in $\mathcal{BV}_c$ that converges to a set $E \subset \mathbb{R}^n$. Then $E \in \mathcal{BV}_c(\mathbb{R}^n)$, and for each $v \in C(\mathbb{R}^n)$,*

$$\lim \int_{\partial_* E_k} v \cdot \nu_{E_k} \, d\mathcal{H}^{n-1} = \int_{\partial_* E} v \cdot \nu_E \, d\mathcal{H}^{n-1}.$$

PROOF. Since each E_k is contained in a fixed compact set $K \subset \mathbb{R}^n$, so is E. By Lemma 2.4.1, the set E is measurable and $\lim \chi_{E_k} = \chi_E$ in $L^1(\mathbb{R}^n)$. Since there is $w \in C_c(\mathbb{R}^n; \mathbb{R}^n)$ such that $w \restriction K = v \restriction K$, the proposition follows from Proposition 5.5.4 and Theorem 6.5.2. $\square$

Proposition 6.7.1 shows that $\overline{\mathcal{DF}} \subset \mathcal{BV}_c(\mathbb{R}^n)$, and that on $\overline{\mathcal{DF}}$ the flux $\widetilde{F}$ defined in Section 2.4 coincides with that defined by formula (6.5.1).

Lemma 6.7.2. *Let Q be a cube, and let $E \subset Q$ be a BV set with $|E| \le |Q|/2$. Then*

$$\mathcal{H}^{n-1}(\partial Q \cap \partial_* E) \le \beta \, \mathcal{H}^{n-1}(\text{int } Q \cap \partial_* E)$$

where $\beta = \beta(n) > 0$.

PROOF. As the lemma follows from Proposition 6.1.1 when $n = 1$, suppose $n \ge 2$. By translation invariance, we may assume $Q = [0, h]^n$. The intersection $S := Q \cap \Pi_n$ is an $(n-1)$-dimensional face of Q, and we estimate $\mathcal{H}^{n-1}(S \cap \partial_* E)$. Letting

$$A := \{u \in S : \mathcal{H}^1[\pi_n^{-1}(u) \cap E \cap Q] = h\},$$
$$B := \{u \in S : 0 < \mathcal{H}^1[\pi_n^{-1}(u) \cap E \cap Q] < h\},$$

it is clear that $S \cap \partial_* E \subset A \cup B$ and $B = \pi_n(\text{int } Q \cap \partial_* E)$. Since $|E| \le |Q - E|$, the relative isoperimetric inequality (Corollary 5.9.13) implies

$$h\mathcal{H}^{n-1}(A) \le |E|^{\frac{1}{n}} |E|^{\frac{n-1}{n}} \le \kappa |Q|^{\frac{1}{n}} \mathcal{H}^{n-1}(\text{int } Q \cap \partial_* E) = \kappa h \mathcal{H}^{n-1}(\text{int } Q \cap \partial_* E).$$

Dividing by $h > 0$ and using $\mathcal{H}^{n-1}(B) \le \mathcal{H}^{n-1}(\text{int } Q \cap \partial_* E)$, we obtain

$$\mathcal{H}^{n-1}(S \cap \partial_* E) \le (1 + \kappa)\mathcal{H}^{n-1}(\text{int } Q \cap \partial_* E).$$

With $\beta = 2n(1 + \kappa)$, the lemma follows by symmetry. $\square$

Proposition 6.7.3. *Let $E \in \mathcal{BV}_c(\mathbb{R}^n)$, and for an integer k, let $\mathcal{A}$ be the collection of all k-cubes Q such that $|Q \cap E| \ge |Q|/2$. If $A := \bigcup \mathcal{A}$ then for $\gamma = \gamma(n) > 0$,*

$$\mathbf{P}(A) \le \gamma \mathbf{P}(E) \quad \text{and} \quad |A \bigtriangleup E| \le 2^{-k} \gamma \mathbf{P}(E).$$

PROOF. We assume $n \geq 2$, since for $n = 1$ the proposition follows from Proposition 6.1.1. Denote by $\mathcal{B}$ the collection of all k-cubes Q such that $0 < |Q \cap E| < |Q|/2$, and let $B = \bigcup \mathcal{B}$. Note $\mathcal{A}$ and $\mathcal{B}$ are finite collections, $\mathcal{A} \cap \mathcal{B} = \emptyset$, and $\mathcal{A} \cup \mathcal{B}$ covers E almost entirely. The relative isoperimetric inequality yields

$$|A - E| = \sum_{Q \in \mathcal{A}} |Q - E| = \sum_{Q \in \mathcal{A}} |Q - E|^{\frac{1}{n}} |Q - E|^{\frac{n-1}{n}}$$

$$\leq \kappa \sum_{Q \in \mathcal{A}} |Q|^{\frac{1}{n}} \mathcal{H}^{n-1}(\text{int } Q \cap \partial_* E) \leq 2^{-k} \kappa \mathbf{P}(E, \text{int } A),$$

$$|E - A| = \sum_{Q \in \mathcal{B}} |Q \cap E| = \sum_{Q \in \mathcal{B}} |Q \cap E|^{\frac{1}{n}} |Q \cap E|^{\frac{n-1}{n}}$$

$$\leq \kappa \sum_{Q \in \mathcal{B}} |Q|^{\frac{1}{n}} \mathcal{H}^{n-1}(\text{int } Q \cap \partial_* E) \leq 2^{-k} \kappa \mathbf{P}(E, \text{int } B).$$

As int $A \cap$ int $B = \emptyset$,

$$|A \triangle E| \leq 2^{-k} \kappa [\mathbf{P}(E, \text{int } A) + \mathbf{P}(E, \text{int } B)] \leq 2^{-k} \kappa \mathbf{P}(E).$$

Now let S be an $(n-1)$-dimensional face of a cube $Q \in \mathcal{A}$ such that $S \subset \partial A$, and let Q' be the other k-cube whose $(n-1)$-dimensional face is S. Clearly $Q' \notin \mathcal{A}$, and

$$S = [S \cap \partial_*(Q - E)] \cup [S \cap \partial_*(Q \cap E)] \subset [\partial Q \cap \partial_*(Q - E)] \cup [S \cap \partial_*(Q \cap E)].$$

If $Q' \notin \mathcal{B}$ then $S \cap \partial_*(Q \cap E) = S \cap \partial_* E \subset \partial_* E$. On the other hand, if $Q' \in \mathcal{B}$ then

$$S \cap \partial_*(Q \cap E) = S \cap \left([\partial_*(Q \cap E) \cap \partial_*(Q' \cap E)] \cup [\partial_*(Q \cap E) - \partial_*(Q' \cap E)] \right)$$

$$= \left(S \cap [\partial_*(Q \cap E) \cap \partial_*(Q' \cap E)] \right) \cup (S \cap \partial_* E)$$

$$\subset [\partial Q' \cap \partial_*(Q' \cap E)] \cup \partial_* E.$$

It follows that

$$\partial A \subset \partial_* E \cup \bigcup_{Q \in \mathcal{A}} [\partial Q \cap \partial_*(Q - E)] \cup \bigcup_{Q \in \mathcal{B}} [\partial Q \cap \partial_*(Q \cap E)]. \qquad (*)$$

From Corollary 4.2.5 we infer

$$\text{int } Q \cap \partial_*(Q - E) = \text{int } Q \cap \partial_*(Q \cap E) = \text{int } Q \cap \partial_* E$$

for each cube $Q \subset \mathbb{R}^n$. Thus $(*)$ and Lemma 6.7.2 imply

$$\mathbf{P}(A) \leq \mathbf{P}(E) + \beta \sum_{Q \in \mathcal{A}} \mathcal{H}^{n-1}(\text{int } Q \cap \partial_* E) + \beta \sum_{Q \in \mathcal{B}} \mathcal{H}^{n-1}(\text{int } Q \cap \partial_* E)$$

$$\leq (1 + \beta) \mathbf{P}(E),$$

and the proposition holds with $\gamma := \max\{\kappa, 1 + \beta\}$. $\qquad \square$

Corollary 6.7.4. $\overline{\mathcal{DF}} = \mathcal{BV}_c(\mathbb{R}^n)$.

Remark 6.7.5. Given the dimension n and the grid of k-cubes, Proposition 6.7.3 asserts that both $\mathbf{P}(A)$ and $|A \triangle E|$ depend only on $\mathbf{P}(E)$. The readers familiar with currents will recognize that Proposition 6.7.3 is a version of the *deformation theorem* for the top dimensional *integral currents*; see [33, Theorem 4.2.9] or [47, Section 5.1].

Theorem 6.7.6. *Let $K \subset \mathbb{R}^n$ be a compact set, and let $c \in \mathbb{R}_+$. The set*

$$\mathcal{X} := \{\chi_E : E \subset K \text{ and } \mathbf{P}(E) \leq c\}$$

is a compact subset of $L^1(\mathbb{R}^n)$.

PROOF. Although the theorem follows immediately from Theorems 5.5.12 and 6.5.5, a different proof based on Proposition 6.6.2 is instructive. From Propositions 4.5.3 and 5.1.7 we infer that $\mathfrak{X}$ is a closed subset of $L^1(\mathbb{R}^n)$. Since $L^1(\mathbb{R}^n)$ is complete, so is $\mathfrak{X}$. Thus it suffices to show that $\mathfrak{X}$ is totally bounded. Choose $\varepsilon > 0$, and recall that

$$|A \bigtriangleup B| = \|\chi_A - \chi_B\|_{L^1(\mathbb{R}^n)}$$

for any pair of measurable sets $A, B \subset \mathbb{R}^n$. If γ is the constant from Proposition 6.7.3, select $k \in \mathbb{N}$ so that $2^{-k}c\gamma < \varepsilon$. Cover K by a finite family $\mathcal{C}$ of k-cubes, and enumerate all nonempty subfamilies of $\mathcal{C}$ as $\mathcal{C}_1, \ldots, \mathcal{C}_p$. Note that the indicator of each $A_i := \bigcup \mathcal{C}_i$ belongs to $L^1(\mathbb{R}^n)$, but not necessarily to $\mathfrak{X}$, since A_i need not be contained in K. Notwithstanding, Proposition 6.7.3 shows that for each $\chi_E \in \mathfrak{X}$, there is $A_{i(E)}$ such that

$$|A_{i(E)} \bigtriangleup E| \leq 2^{-k}c\gamma < \varepsilon.$$

Hence $\mathfrak{X}$ is covered by the sets $\{\chi_E \in \mathfrak{X} : |E \bigtriangleup A_i| < \varepsilon\}$, $i = 1, \ldots, p$, whose diameters are smaller than 2ε. As ε is arbitrary, $\mathfrak{X}$ is totally bounded. $\qquad\square$

We have shown that bounded BV sets can be approximated by dyadic figures. Since BV functions can be approximated by C^∞ functions (Theorem 5.6.3), one expects that BV sets can be also approximated by sets whose boundaries are C^∞ manifolds. This is indeed true, but we do not need it for our exposition. The interested reader is referred to [36, Theorem 1.24].

Part 3

The divergence theorem

Chapter 7

Bounded vector fields

We extend the divergence theorem for admissible vector fields from dyadic figures to BV sets. As BV sets need not be compact, we limit our attention to bounded admissible vector fields. While the proof for BV sets is conceptually similar to that for figures, its technical aspects are more involved and require some preparatory results.

7.1. Approximating from inside

It follows from Proposition 6.7.3 that each bounded BV set is the L^1 limit of a sequence of figures. However, for a bounded BV set A, it is generally not possible to find a sequence $\{A_k\}$ of figures which L^1 converges to A and satisfies either $A_k \subset A$ for all $k \in \mathbb{N}$ — a convergence from *inside*, or $A \subset A_k$ for all $k \in \mathbb{N}$ — a convergence from *outside*. Indeed, in Example 5.2.6 no sequence of figures L^1 converges to "caviar" from outside, to "Swiss cheese" from inside, and to the union of "caviar" and "Swiss cheese" from either side.

On the other hand, I. Tamanini and C. Giacomelli proved that for a bounded BV set A there exists a sequence $\{A_k\}$ of BV sets with additional regularity properties such that $A_k \subset A$ for all $k \in \mathbb{N}$ and $\lim \mathbf{P}(A - A_k) = 0$. In particular, such a sequence L^1 converges to A from inside: by Proposition 6.1.1 if $n = 1$, and by the isoperimetric inequality if $n \geq 2$.

Throughout this section, we fix real numbers $0 < \alpha < 1$ and $\beta > 0$, as well as a bounded set $E \in \mathcal{BV}(\mathbb{R}^n)$ such that $\mathrm{cl}_* E = E$. We let

$$\mathcal{E} := \{A \in \mathcal{BV}(\mathbb{R}^n) : A \subset E\},$$

and for each $A \in \mathcal{E}$, define

$$F(A) := \beta|E - A| + \mathbf{P}(A) - \alpha\mathcal{H}^{n-1}(\partial_* A \cap \partial_* E).$$

Note that $F(A)$ depends only on the equivalence class of A. The inequality $\alpha\mathcal{H}^{n-1}(\partial_* A \cap \partial_* E) \leq \mathbf{P}(A)$ implies that the functional

$$F : A \mapsto F(A) : \mathcal{E} \to \mathbb{R}$$

is nonnegative.

Lemma 7.1.1. *If $\{A_k\}$ is a sequence in $\mathcal{E}$ that L^1 converges to a set $A \in \mathcal{E}$, then $F(A) \le \liminf F(A_k)$.*

PROOF. For $B \in \mathcal{E}$, let $G(B) := \mathcal{H}^{n-1}(\partial_* B - \partial_* E)$ and observe that

$$F(B) = \beta |E - B| + (1 - \alpha)\mathbf{P}(B) + \alpha G(B).$$

In view of Proposition 5.1.7, it suffices to show $G(A) \le \liminf G(A_k)$. To this end, choose $\varepsilon > 0$ and find a compact set $K \subset \partial_* E$ with $\mathcal{H}^{n-1}(\partial_* E - K) < \varepsilon$. If $U = \mathbb{R}^n - K$ and $B \in \mathcal{E}$, then

$$\partial_* B - \partial_* E \subset \partial_* B \cap U \subset (\partial_* B - \partial_* E) \cup (\partial_* E - K)$$

for each $B \in \mathcal{E}$, and consequently $G(B) \le \mathbf{P}(B, U) \le G(B) + \varepsilon$. In view of Theorem 6.5.5 and Proposition 5.1.7,

$$G(A) \le \mathbf{P}(A, U) \le \liminf \mathbf{P}(A_k, U) \le \liminf G(A_k) + \varepsilon$$

and the lemma follows from the arbitrariness of ε.   □

We say that a set $A \in \mathcal{E}$ is a *minimizer* of F, or *minimizes* F, if

$$F(A) = \inf\{F(B) : B \in \mathcal{E}\}.$$

If A minimizes F, then so does any set $B \in \mathcal{E}$ that is equivalent to A. In particular, replacing A by $\mathrm{int}_* A$ or $\mathrm{cl}_* A$, we may assume that a minimizer of F satisfies $A = \mathrm{int}_* A$ or $A = \mathrm{cl}_* A$, respectively.

Lemma 7.1.2. *The functional F has a minimizer.*

PROOF. Let $c := \inf\{F_k(B) : B \in \mathcal{E}\}$. As $\emptyset \in \mathcal{E}$ and $F(\emptyset) = \beta |E|$, we see that $c \le \beta |E|$. Thus there is a sequence $\{A_k\}$ in $\mathcal{E}$ such that

$$c \le F(A_k) < c + 1/k, \quad k = 1, 2, \ldots.$$

From $F(A_k) \le \beta |E| + 1$ we obtain $\mathbf{P}(A_k) \le \beta |E| + 1 + \alpha \mathbf{P}(E)$. Theorem 5.5.12 implies that there are a set $A \in \mathcal{E}$ and a subsequence of $\{A_k\}$, still denoted by $\{A_k\}$, which L^1 converges to A. By Lemma 7.1.1,

$$c \le F(A) \le \lim F(A_k) = c.$$ □

Lemma 7.1.3. *Let A be a minimizer of F such that $A = \mathrm{int}_* A$, and let $x \in \mathbb{R}^n$. If $\mathcal{H}^{n-1}(\partial_* A \cap \partial B_r) = 0$ for $B_r = B(x, r)$, then*

$$\mathcal{H}^{n-1}(\partial_* A \cap B_r) - \alpha \mathcal{H}^{n-1}(\partial_* E \cap \partial_* A \cap B_r)$$
$$\le \mathcal{H}^{n-1}(A \cap \partial B_r) + \beta |A \cap B_r|.$$

PROOF. Fix $r > 0$ so that $\mathcal{H}^{n-1}(\partial_* A \cap \partial B_r) = 0$, and define a measure

$$\mu := \mathcal{H}^{n-1} - \alpha \mathcal{H}^{n-1} \llcorner \partial_* E.$$

Observe that $F(B) = \beta|E - B| + \mu(\partial_* B)$ for each $B \in \mathcal{E}$, and that the inequality we wish to prove is transformed to

$$\mu(\partial_* A \cap B_r) \leq \mathcal{H}^{n-1}(A \cap \partial B_r) + \beta|A \cap B_r|. \qquad (*)$$

By Lemma 4.5.5, there is an increasing sequence $\{s_k\}$ of positive numbers such that $\lim s_k = r$ and $\mathcal{H}^{n-1}(\partial_* A \cap \partial B_{s_k}) = 0$ for $k = 1, 2, \ldots$. As A minimizes F, for $C_k := A - B_{s_k}$ we obtain

$$\beta|E - A| + \mu(\partial_* A) = F(A) \leq F(C_k) = \beta|E - C_k| + \mu(\partial_* C_k),$$

and consequently

$$\mu(\partial_* A) \leq \mu(\partial_* C_k) + \beta|A - C_k|$$
$$= \mu(\partial_* C_k) + \beta|A \cap B_{s_k}| \leq \mu(\partial_* C_k) + \beta|A \cap B_r|.$$

In addition $\partial_* A - B_r = \partial_* C_k - B_r$, since $B_{s_k} \subset \text{int } B_r$. Hence

$$\mu(\partial_* A \cap B_r) = \mu(\partial_* A) - \mu(\partial_* A - B_r)$$
$$\leq \mu(\partial_* C_k) + \beta|A \cap B_r| - \mu(\partial_* C_k - B_r)$$
$$= \mu(\partial_* C_k \cap B_r) + \beta|A \cap B_r|$$

for $k = 1, 2, \ldots$. Thus to prove $(*)$, it suffices to verify

$$\lim \mu(\partial_* C_k \cap B_r) = \mathcal{H}^{n-1}(A \cap \partial B_r). \qquad (**)$$

To this end, let $D_k := B_r - B_{s_k}$, and note that $\mathcal{H}^{n-1}(\partial_* A \cap \partial D_k) = 0$ by the choice of r and s_k. In particular,

$$\lim \mathcal{H}^{n-1}(\partial_* A \cap D_k) = \lim \mathcal{H}^{n-1}(\partial_* A \cap \text{int } D_k) = 0. \qquad (\dagger)$$

As $A = \text{int}_* A$ and $C_k = A \cap (\mathbb{R}^n - B_{s_k})$, Corollary 4.2.5 shows that the pairs

$$\left[\partial_* (A \cap D_k), (\partial_* A \cap D_k) \cup (A \cap \partial D_k) \right],$$
$$\left[\partial_* C_k, (\partial_* A - B_{s_k}) \cup (A \cap \partial B_{s_k}) \right] \qquad (\dagger\dagger)$$

consist of $\mathcal{H}^{n-1}$ equivalent sets. Intersecting the sets of the second pair by B_r produces a pair $\left[\partial_* C_k \cap B_r, (\partial_* A \cap D_k) \cup (A \cap \partial B_{s_k}) \right]$ of $\mathcal{H}^{n-1}$ equivalent sets. In addition, $A = \text{int}_* A$ and $A \subset E$ imply $A \cap \partial_* E = \emptyset$. Thus

$$\mu(\partial_* C_k \cap B_r) = \mu(\partial_* A \cap D_k) + \mu(A \cap \partial B_{s_k})$$
$$= \mu(\partial_* A \cap D_k) + \mathcal{H}^{n-1}(A \cap \partial B_{s_k}).$$

In view of this equality and $(\dagger)$, we prove $(**)$ by establishing

$$\lim \mathcal{H}^{n-1}(A \cap \partial B_{s_k}) = \mathcal{H}^{n-1}(A \cap \partial B_r). \qquad (\diamond)$$

Via mollification, define $v \in C_c^1(\mathbb{R}^n; \mathbb{R}^n)$ so that $v(y) = (y - x)|y - x|^{-1}$ for every $y \in \text{cl } D_k$, and observe that

$$v(y) = \begin{cases} \nu_{A \cap D_k}(y) & \text{if } y \in A \cap \partial B_r, \\ -\nu_{A \cap D_k}(y) & \text{if } y \in A \cap \partial B_{s_k}. \end{cases}$$

Theorem 6.5.4 and (††) imply

$$\int_{A \cap D_k} \operatorname{div} v = \int_{\partial_*(A \cap D_k)} v \cdot \nu_{A \cap D_k} \, d\mathcal{H}^{n-1}$$

$$= \int_{\partial_* A \cap D_k} v \cdot \nu_{A \cap D_k} \, d\mathcal{H}^{n-1} + \int_{A \cap \partial D_k} v \cdot \nu_{A \cap D_k} \, d\mathcal{H}^{n-1}$$

$$= \int_{\partial_* A \cap D_k} v \cdot \nu_{A \cap D_k} \, d\mathcal{H}^{n-1} +$$

$$\mathcal{H}^{n-1}(A \cap \partial B_r) - \mathcal{H}^{n-1}(A \cap \partial B_{s_k})$$

for $k = 1, 2, \ldots$. Since $\lim \int_{A \cap D_k} \operatorname{div} v = 0$, and since

$$\lim \left| \int_{\partial_* A \cap D_k} v \cdot \nu_{A \cap D_k} \, d\mathcal{H}^{n-1} \right| \leq \lim \mathcal{H}^{n-1}(\partial_* A \cap D_k) = 0$$

by (†), the desired equality ($\Diamond$) is established.   □

Lemma 7.1.4. *If* $A = \operatorname{int}_* A$ *minimizes* F, *then* $\operatorname{cl}_* A = \operatorname{cl} A$.

PROOF. Since Proposition 6.1.1 shows that $\operatorname{cl}_* A = \operatorname{cl} A$ for any one-dimensional BV set A, we assume $n \geq 2$ and prove that $\operatorname{cl} A \subset \operatorname{cl}_* A$. Choose $x \in \operatorname{cl} A$, and let $B_s := B(x, s)$ for each $s > 0$. Select $R > r > 0$ so that $|A \cap B_R| \leq \frac{1}{2}|A|$ and $\mathcal{H}^{n-1}(\partial_* A \cap \partial B_r) = 0$. By Lemma 4.5.5, the latter condition is satisfied for all but countably many $0 < r < R$. The relative isoperimetric inequality and Lemma 7.1.3 yield

$$(1 - \alpha)\kappa |A \cap B_r|^{\frac{n-1}{n}} \leq (1 - \alpha)\mathcal{H}^{n-1}(\partial_* A \cap B_r)$$

$$\leq \mathcal{H}^{n-1}(\partial_* A \cap B_r) - \alpha \mathcal{H}^{n-1}(\partial_* A \cap B_r \cap \partial_* E)$$

$$\leq \mathcal{H}^{n-1}(A \cap \partial B_r) + \beta |A \cap B_r|$$

$$\leq \mathcal{H}^{n-1}(A \cap \partial B_r) + \beta |B_r|^{\frac{1}{n}} |A \cap B_r|^{\frac{n-1}{n}}$$

where $\kappa = \kappa(n) > 0$. Observe that $A = \operatorname{int}_* A$ and $A \cap U(x, r) \neq \emptyset$ implies $|A \cap B_r| > 0$. Dividing by $|A \cap B_r|^{\frac{n-1}{n}}$, we obtain

$$(1 - \alpha)\kappa - \beta |B_r|^{\frac{1}{n}} \leq |A \cap B_r|^{\frac{1}{n} - 1} \mathcal{H}^{n-1}(A \cap \partial B_r).$$

Let $\gamma := \frac{1}{2}(1 - \alpha)\kappa$, and make R so small that $\beta |B_R|^{\frac{1}{n}} \leq \gamma$. Then

$$\gamma \leq |A \cap B_r|^{\frac{1}{n} - 1} \mathcal{H}^{n-1}(A \cap \partial B_r)$$

for all but countably many $0 < r < R$. Since

$$|A \cap B_r| = \int_0^r \mathcal{H}^{n-1}(A \cap \partial B_s) \, ds$$

[29, Section 3.4.4, Proposition 1], the previous inequality transforms to

$$\gamma \leq |A \cap B_r|^{\frac{1}{n} - 1} \frac{d}{dr} |A \cap B_r| = n \frac{d}{dr} \left(|A \cap B_r|^{\frac{1}{n}} \right)$$

which holds for $\mathcal{L}^1$ almost all $0 < r < R$. Integrating over the interval $[0, r] \subset [0, R]$ gives $\gamma r \leq n|A \cap B_r|^{\frac{1}{n}}$ for all $0 < r < R$. Thus

$$0 < \frac{1}{\alpha(n)} \left(\frac{\gamma}{n} \right)^n \leq \frac{|A \cap B_r|}{|B_r|}$$

for all sufficiently small $r > 0$. It follows that $x \in \mathrm{cl}_* A$. $\qquad\square$

Combining Lemmas 7.1.2 and 7.1.4, we obtain the aforementioned result of Tamanini and Giacomelli [72].

Theorem 7.1.5. *Let $E \in \mathcal{BV}(\mathbb{R}^n)$. There exists a sequence $\{E_k\}$ in $\mathcal{BV}(\mathbb{R}^n)$ such that $E_k \subset E$ and $\mathrm{cl}\, E_k = \mathrm{cl}_* E_k$ for $k = 1, 2, \ldots$, and*

$$\lim |E - E_k| = \lim \mathbf{P}(E - E_k) = 0.$$

PROOF. Assume $E = \mathrm{cl}_* E$. For $k = 1, 2, \ldots$ and $A \in \mathcal{E}$, define

$$F_k(A) := k|E - A| + P(A) - \alpha \mathcal{H}^{n-1}(\partial_* A \cap \partial_* E).$$

By Lemmas 7.1.2 and 7.1.4, each F_k has a minimizer E_k which satisfies $\mathrm{cl}_* E_k = \mathrm{cl}\, E_k$. Since $\alpha \mathcal{H}^{n-1}(\partial_* A \cap \partial_* E) \leq \mathbf{P}(A)$ for every $A \in \mathcal{E}$, and since E_k minimizes F_k, we obtain

$$k|E - E_k| \leq F_k(E_k) \leq F_k(E) \leq (1 - \alpha)\mathbf{P}(E). \qquad (*)$$

Hence $\lim |E - E_k| = 0$, and consequently $\mathbf{P}(E) \leq \liminf \mathbf{P}(E_k)$ by Proposition 5.1.7. On the other hand, inequality $(*)$ yields

$$\mathbf{P}(E_k) - \alpha \mathbf{P}(E) \leq \mathbf{P}(E_k) - \alpha \mathcal{H}^{n-1}(\partial_* E_k \cap \partial_* E)$$
$$\leq F(E_k) \leq (1 - \alpha)\mathbf{P}(E).$$

Now deduce first that $\lim \mathbf{P}(E_k) = \mathbf{P}(E)$, and then

$$\lim \mathcal{H}^{n-1}(\partial_* E_k \cap \partial_* E) = \mathbf{P}(E).$$

This and Proposition 6.6.3 imply $\lim \mathbf{P}(E - E_k) = 0$. $\qquad\square$

7.2. Relative derivatives

Let $E \subset \mathbb{R}^n$. A map $\phi : E \to \mathbb{R}^m$ is *relatively differentiable* at a point $x \in E \cap \mathrm{int}_* E$ if there is a linear map $L : \mathbb{R}^n \to \mathbb{R}^m$ such that

$$\lim_{\substack{y \to x \\ y \in E}} \frac{|\phi(y) - \phi(x) - L(y - x)|}{|y - x|} = 0.$$

The linear map L, which is unique by Proposition 7.2.1 below, is called the *relative derivative* of ϕ at x, denoted by $D_E \phi(x)$. Note that the concepts of relative differentiability and differentiability coincide whenever $x \in \mathrm{int}\, E$; in particular $D_E\phi(x) = D\phi(x)$ for each $x \in \operatorname{int} E$. If $A \subset E$ and ϕ is differentiable at $x \in A \cap \operatorname{int}_* A$ relative to E, then the restriction $\phi \upharpoonright A$ is differentiable at x relative to A and

$$D_A(\phi \upharpoonright A)(x) = D_E\phi(x). \tag{7.2.1}$$

Proposition 7.2.1. *If the relative derivative exists, it is unique.*

PROOF. Let $x \in E \cap \operatorname{int}_* E$, and assume that there exist distinct linear maps L_1 and L_2 satisfying the following condition: given $\varepsilon > 0$, there is $\delta > 0$ such that for each $y \in E \cap U(x,\delta)$ and $k = 1, 2$,

$$\left|\phi(y) - \phi(x) - L_k(y - x)\right| < \varepsilon|y - x|. \tag{$*$}$$

As $L_1 \neq L_2$, the linear map $L := L_1 - L_2$ has a positive norm N. Let $\varepsilon := N/4$, and find $\delta > 0$ so that $(*)$ holds for each $y \in E \cap U(x,\delta)$. Define open sets

$$S := \left\{y \in \mathbb{R}^n : \left|L(y)\right| > 2\varepsilon|y|\right\} \quad \text{and} \quad T := x + S.$$

If $S = \emptyset$ then $N \leq 2\varepsilon = N/2$, a contradiction. Being open, the set S has positive measure, and as $0 \in \operatorname{cl} S$, so does the intersection $S \cap B(0,1)$. As $rS = S$ for each $r > 0$, we obtain

$$\lim_{r \to 0} \frac{\left|S \cap B(0,r)\right|}{\left|B(0,r)\right|} = \frac{\left|S \cap B(0,1)\right|}{\left|B(0,1)\right|} > 0,$$

which means $0 \in \operatorname{cl}_* S$. Consequently $x \in \operatorname{cl}_* T$, and Proposition 4.2.4 yields $x \in \operatorname{cl}_*(T \cap E)$; in particular, $T \cap E \cap U(x,\delta) \neq \emptyset$. However,

$$\begin{aligned} 2\varepsilon|y - x| < \left|L(y - x)\right| &= \left|L_1(y - x) - L_2(y - x)\right| \\ &\leq \left|L_1(y - x) - \phi(y) + \phi(x)\right| + \left|\phi(y) - \phi(x) - L_2(y - x)\right| \\ &< 2\varepsilon|y - x| \end{aligned}$$

for each y in $T \cap E \cap U(x,\delta)$ — a contradiction. $\square$

Using relatively differentiable maps, we extend Stepanoff's theorem (Theorem 1.3.3) to pointwise Lipschitz maps defined on arbitrary subsets of $\mathbb{R}^n$. The proof is almost identical to that of the original Stepanoff's theorem [33, Theorem 3.1.9]. We begin with a preparatory lemma.

Lemma 7.2.2. *Let $E \subset \mathbb{R}^n$, $C \subset E \cap \operatorname{int}_* E$, and let $\phi : E \to \mathbb{R}^m$ satisfy the following conditions:*

   (i) *there are positive numbers c and δ such that*

$$\left|\phi(x) - \phi(y)\right| \leq c|x - y|$$

   for each $y \in C$ and each x in $E \cap U(y,\delta)$;

   (ii) *there is a Lipschitz map $\psi : \mathbb{R}^n \to \mathbb{R}^m$ such that $\psi \upharpoonright C = \phi \upharpoonright C$.*

If ψ is differentiable at a point $z \in C \cap \operatorname{int}_* C$, then ϕ is relatively differentiable at z and $D_E\phi(z) = D\psi(z)$.

PROOF. Suppose ψ is differentiable at $z \in C \cap \text{int}_* C$, and choose a positive $\varepsilon \leq 1$. Making δ smaller, we may assume that for all $x \in U(z,\delta)$,

$$\left| \psi(x) - \psi(z) - D\psi(z)(x - z) \right| \leq \varepsilon |x - z|.$$

If $x \in U(z,\delta)$ and $r_x := |x - z| > 0$, then $d\left[U(x,\varepsilon r_x) \cup \{z\}\right] \leq 2r_x$ and $s\left[U(x,\varepsilon r_x) \cup \{z\}\right] = \alpha(n)(\varepsilon/2)^n$. By Lemma 4.2.3,

$$\lim_{x \to z} \frac{|C \cap U(x,\varepsilon r_x)|}{|U(x,\varepsilon r_x)|} = 1.$$

Thus making δ still smaller, we may assume that $C \cap U(x,\varepsilon r_x) \neq \emptyset$ for each $x \in U(z,\delta)$ with $x \neq z$. Select $x \in E \cap \left[U(z,\delta) - \{z\}\right]$, and choose a point $y \in C \cap U(x,\varepsilon r_x)$. Observing that $|x - y| < \varepsilon |x - z| < \delta$, and that $\psi(z) = \phi(z)$ and $\psi(y) = \phi(y)$, we obtain

$$\begin{aligned}
\left| \phi(x) - \phi(z) - D\psi(z)(x - z) \right| &= \left| \phi(x) - \psi(z) - D\psi(z)(x - z) \right| \\
&\leq \left| \phi(x) - \psi(x) \right| + \left| \psi(x) - \psi(z) - D\psi(z)(z - z) \right| \\
&\leq \left| \phi(x) - \phi(y) \right| + \left| \psi(y) - \psi(x) \right| + \varepsilon |x - z| \\
&\leq (c + \text{Lip}\,\psi)|x - y| + \varepsilon |x - z| \leq \varepsilon(c + \text{Lip}\,\psi + 1)|x - z|. \qquad \square
\end{aligned}$$

Theorem 7.2.3. *Let* $E \subset \mathbb{R}^n$. *If* $\phi : E \to \mathbb{R}^m$ *is pointwise Lipschitz in* $C \subset E \cap \text{int}_* E$, *then* ϕ *is relatively differentiable at almost all* $x \in C$.

PROOF. For $j = 1, 2, \ldots$, denote by C_j the set of all $x \in C$ such that

$$\left| \phi(x) - \phi(y) \right| \leq j|x - y| \qquad (*)$$

for each $y \in E \cap U(x, 1/j)$. Then $C = \bigcup_{j=1}^{\infty} C_j$, and each C_j is the union of sets $C_{j,k}$, $k = 1, 2, \ldots$, of diameters smaller than $1/j$. Observe that the sets C_j satisfy condition (i) of Lemma 7.2.2, and hence so do the sets $C_{j,k}$. In addition, inequality $(*)$ implies that the restrictions $\phi \upharpoonright C_{i,j}$ are Lipschitz maps. Using Proposition 1.5.2, extend $\phi \upharpoonright C_{i,j}$ to a Lipschitz map $\psi_{i,j} : \mathbb{R}^n \to \mathbb{R}^m$. By Rademacher's theorem (Theorem 1.5.3), each $\psi_{i,j}$ is differentiable at almost all $x \in \mathbb{R}^n$. Lemma 7.2.2 implies that ϕ is relatively differentiable almost everywhere in every $C_{i,j} \cap \text{int}_* C_{i,j}$, and hence almost everywhere in C by Corollary 4.4.4. $\qquad \square$

Remark 7.2.4. If the set E in Theorem 7.2.3 is open, we obtain Stepanoff's theorem quoted without proof in Theorem 1.5.4.

Let $E \subset \mathbb{R}^n$, and let $f : E \to \mathbb{R}$ be relatively differentiable at a point $x \in E \cap \text{int}_* E$. The real numbers

$$D_{E,i} f(x) := D_i \left[D_E f(x) \right], \quad i = 1, \ldots, n,$$

are called the *relative partial derivatives* of f at x. In other words, the relative partial derivatives $D_{E,i} f(x)$ are defined as the usual partial derivatives of the linear map $D_E f(x) : \mathbb{R}^n \to \mathbb{R}$. The usual partial derivatives $D_i f$ of f may not exist unless $x \in \mathrm{int}\, E$, in which case

$$D_i f(x) = D_{E,i} f(x).$$

If $v = (v_1, \ldots, v_n)$ is a vector field defined on a set $E \subset \mathbb{R}^n$ that is relatively differentiable at $x \in E \cap \mathrm{int}_* E$, then

$$\mathrm{div}_E v(x) := \sum_{i=1}^{n} D_{E,i} v_i(x)$$

is called the *relative divergence* of v at x. Clearly, $\mathrm{div}_E v(x)$ is the *trace* of the linear map $D_E v(x) : \mathbb{R}^n \to \mathbb{R}^n$, and $\mathrm{div}_E v(x) = \mathrm{div}\, v(x)$ whenever $x \in \mathrm{int}\, E$.

7.3. The critical interior

The *critical interior* of a set $E \subset \mathbb{R}^n$ is the set

$$\mathrm{int}_c E := \left\{ x \in \mathrm{int}_* E : \lim_{r \to 0} \frac{\mathcal{H}^{n-1}\big[\partial_* E \cap B(x,r)\big]}{r^{n-1}} = 0 \right\}.$$

In general, $\mathrm{int}_c E$ is a proper subset of $\mathrm{int}_* E$ even when E is a compact BV set; see Example 7.3.3 below. However, for locally BV sets, the difference between the essential and critical interiors is small.

Proposition 7.3.1. *If $E \in \mathcal{BV}_{\mathrm{loc}}(\mathbb{R}^n)$ then $\mathcal{H}^{n-1}(\mathrm{int}_* E - \mathrm{int}_c E) = 0$.*

PROOF. It suffices to choose $t > 0$ and show that the set

$$A := \left\{ x \in \mathrm{int}_* E : \limsup_{r \to 0+} \frac{\mathcal{H}^{n-1}\big[\partial_* E \cap B(x,r)\big]}{r^{n-1}} > t \right\}$$

is $\mathcal{H}^{n-1}$ negligible. To this end, choose positive ε and δ, and recall that the reduced measure $\mathcal{H}^{n-1} \llcorner \partial_* E$ is Radon (Theorems 6.5.2). As $A \cap \partial_* E = \emptyset$, there is an open set U with $A \subset U$ and $\mathcal{H}^{n-1}(\partial_* E \cap U) \le \varepsilon$. For each $x \in A$ select $0 < r_x < \delta\,/10$ so that

$$\frac{\mathcal{H}^{n-1}\big[\partial_* E \cap B(x, r_x)\big]}{r_x^{n-1}} > t$$

and $B(x, r_x) \subset U$. By Vitali's theorem (Theorem 4.3.2), there is a countable set $C \subset A$ such that $\big\{ B(x, r_x) : x \in C \big\}$ is a disjoint family and $A \subset \bigcup_{x \in C} B(x, 5r_x)$. We calculate:

$$\mathcal{H}_\delta^{n-1}(A) \le \sum_{x \in C} \alpha(n-1)(5r_x)^{n-1}$$

$$\le \alpha(n-1)5^{n-1}t^{-1} \sum_{x \in C} \mathcal{H}^{n-1}\big[\partial_* E \cap B(x, r_x)\big]$$

$$\le \alpha(n-1)5^{n-1}t^{-1}\mathcal{H}^{n-1}(\partial_* E \cap U) \le \alpha(n-1)5^{n-1}t^{-1}\varepsilon.$$

The arbitrariness of ε implies $\mathcal{H}^s_\delta(A) = 0$. As δ is also arbitrary, the proposition follows. $\square$

Throughout the remainder of this book, $\rho(n) := \frac{1}{8}n^{-3/2}$ is a fixed number. A bounded set $E \in \mathbf{BV}(\mathbb{R}^n)$ is called *regular* if $|E| > 0$ and

$$\frac{|E|}{d(E)\mathbf{P}(E)} > \rho(n).$$

A direct calculation shows that each cube $C \subset \mathbb{R}^n$ is a regular BV set.

We note that the value of $\rho(n)$ is to a large extent arbitrary. Our choice simplifies the proof of Lemma 7.3.2 below, but any positive value of $\rho(n)$ which makes cubes regular can be used; cf. Example 7.3.3 below.

Recall from Section 4.2 that the shape $s(E)$ of a bounded set $E \subset \mathbb{R}^n$ of positive diameter is defined as the ratio $|E|/d(E)^n$. Suppose that a bounded set $E \in \mathbf{BV}(\mathbb{R}^n)$ is regular. If $n = 1$ then $s(E) > \rho(n)$, since $\mathbf{P}(E) \geq 2$. If $n \geq 2$, then the isoperimetric inequality implies $|E|^{n-1} \leq \gamma^n \mathbf{P}(E)^n$ where $\gamma = \gamma(n) > 0$. Thus

$$s(E) \geq \gamma^{-n} \frac{|E|^{n-1}}{\mathbf{P}(E)^n} \cdot \frac{|E|}{d(E)^n} = \gamma^{-n} \left[\frac{|E|}{d(E)\mathbf{P}(E)} \right]^n > \left[\frac{\rho(n)}{\gamma} \right]^n,$$

and we see that the shapes of regular bounded BV sets are bounded away from zero.

Lemma 7.3.2. *Let $A \in \mathbf{BV}_{\mathrm{loc}}(\mathbb{R}^n)$ and $x \in \mathrm{int}_c A$. If a cube C containing x is sufficiently small, then $A \cap C$ is a regular BV set.*

PROOF. If $r := d(C)$, then Corollary 4.2.5 implies

$$\mathbf{P}(A \cap C) \leq \mathcal{H}^{n-1}\big(\partial_* A \cap B[x,r]\big) + \mathbf{P}(C),$$

and we calculate:

$$\frac{r\mathbf{P}(A \cap C)}{|C|} \leq \frac{r^n}{|C|} \cdot \frac{\mathcal{H}^{n-1}\big[\partial_* A \cap B(x,r)\big]}{r^{n-1}} + \frac{d(C)\mathbf{P}(C)}{|C|}$$

$$= n^{n/2} \frac{\mathcal{H}^{n-1}\big[\partial_* A \cap B(x,r)\big]}{r^{n-1}} + \frac{1}{4\rho(n)}.$$

By the definition of $\mathrm{int}_c E$ and Lemma 4.2.3,

$$\lim_{r \to 0} \frac{\mathcal{H}^{n-1}\big[\partial_* A \cap B(x,r)\big]}{r^{n-1}} = 0 \quad \text{and} \quad \lim_{r \to 0} \frac{|A \cap C|}{|C|} = 1.$$

Hence for each sufficiently small r,

$$\frac{d(A \cap C)\mathbf{P}(A \cap C)}{|A \cap C|} \leq \frac{r\mathbf{P}(A \cap C)}{|C|} \cdot \frac{|C|}{|A \cap C|} < \frac{1}{2\rho(n)} \cdot 2 = \frac{1}{\rho(n)}$$

and the lemma follows. $\square$

Example 7.3.3. Assume that $n = 2$, and choose an integer $k \geq 8$. For $i = 1, \ldots, k$, let U_i be the set

$$\left\{ (x,y) \in \mathbb{R}^2 : 0 < y < x^2 \tan \tfrac{2\pi}{k} \text{ and } x > 0 \right\}$$

rotated counterclockwise by $2\pi(i-1)/k$. Then $A_k := B(0,1) - \bigcup_{i=1}^{k} U_i$ is a compact BV set, and a direct calculation reveals that the origin 0 of $\mathbb{R}^2$ is a point of $\mathrm{int}_* A_k$ which does not belong to $\mathrm{int}_c A_k$. If $C_r := [-r, r]^2$, then it is easy to see that

$$\lim_{r \to 0} \frac{|A_k \cap C_r|}{d(A_k \cap C_r) \mathbf{P}(A_k \cap C_r)} = \frac{1}{(4+k)\sqrt{2}} \to 0 \text{ as } k \to \infty.$$

Thus if k is sufficiently large, $A_k \cap C_r$ is not a regular set for all sufficiently small $r > 0$. It follows that the assumption $x \in \mathrm{int}_c A$ of Lemma 7.3.2 cannot be omitted no matter how small the constant $\rho(n) > 0$ we select.

7.4. The divergence theorem

Lemma 7.4.1. *Let $E \subset \mathbb{R}^n$, $x \in \mathrm{cl}_* E$, and $v \in L^\infty(\mathrm{cl}_* E, \mathcal{H}^{n-1}; \mathbb{R}^n)$. Assume $\{B_k\}$ is a sequence in $\boldsymbol{BV}(\mathbb{R}^n)$ such that $B_k \subset \mathrm{cl}_* E$ and $x \in B_k$ for $k = 1, 2, \ldots$, and $\lim d(B_k) = 0$. Then*

$$\limsup \frac{1}{\mathbf{P}(B_k)} \int_{\partial_* B_k} v \cdot \nu_{B_k} \, d\mathcal{H}^{n-1} \leq H_0 v(x). \tag{1}$$

Assume in addition that each B_k is regular. For $0 \leq s \leq 1$,

$$\limsup \frac{1}{d(B_k)^{n-1+s}} \left| \int_{\partial_* B_k} v \cdot \nu_{B_k} \, d\mathcal{H}^{n-1} \right| \leq \beta H_s v(x) \tag{2}$$

where $\beta = \beta(n) > 0$. If $x \in \mathrm{int}_* E$ and v is relatively differentiable at x, then

$$\lim \frac{1}{|B_k|} \int_{\partial_* B_k} v \cdot \nu_{B_k} \, d\mathcal{H}^{n-1} = \mathrm{div}_E v(x). \tag{3}$$

PROOF. The proof is similar to that of Corollary 2.1.3. We may assume $H_s v(x) < \infty$. Choose $\varepsilon > 0$, and for $B \in \boldsymbol{BV}(\mathbb{R}^n)$ with $B \subset \mathrm{cl}_* E$, let

$$F(B) := \int_{\partial_* B} v \cdot \nu_B \, d\mathcal{H}^{n-1}.$$

There is $\delta > 0$ such that $|v(y) - v(x)| \leq [H_s v(x) + \varepsilon]|y - x|^s$ for every $y \in E \cap U(x, \delta)$. By Theorem 6.5.4,

$$F(B_k) = \int_{\partial_* B_k} [v(y) - v(x)] \cdot \nu_{B_k} \, d\mathcal{H}^{n-1}(y)$$

$$\leq [H_s v(x) + \varepsilon] \int_{\partial_* B_k} |y - x|^s \, d\mathcal{H}^{n-1}(y) \tag{*}$$

$$\leq [H_s v(x) + \varepsilon] d(B_k)^s \mathbf{P}(B_k)$$

for all sufficiently large k. If $s = 0$, inequality (1) follows from the arbitrariness of ε. If B_k is regular, then $(*)$ and the isodiametric inequality (1.4.2) imply

$$F(B_k) \le \left[H_s v(x) + \varepsilon\right] d(B_k)^{s-1} \frac{1}{\rho(n)} |B_k|$$

$$\le \left[H_s v(x) + \varepsilon\right] \frac{\alpha(n)}{2^n \rho(n)} d(B_k)^{n-1+s}.$$

As ε is arbitrary, inequality (2) holds with $\beta = \alpha(n)/\left[2^n \rho(n)\right]$. Finally, assume that in addition $x \in \mathrm{int}_* E$ and v is relatively differentiable at x. Letting

$$w : y \mapsto v(x) + \left[D_E v(x)\right](y - x) : \mathbb{R}^n \to \mathbb{R}^n,$$

we have $\mathrm{div}\, w(y) = \mathrm{div}_E v(x)$ for each $y \in \mathbb{R}^n$, and there is $\eta > 0$ such that

$$\left|v(y) - w(y)\right| < \varepsilon |y - x|$$

for every $y \in E \cap U(x,\eta)$. By Theorem 6.5.4,

$$\left|F(B_k) - \mathrm{div}_E v(x)|B_k|\right| = \left|\int_{\partial_* B_k} \left[v(y) - w(y)\right] \cdot \nu_{B_k}(y) \, d\mathcal{H}^{n-1}(y)\right|$$

$$\le \varepsilon \int_{\partial_* B_k} |y - x| \, d\mathcal{H}^{n-1}(y)$$

$$\le \varepsilon d(B_k)\mathbf{P}(B_k) \le \frac{\varepsilon}{\rho(n)}|B_k|$$

for all sufficiently large k. Equality (3) follows. □

Lemma 7.4.2. *Let $A \in \boldsymbol{BV}(\mathbb{R}^n)$ be a compact set such that $A = \mathrm{cl}_* A$, and let $v \in Adm(A; \mathbb{R}^n)$. Define $f : A \to \mathbb{R}$ by*

$$f(x) := \begin{cases} \mathrm{div}_A v(x) & \text{if } x \in \mathrm{int}_* A \text{ and } v \text{ is relatively differentiable at } x, \\ 0 & \text{otherwise.} \end{cases}$$

For each $\varepsilon > 0$ and each $\delta : A \to \mathbb{R}_+$, there is a δ-fine partition

$$P = \left\{(C_1, x_1), \ldots, (C_p, x_p)\right\}$$

such that $C_i \in \boldsymbol{BV}(\mathbb{R}^n)$ for $i = 1, \ldots, p$, $[P] = A$, and

$$\left|\sum_{i=1}^p f(x_i)|C_i| - \int_{\partial_* A} v \cdot \nu_A \, d\mathcal{H}^{n-1}\right| < \varepsilon.$$

PROOF. We recall from Section 2.2 that $[P] = \bigcup_{i=1}^p C_i$. According to Remark 2.3.3, the admissible vector field v is bounded and $\mathcal{H}^{n-1}$ measurable. Let $\mathcal{C} := \left\{C \in \boldsymbol{BV}(\mathbb{R}^n) : C \subset A\right\}$ and define

$$F : C \mapsto \int_{\partial_* C} v \cdot \nu_C \, d\mathcal{H}^{n-1} : \mathcal{C} \to \mathbb{R}.$$

In view of Remark 2.3.2, there are $0 \le s_k < 1$ and disjoint, possibly empty, sets $B_k \subset A$ such that v is pointwise Lipschitz in $A - \bigcup_{k=1}^{\infty} B_k$, and for $k = 1, 2, \dots$, the following conditions hold:

(i*) $\mathcal{H}^{n-1+s_k}(B_k) < \infty$, and $H_{s_k} v(x) < \infty$ for each $x \in B_k$;

(ii) $\mathcal{H}^{n-1+s_k}(B_k) > 0$ implies $H_{s_k} v(x) = 0$ for each $x \in B_k$.

We apply separate arguments to the sets $\mathrm{int}_c A$ and $A - \mathrm{int}_c A$. While the proof for $\mathrm{int}_c A$ is similar to that of Lemma 2.3.6, new techniques are needed to deal with $A - \mathrm{int}_c A$.

The intersections $E_k = B_k \cap \mathrm{int}_c A$, $k = 1, 2, \dots$, satisfy conditions (i*) and (ii). Theorem 7.2.3 shows that $\mathrm{int}_c A - \bigcup_{k=1}^{\infty} E_k$ is the union of disjoint sets E_0 and D such that $\mathcal{H}^n(E_0) = 0$ and v is relatively differentiable at each $x \in D$. Thus $\mathrm{int}_c A$ is the union of disjoint sets $D, E_0, E_1, \dots$, and we let $s_0 = 1$. The family $\{(E_k, s_k) : k = 0, 1, \dots\}$ is the disjoint union of

$$\{(E_k, s_k) : \mathcal{H}^{n-1+s_k}(E_k) > 0\} \quad \text{and} \quad \{(E_k, s_k) : \mathcal{H}^{n-1+s_k}(E_k) = 0\}.$$

We enumerate these subfamilies as $\{(E_i^+, s_i^+) : i \ge 1\}$ and $\{(E_i^0, s_i^0) : i \ge 1\}$, respectively. For $j \in \mathbb{N}$, let

$$E_{i,j}^0 := \{x \in E_i^0 : j - 1 \le H_{s_i^0} v(x) < j\}$$

and define $t_i^+ = n - 1 + s_i^+$ and $t_i^0 = n - 1 + s_i^0$. Now $E = \mathrm{int}_c A - D$ is the union of disjoint sets E_i^+ and $E_{i,j}^0$. Choose $\varepsilon > 0$, and select $c > \|v\|_{L^\infty(A;\mathbb{R}^n)}$ and $c_i > \mathcal{H}^{t_i^+}(E_i^+)$, $i = 1, 2, \dots$. By Lemma 7.4.1, there are $\beta = \beta(n) > 0$ and $\gamma : A \to \mathbb{R}_+$ such that for each regular set $C \in \mathcal{C}$, the following is true:

(1) $\big| f(x)|C| - F(C) \big| \le \varepsilon |C|$ if $d(C) < \gamma(x)$ for some $x \in D \cap C$,

(2) $\big| F(C) \big| \le \varepsilon 2^{-i} c_i^{-1} d(C)^{t_i^+}$ if $d(C) < \gamma(x)$ for some $x \in E_i^+ \cap C$,

(3) $\big| F(C) \big| \le \beta j \, d(C)^{t_i^0}$ if $d(C) < \gamma(x)$ for some $x \in E_{i,j}^0 \cap C$.

Making γ smaller, we may assume that the intersection $A \cap Q$ is regular for every cube Q such that $d(Q) < \gamma(x)$ for some $x \in \mathrm{int}_c A \cap Q$; see Lemma 7.3.2.

Turning our attention to the set $A - \mathrm{int}_c A$, let

$$N := (A - \mathrm{int}_c A) \cap \bigcup_{k=1}^{\infty} \{B_k : s_k = 0 \text{ and } \mathcal{H}^{n-1}(B_k) = 0\}$$

and $M = (A - \mathrm{int}_c A) - N$. Clearly $\mathcal{H}^{n-1}(N) = 0$, and condition (ii) implies that $H_0 v(x) = 0$ for each $x \in M$. By Proposition 6.6.6, there is $\eta > 0$ such that $\mathbf{P}(A \cap S) \le \varepsilon/c$ for each set $S \in \mathcal{BV}(\mathbb{R}^n)$ with $\mathbf{P}(S) \le \eta$. By making c larger, we achieve $\mathcal{H}^{n-1}(M) < c$. Moreover, in view of Lemma 7.4.1, making γ smaller, we may assume that

$$\big| F(C) \big| \le \varepsilon c^{-1} \mathbf{P}(C) \tag{$*$}$$

for each $C \in \mathcal{C}$ with $d(C) < \gamma(x)$ for some $x \in M \cap C$.

Select a dyadic figure T containing A, and define $\delta : T \to \mathbb{R}_+$ by letting $\delta(x) = \gamma(x)$ if $x \in A$, and $\delta(x) = \text{dist}(x, A)$ if $x \in T - A$. Proposition 2.2.2 shows that there is a δ-fine dyadic partition

$$P = \{(Q_1, x_1), \ldots, (Q_p, x_p)\}$$

such that $[P] = T$, and for a fixed $\kappa = \kappa(n) > 0$, the following is true:

(a) $\sum_{x_k \in N} d(Q_k)^{n-1} \leq (2n)^{-1}\eta$,

(b) $\sum_{x_k \in M} d(Q_k)^{n-1} \leq \kappa c$,

(c) $\sum_{x_k \in B_i^+} d(Q_k)^{t_i^+} \leq \kappa c_i$,

(d) $\sum_{x_k \in B_{i,j}^0} d(Q_k)^{t_i^0} < \varepsilon \, j^{-1} 2^{-i-j}$.

If $C_k := A \cap Q_k$, then $R := \{(C_k, x_k) : x_k \in A\}$ is a δ-fine partition such that $[R] = A$ and C_k is regular whenever $x_k \in \text{int}_c A$. We distinguish four cases.

Case 1. If $C = \bigcup_{x_k \in N} C_k$ and $Q = \bigcup_{x_k \in N} Q_k$, then $C = A \cap Q$. By inequality (a),

$$\mathbf{P}(Q) \leq \sum_{x_k \in N} \mathbf{P}(Q_k) = 2n \sum_{x_k \in N} d(Q_k)^{n-1} \leq \eta$$

and our choice of η implies $\mathbf{P}(C) < \varepsilon/c$. Since $\|v\|_{L^\infty(A;\mathbb{R}^n)} < c$, we obtain $|F(C)| \leq c\mathbf{P}(C) \leq \varepsilon$.

Case 2. By $(*)$, Corollary 4.2.5, and inequality (b),

$$\sum_{x_k \in M} |F(C_k)| \leq \varepsilon c^{-1} \sum_{x_k \in M} \mathbf{P}(C_k)$$

$$\leq \varepsilon c^{-1} \sum_{x_k \in M} \left[\mathcal{H}^{n-1}(\partial_* A \cap \text{int } Q_k) + \mathbf{P}(Q_k) \right]$$

$$\leq \varepsilon c^{-1} \mathbf{P}(A) + 2n\varepsilon c^{-1} \sum_{x_k \in M} d(Q_k)^{n-1} \leq \varepsilon \left[P(A) + 2n\kappa \right].$$

Case 3. Condition (1) yields

$$\sum_{x_k \in D} \left| f(x_k) |C_k| - F(C_k) \right| \leq \varepsilon \sum_{x_k \in D} |C_k| \leq \varepsilon |A|.$$

Case 4. Conditions (2) and (3) and inequalities (c) and (d) imply

$$\sum_{x_k \in E} |F(C_k)| \leq \sum_{i \geq 1} \left[\sum_{x_k \in E_i^+} |F(C_k)| + \sum_{j=1}^{\infty} \sum_{x_k \in E_{i,j}^0} |F(C_k)| \right]$$

$$\leq \varepsilon \sum_{i \geq 1} \left[2^{-i} c_i^{-1} \sum_{x_k \in E_i^+} d(C_k)^{t_i^+} + \beta \sum_{j=1}^{\infty} j \sum_{x_k \in E_{i,j}^0} d(C_k)^{t_i^0} \right]$$

$$\leq \varepsilon\kappa \sum_{i=1}^{\infty} 2^{-i} + \varepsilon\beta \sum_{i,j=1}^{\infty} 2^{-i-j} = \varepsilon(\kappa + \beta).$$

Since A is the disjoint union of the sets N, M, D, and E, and since $f(x) = 0$ for each $x \in A - D$, the lemma follows by summing up the inequalities established in Cases 1–4. $\square$

Proposition 7.4.3. *Let* $A \in \mathcal{BV}(\mathbb{R}^n)$, *and let* $v : \mathrm{cl}_* A \to \mathbb{R}^n$ *be a bounded admissible vector field. If* $\mathrm{div}_{\mathrm{cl}_* A} v$ *belongs to* $L^1(A)$, *then*

$$\int_A \mathrm{div}_{\mathrm{cl}_* A} v(x)\, dx = \int_{\partial_* A} v \cdot \nu_A \, d\mathcal{H}^{n-1}.$$

PROOF. As Theorem 4.4.2 implies $\partial_*(\mathrm{cl}_* A) = \partial_* A$, no generality is lost by assuming $A = \mathrm{cl}_* A$. Define $f : A \to \mathbb{R}$ as in Lemma 7.4.2, and observe that by our assumptions, $f \in L^1(A)$ and

$$\int_A f(x)\, dx = \int_A \mathrm{div}_A v(x)\, dx.$$

   Case 1. Let $A = \mathrm{cl}_* A$ be a compact BV set. Choose $\varepsilon > 0$, and find $\delta : A \to \mathbb{R}_+$ associated with f and ε according to the Henstock lemma (Lemma 1.2.3). If $P := \{(B_1, x_1), \dots, (B_p, x_p)\}$ is a δ-fine partition for which Lemma 7.4.2 holds, then

$$\left| \int_A \mathrm{div}_A v(x)\, dx - \int_{\partial_* A} v \cdot \nu_A \, d\mathcal{H}^{n-1} \right| \le \left| \int_A f(x)\, dx - \sum_{i=1}^p f(x_i)|B_i| \right|$$

$$+ \left| \sum_{i=1}^p f(x_i)|B_i| - \int_{\partial_* A} v \cdot \nu_A \, d\mathcal{H}^{n-1} \right| < 2\varepsilon.$$

The arbitrariness of ε implies the desired equality.

   Case 2. Let $A = \mathrm{cl}_* A$ be a bounded BV set. Use Theorem 7.1.5 to find a sequence $\{A_k\}$ of compact BV sets such that $A_k \subset A$ and $A_k = \mathrm{cl}_* A_k$ for $k = 1, 2, \dots$, and

$$\lim |A - A_k| = \lim \mathbf{P}(A - A_k) = 0.$$

As $\mathrm{div}_{A_k}(v \restriction A_k)(x) = \mathrm{div}_A v(x)$ for almost all $x \in \mathrm{int}\, A_k$, Case 1 yields

$$\int_{A_k} \mathrm{div}_A v(x)\, dx = \int_{\partial_* A_k} v \cdot \nu_{A_k} \, d\mathcal{H}^{n-1}$$

for $k = 1, 2, \dots$. From Theorem 6.6.1, we obtain

$$\lim \left| \int_{\partial_* A} v \cdot \nu_A \, d\mathcal{H}^{n-1} - \int_{\partial_* A_k} v \cdot \nu_{A_k} \, d\mathcal{H}^{n-1} \right|$$

$$= \lim \left| \int_{\partial_*(A - A_k)} v \cdot \nu_{A - A_k} \, d\mathcal{H}^{n-1} \right|$$

$$\le \|v\|_{L^\infty(A;\mathbb{R}^n)} \lim \mathbf{P}(A - A_k) = 0.$$

The proposition follows from the equality

$$\lim \int_{A_k} \operatorname{div} A(x)\, dx = \int_A \operatorname{div} A(x)\, dx.$$

Case 3. Let $A = \operatorname{cl}_* A$ be an arbitrary BV set. If $C_k := [-k, k]^n$ for $k = 1, 2, \ldots$, then

$$\lim \left| \int_{\partial_*(A-C_k)} v \cdot \nu_{A-C_k}\, d\mathcal{H}^{n-1} \right| \leq \|v\|_{L^\infty(A;\mathbb{R}^n)} \lim \mathbf{P}(A - C_k) = 0$$

by Proposition 4.6.11. Since

$$\lim \int_{A \cap C_k} \operatorname{div}_A v(x)\, dx = \int_A \operatorname{div}_A v(x)\, dx,$$

an application of Case 2 and Theorem 6.6.1 completes the proof. $\qquad\square$

Proposition 7.4.4 (Partition of unity). *Let $\mathcal{O}$ be a family of open subsets of $\mathbb{R}^n$. There are nonnegative functions $\varphi_k \in C_c^\infty(\mathbb{R}^n)$ such that*

   (1) *for each $k \in \mathbb{N}$ there is $O \in \mathcal{O}$ with $\operatorname{spt} \varphi_k \subset O$,*

   (2) $\sum_{k \in \mathbb{N}} \varphi_k(x) = 1$ *for each $x \in \bigcup \mathcal{O}$,*

   (3) *the family $\{\operatorname{spt} \varphi_k : k \in \mathbb{N}\}$ is locally finite.*

PROOF. Let $\Omega = \bigcup \mathcal{O}$. Denote by $\mathcal{U}$ the family of all open balls $U(x, r)$ such that $x \in \Omega \cap \mathbb{Q}^n$, $r \in \mathbb{Q}_+$, and $U(x, r) \subset O$ for some $O \in \mathcal{O}$. Enumerate $\mathcal{U}$ as $\{U_k : k \in \mathbb{N}\}$. If $U_k = U(x, r)$, let $V_k := U(x, r/2)$ and $B_k := B(x, 2r/3)$. Mollifying the indicator χ_{B_k} of B_k, we obtain a function $\psi_k \in C_c^\infty(\mathbb{R}^n)$ such that $\chi_{V_k} \leq \psi_k \leq 1$ and $\operatorname{spt} \psi_k \subset U_k$. Let $\varphi_1 := \psi_1$, and for $k = 1, 2, \ldots$, define $\varphi_{k+1} \in C_c^\infty(\mathbb{R}^n)$ by the formula

$$\varphi_{k+1} := (1 - \psi_1) \cdots (1 - \psi_k) \psi_{k+1}. \qquad (*)$$

Since $\operatorname{spt} \varphi_k \subset U_k$ assertion (1) is proved. Moreover, for each $k \in \mathbb{N}$,

$$\varphi_1 + \cdots + \varphi_k = 1 - (1 - \psi_1) \cdots (1 - \psi_k). \qquad (**)$$

Indeed $(**)$ is true for $k = 1$, and assuming it holds for $k \in \mathbb{N}$, then adding $(*)$ and $(**)$ shows that $(**)$ holds for $k + 1$. If $x \in V_k$ then $\psi_k(x) = 1$, and

$$\varphi_1(x) + \cdots + \varphi_j(x) = 1 \qquad (\dagger)$$

for each integer $j \geq k$ by $(**)$. In particular, $\varphi_j(x) = 0$ when $x \in V_k$ and $j > k$; this fact is also implied by $(*)$. Since V_k is an open set,

$$\operatorname{spt} \varphi_j \cap V_k = \emptyset \quad \text{for each } j > k. \qquad (\dagger\dagger)$$

Let $x \in \Omega$. Then $x \in O$ for some $O \in \mathcal{O}$, and there is $r \in \mathbb{Q}_+$ with $B(x, r) \subset O$. Find $y \in U(x, r/3) \cap \mathbb{Q}^n$, and note $U(y, 2r/3) \subset O$. Thus $U(y, 2r/3) = U_{k_x}$ for an integer $k_x \geq 1$. As $V_{k_x} = U(y, r/3)$, we see that $x \in V_{k_x}$, and $(\dagger)$ implies $\sum_{k \in \mathbb{N}} \varphi_k(x) = 1$. Since V_{k_x} is a neighborhood of x, assertion (3) follows from identity $(\dagger\dagger)$. $\qquad\square$

A family $\{\varphi_k : k \in \mathbb{N}\}$ of nonnegative functions defined on $\mathbb{R}^n$ which satisfies conditions (1)–(3) of Proposition 7.4.4 is called a *partition of unity* subordinated to $\mathcal{O}$. Our proof follows that of [64, Theorem 6.20]; for Lipschitz partitions of unity, it applies to any separable metric space.

Theorem 7.4.5. *Let $A \in \mathcal{BV}_{\mathrm{loc}}(\mathbb{R}^n)$, and let $v : \mathrm{cl}_* A \to \mathbb{R}^n$ be a bounded admissible vector field. Then*

$$\int_A \mathrm{div}_{\mathrm{cl}_* A} v(x) \, dx = \int_{\partial_* A} v \cdot \nu_A \, d\mathcal{H}^{n-1} \tag{7.4.1}$$

whenever $\mathrm{div}_{\mathrm{cl}_* A} v \in L^1(A)$ and $v \restriction \partial_* A \in L^1(\partial_* A, \mathcal{H}^{n-1}; \mathbb{R}^n)$.

PROOF. If $\mathrm{spt}\, v$ is compact, choose an open ball U with $\mathrm{spt}\, v \subset U$. Equality (7.4.1) holds trivially when A is replaced by $A - U$, and when A is replaced by $A \cap U$, it holds according to Proposition 7.4.3. The theorem follows from Theorem 6.6.1.

If v is arbitrary, cover $\mathrm{cl}_* A$ by a family $\mathcal{U}$ of open balls, and choose a partition of unity $\{\varphi_k : k \in \mathbb{N}\} \subset C_c^\infty(\mathbb{R}^n)$ subordinated to $\mathcal{U}$. If $k \in \mathbb{N}$ and $v_k := (\varphi_k \restriction \mathrm{cl}_* A)v$, then

$$\int_A \mathrm{div}_{\mathrm{cl}_* A} v_k(x) \, dx = \int_{\partial_* A} v_k \cdot \nu_A \, d\mathcal{H}^{n-1} \tag{$*$}$$

by Proposition 2.3.5 and the first part of the proof. Since the family $\{\mathrm{spt}\, \varphi_k\}$ is locally finite and $\sum_{k \in \mathbb{N}} v_k = v$, we have

$$\mathrm{div}_{\mathrm{cl}_* A} v(x) = \mathrm{div}_{\mathrm{cl}_* A}\left[\sum_{k \in \mathbb{N}} v_k(x)\right] = \sum_{k \in \mathbb{N}} \mathrm{div}_{\mathrm{cl}_* A} v_k(x)$$

at each $x \in \mathrm{int}_* A$ at which v is relatively differentiable. If $\mathrm{div}_{\mathrm{cl}_* A} v$ and $v \restriction \partial_* A$ belong, respectively, to $L^1(A)$ and $L^1(\partial_* A, \mathcal{H}^{n-1}; \mathbb{R}^n)$, the theorem follows by summing up equalities $(*)$.   □

7.5. Lipschitz domains

We show that BV functions in Lipschitz domains have boundary values. Using this result, we prove the divergence theorem for vector fields whose components are BV functions.

Proposition 7.5.1. *Let $\Omega \subset \mathbb{R}^n$ be an open set, and let $f, f_k \in BV(\Omega)$, $k = 1, 2, \ldots$. If $\lim \|f_k - f\|_{L^1(\Omega)} = 0$ and $\lim \|Df_k\|(\Omega) = \|Df\|(\Omega)$, then*

$$\lim \int_\Omega v \cdot Df_k = \int_\Omega v \cdot d(Df)$$

for each bounded $v \in C(\Omega; \mathbb{R}^n)$.

PROOF. Choose $\varepsilon > 0$, and find an open set $U \Subset \Omega$ so that $\|Df\|(B) \leq \varepsilon$ for $B := \Omega - U$. Then $\limsup \|Df_k\|(B) \leq \|Df\|(B) < \varepsilon$ by Proposition 5.5.4.

Assume first that $v \in C^1(\Omega; \mathbb{R}^n)$ and $\beta = \|v\|_{L^\infty(\Omega;\mathbb{R}^n)} < \infty$. Choose a function $\varphi \in C_c^1(\mathbb{R}^n)$ so that $\chi_U \leq \varphi \leq 1$ and $\mathrm{spt}\,\varphi \subset \Omega$, and calculate:

$$\int_\Omega v \cdot Df_k = \int_\Omega \varphi v \cdot Df_k + \int_\Omega (1 - \varphi) v \cdot Df_k$$

$$= -\int_\Omega f_k \,\mathrm{div}\,(\varphi v) + \int_B (1 - \varphi) v \cdot Df_k$$

$$\leq -\int_\Omega f_k \,\mathrm{div}\,(\varphi v) + \beta \|Df_k\|(B),$$

$$-\lim \int_\Omega f_k \,\mathrm{div}\,(\varphi v) = -\int_\Omega f \,\mathrm{div}\,(\varphi w) = \int_\Omega \varphi w \cdot d(Df)$$

$$= \int_\Omega w \cdot d(Df) + \int_B (\varphi - 1) v \cdot d(Df)$$

$$\leq \int_\Omega v \cdot d(Df) + \beta \|Df\|(B).$$

Combining these inequalities with $(*)$ and $(**)$ implies

$$\lim \int_\Omega v \cdot Df_k \leq \int_\Omega v \cdot d(Df) + 2\beta\varepsilon.$$

As the same inequality holds for $-v$, the arbitrariness of ε yields

$$\lim \int_\Omega v \cdot Df_k = \int_\Omega v \cdot d(Df).$$

If $v \in C(\Omega; \mathbb{R}^n)$ is bounded, use Corollary 1.1.2 to extend $v \upharpoonright \mathrm{cl}\,U$ to a vector field $u \in C_c(\mathbb{R}^n; \mathbb{R}^n)$, and let

$$\beta = \max\{\|v\|_{L^\infty(\Omega;\mathbb{R}^n)}, \|u\|_{L^\infty(\mathbb{R}^n;\mathbb{R}^n)}\}.$$

A mollification of u provides $w \in C^1(\mathbb{R}^n; \mathbb{R}^n)$ such that $\|w\|_{L^\infty(\mathbb{R}^n;\mathbb{R}^n)} \leq \beta$ and $\|v - w\|_{L^\infty(U;\mathbb{R}^n)} < \varepsilon$. We obtain

$$\left| \int_\Omega v \cdot d(Df) - \int_\Omega w \cdot d(Df) \right|$$

$$\leq \int_U |v - w| \, d(\|Df\|) + \int_B |v - w| \, d(\|Df\|)$$

$$\leq \varepsilon \|Df\|(\Omega) + 2\beta \|Df\|(B) \leq \varepsilon (\|Df\|(\Omega) + 2\beta),$$

and by a similar calculation,

$$\lim \left| \int_\Omega v \cdot Df_k - \int_\Omega w \cdot Df_k \right| \leq \varepsilon (\|Df\|(\Omega) + 2\beta).$$

The proposition follows from the arbitrariness of ε, since

$$\lim \int_\Omega w \cdot Df_k = \int_\Omega w \cdot d(Df)$$

by the first part of the proof applied to $w \upharpoonright \Omega$.   □

Lemma 7.5.2. *Let $\Omega \subset \mathbb{R}^n$ be an open set. If $f \in BV(\Omega)$, then*

$$\lim_{r \to 0} \frac{\|Df\|[\Omega \cap B(x,r)]}{r^{n-1}} = 0$$

for $\mathcal{H}^{n-1}$ almost all $x \in \partial\Omega$.

PROOF. For $\beta > 0$, let A_β be the set of all $x \in \partial\Omega$ such that

$$\limsup_{r \to 0} \frac{\|Df\|[\Omega \cap B(x,r)]}{r^{n-1}} > \beta \,.$$

Given $\delta > 0$ and $x \in A_\beta$, find $0 < r_x < \delta$ with

$$\|Df\|[\Omega \cap B(x, r_x)] > \beta \, r_x^{n-1}.$$

By Vitali's theorem there are $x_i \in A_\beta$ such that the balls $B(x_i, r_{x_i})$ are disjoint and $A_\beta \subset \bigcup_i B(x_i, 5r_{x_i})$. Thus

$$\mathcal{H}_{10\delta}^{n-1}(A_\beta) \leq \sum_i \alpha(n-1)(5r_{x_i})^{n-1}$$

$$\leq 5^{n-1}\beta^{-1}\alpha(n-1) \sum_i \|Df\|[\Omega \cap B(x_i, r_{x_i})] \leq \gamma \|Df\|(\Omega_\delta)$$

where $\gamma = 5^{n-1}\beta^{-1}\alpha(n-1)$ and$\Omega_\delta = \Omega \cap B(\partial\Omega, \delta)$. As $\|Df\|(\Omega) < \infty$ and $\bigcap_{\delta > 0} \Omega_\delta = \emptyset$, we infer

$$\mathcal{H}^{n-1}(A_\beta) = \lim_{\delta \to 0} \mathcal{H}_{10\delta}^{n-1}(A_\beta) \leq \gamma \lim_{\delta \to 0} \|Df\|(\Omega_\delta) = 0.$$

The lemma follows, since

$$\left\{ x \in \partial\Omega : \limsup_{r \to 0} \frac{\|Df\|[\Omega \cap B(x,r)]}{r^{n-1}} > 0 \right\} = \bigcup_{k=1}^{\infty} A_{1/k}. \qquad \square$$

Theorem 7.5.3. *Let $\Omega \subset \mathbb{R}^n$ be a Lipschitz domain. If $f \in BV_{\mathrm{loc}}(\Omega)$, then*

$$Tf : x \mapsto \lim_{r \to 0} \frac{1}{|\Omega \cap B(x,r)|} \int_{\Omega \cap B(x,r)} f(y) \, dy : \partial\Omega \to \mathbb{R} \qquad (7.5.1)$$

is defined for $\mathcal{H}^{n-1}$ almost all $x \in \partial\Omega$ and belongs to $L^1_{\mathrm{loc}}(\partial\Omega, \mathcal{H}^{n-1})$. If $f \in BV(\Omega)$ and $\partial\Omega$ is compact, then

$$\int_{\partial\Omega} (Tf)v \cdot \nu_\Omega \, d\mathcal{H}^{n-1} = \int_\Omega f(x) \operatorname{div} v(x) \, dx + \int_\Omega v \cdot d(Df) \qquad (7.5.2)$$

for each continuous $v \in Adm(\mathrm{cl}\,\Omega; \mathbb{R}^n)$ with $\operatorname{div} v \in L^\infty(\Omega)$.

PROOF. For $z \in \mathbb{R}^n$, let $z' = \pi_n(z)$ and $z_n = z \cdot e_n$. Fix $x \in \partial\Omega$. After a suitable rotation about x, there are an open cylinder

$$C(x; r, h) = \pi_n\big[U(x, r)\big] \times (x_n - h, x_n + h)$$

and a Lipschitz function $g : \mathbb{R}^{n-1} \to \mathbb{R}$ such that

$$\Omega \cap C(x; r, h) = \big\{y \in C(x; r, h) : g(y') < y_n\big\}.$$

Letting $L := \max\{1, \operatorname{Lip} g\}$, and making r smaller, we may assume $h = 2Lr$. If $C_{x,r} := C(x; r, 2Lr)$ and $U_{x,r} := \pi_n\big[U(x, r)\big]$, then

$$\big|g(y') - x_n\big| = \big|g(y') - g(x')\big| \le (\operatorname{Lip} g)|y' - x'| < Lr$$

for each $y' \in U_{x,r}$. It follows that for each $0 \le s < t \le r$, the intersection $\Omega \cap C_{x,r}$ contains the open set

$$C_{x,r}(s, t) := \big\{y \in C_{x,r} : g(y') + s < y_n < g(y') + t\big\}.$$

We employ the Lipschitz maps

$$\phi : y' \mapsto (y', g(y')) : U_{x,r} \to \partial\Omega \cap C_{x,r},$$

$$\psi : (y', t) \mapsto (y', g(y') + t) : U_{x,r} \times (0, r) \to \Omega \cap C_{x,r}(0, r),$$

$$\sigma : (y, t) \mapsto (y', y_n + t) : (\partial\Omega \cap C_{x,r}) \times (0, r) \to \Omega \cap C_{x,r}(0, r),$$

and for $0 < t \le r$ define a translation

$$\sigma^t : y \mapsto (y', y_n + t) : \partial\Omega \cap C_{x,r} \to \Omega \cap C_{x,r}(0, r).$$

Note that ϕ and ψ are surjective lipeomorphisms whose Jacobians satisfy $1 \le J_\phi \le \beta$ where $\beta = \sqrt{1 + (\operatorname{Lip} g)^2}$, and $J_\psi = J_{\psi^{-1}} = 1$. The diagram

$$
\begin{array}{ccc}
U_{x,r} \times (0, r) & & \\
{\scriptstyle \phi \times \mathrm{id}}\big\downarrow & \searrow^{\psi} & \\
(\partial\Omega \cap C_{x,r}) \times (0, r) & \xrightarrow{\;\;\sigma\;\;} & \Omega \cap C_{x,r}(0, r)
\end{array}
$$

commutes. Given $u : \Omega \to \mathbb{R}$ and $0 < t < r$, let $u^t := u \circ \sigma^t$; explicitly

$$u^t : y \mapsto u\big[y', g(y') + t\big] : \partial\Omega \cap C_{x,r} \to \mathbb{R}.$$

PART 1. $f \in C^1(\Omega) \cap BV(\Omega)$.

Case 1a. Let $0 < s < t < r$, and observe that

$$
\begin{aligned}
\big|f^s(y) - f^t(y)\big| &= \big|f\big[y', g(y') + s\big] - f\big[y', g(y') + t\big]\big| \\
&= \left|\int_s^t \frac{\partial f}{\partial x_n}\big[y', g(y') + \tau\big]\, d\tau\right| \\
&\le \int_s^t \big|Df\big[y', g(y') + \tau\big]\big|\, d\tau = \int_s^t \big|(Df)^\tau(y)\big|\, d\tau
\end{aligned}
$$

for each $y \in \partial\Omega \cap C_{x,r}$. In view of [29, Section 3.3.3, Theorem 2] and Fubini's theorem, the preceding inequality implies

$$
\begin{aligned}
\|f^s - f^t\|_{L^1(\partial\Omega \cap C_{x,r}, \mathcal{H}^{n-1})} & \\
= \int_{\partial\Omega \cap C_{x,r}} & \left| f^s(y) - f^t(y) \right| d\mathcal{H}^{n-1}(y) \\
\leq \int_{\partial\Omega \cap C_{x,r}} & \left(\int_s^t \left| (Df)^\tau(y) \right| d\tau \right) d\mathcal{H}^{n-1}(y) \\
= \int_{U_{x,r}} & \left(\int_s^t \left| (Df)^\tau [\phi(y')] \right| d\tau \right) J_\phi(y') \, dy' \qquad (*) \\
\leq \beta \int_{U_{x,r} \times (s,t)} & \left| Df[\psi(y', \tau)] \right| dy' \, d\tau \\
= \beta \int_{C_{x,r}(s,t)} & \left| Df(y) \right| d(y) = \beta \|Df\| \left[C_{x,r}(s,t) \right].
\end{aligned}
$$

Since $\bigcap_{0<s<t<r} C_{x,r}(s,t) = \emptyset$, and since $L^1(\partial\Omega \cap C_{x,r}, \mathcal{H}^{n-1})$ is a complete space, there is $T_x f \in L^1(\partial\Omega \cap C_{x,r}, \mathcal{H}^{n-1})$ which satisfies

$$
\lim_{t \to 0} \|T_x f - f^t\|_{L^1(\partial\Omega \cap C_{x,r}, \mathcal{H}^{n-1})} = 0. \qquad (**)
$$

Moreover, fixing $0 < t < r$ and letting $s \to 0$ in $(*)$, we obtain

$$
\|T_x f - f^t\|_{L^1(\partial\Omega \cap C_{x,r}, \mathcal{H}^{n-1})} \leq \beta \|Df\| \left[C_{x,r}(0,t) \right]. \qquad (* * *)
$$

Case 1b. Given $\rho > 0$, define

$$
\rho' = \rho\sqrt{1 + 4L^2} \quad \text{and} \quad \rho'' = \frac{\rho}{L+1}.
$$

Fix a point $z \in C_{x,r} \cap \partial\Omega$, and for $s > 0$ let $B_s = B(z, s)$. Select $\rho > 0$ so that $B_{\rho'} \subset C_{x,r}$. Note ρ' is the least radius such that $C_{z,\rho} \subset B_{\rho'}$, and observe

$$
\begin{aligned}
\Omega \cap B_{\rho''} \subset C_{z,\rho}(0, \rho) \subset \Omega \cap C_{z,\rho} \subset \Omega \cap B_{\rho'}, & \\
\partial\Omega \cap C_{z,\rho} \subset \partial\Omega \cap B_{\rho'}. & \qquad (\dagger)
\end{aligned}
$$

If $0 < s < t < \rho$, then calculating as in $(*)$,

$$
\|f^s - f^t\|_{L^1(\partial\Omega \cap C_{z,\rho}, \mathcal{H}^{n-1})} \leq \beta \|Df\| \left[C_{z,\rho}(s,t) \right].
$$

In view of $(\dagger)$, fixing t so that $0 < t < \rho$, and letting $s \to 0$, we obtain

$$
\|T_x f - f^t\|_{L^1(\partial\Omega \cap C_{z,\rho}, \mathcal{H}^{n-1})} \leq \beta \|Df\| (\Omega \cap B_{\rho'}). \qquad (\dagger\dagger)
$$

Combining $(\dagger)$, Fubini's theorem, [29, Section 3.3.3, Theorem 2], and $(\dagger\dagger)$,

$$\int_{\Omega \cap B_{\rho''}} \left| T_x f(z) - f(y) \right| dy \le \int_{C_{z,\rho}(0,\rho)} \left| T_x f(z) - f(y) \right| dy$$

$$= \int_{U_{z,\rho}} \left(\int_{g(y')}^{g(y')+\rho} \left| T_x f(z) - f(y',t) \right| dt \right) dy'$$

$$= \int_{U_{z,\rho}} \left(\int_0^\rho \left| T_x f(z) - f\left[y',g(y')+t\right] \right| dt \right) dy'$$

$$\le \int_0^\rho \left(\int_{U_{z,\rho}} \left| T_x f(z) - f^t\left[\phi(y')\right] \right| J_\phi(y') \, dy' \right) dt$$

$$= \int_0^\rho \left(\int_{\partial\Omega \cap C_{z,\rho}} \left| T_x f(z) - f^t(y) \right| d\mathcal{H}^{n-1}(y) \right) dt$$

$$\le \int_0^\rho \left(\int_{\partial\Omega \cap C_{z,\rho}} \left| T_x f(z) - T_x f(y) \right| d\mathcal{H}^{n-1}(y) \right) dt$$

$$+ \int_0^\rho \| T_x f - f^t \|_{L^1(\partial\Omega \cap C_{z,\rho}, \mathcal{H}^{n-1})} \, dt$$

$$\le \rho \int_{\partial\Omega \cap B_{\rho'}} \left| T_x f(z) - T_x f(y) \right| d\mathcal{H}^{n-1}(y)$$

$$+ \rho\beta \| Df \| (\Omega \cap B_{\rho'}).$$

Case 1c. Select a continuous $v \in Adm(\mathrm{cl}\,\Omega; \mathbb{R}^n)$ so that $\mathrm{div}\, v$ belongs to $L^\infty(\Omega)$ and $\mathrm{spt}\, v \subset C_{x,r}$. Choose $0 < t < r$ and let

$$C^t = \Omega \cap C_{x,r} - C_{x,r}(0,t).$$

Now $fv \upharpoonright C^t$ is continuous, bounded, and admissible, and $\mathrm{div}(fv)$ belongs to $L^1(\Omega)$. By Proposition 7.4.3,

$$\int_{C^t} \mathrm{div}(fv) = \int_{\partial C^t} (fv) \cdot \nu_{C^t} \, d\mathcal{H}^{n-1} = \int_{\sigma^t(\partial\Omega \cap C_{x,r})} (fv) \cdot \nu_{C^t} \, d\mathcal{H}^{n-1}$$

$$= \int_{\partial\Omega \cap C_{x,r}} (fv)^t \nu_\Omega \, d\mathcal{H}^{n-1}.$$

The last equality holds, since $(fv)^t = (fv) \circ \sigma^t$ where σ^t is the translation by t in the direction e_n. Because v is continuous, $\lim_{t\to 0} v^t(y) = v(y)$ for all $y \in \partial\Omega \cap C_{x,r}$. Letting $t \to 0$, equality $(**)$ and the generalized dominated convergence theorem [29, Section 1.3, Theorem 4] yield

$$\int_{\Omega \cap C_{x,r}} \mathrm{div}(fv) = \int_{\partial\Omega \cap C_{x,r}} (T_x f) v \cdot \nu_\Omega \, d\mathcal{H}^{n-1}.$$

As $\mathrm{spt}\, v \subset C_{x,r}$, the previous equality expands to

$$\int_\Omega f \, \mathrm{div}\, v + \int_\Omega v \cdot Df = \int_{\partial\Omega} (T_x f) v \cdot \nu_\Omega \, d\mathcal{H}^{n-1}.$$

PART 2. $f \in BV_{\mathrm{loc}}(\Omega)$.

Case 2a. By Theorem 5.6.3 there is a sequence $\{f_k\}$ in $C^1(\Omega)$ such that $\lim \|f_k - f\|_{L^1(\Omega)} = 0$ and $\lim \|Df_k\|(\Omega) = \|Df\|(\Omega)$. Fix $0 < \varepsilon < r$, and for each $y \in \partial\Omega \cap C_{x,r}$, define

$$F_k(y) := \frac{1}{\varepsilon} \int_0^\varepsilon (f_k)^t(y)\, dt.$$

By Fubini's theorem and inequality $(\ast\ast\ast)$,

$$\int_{\partial\Omega \cap C_{x,r}} \left| T_x f_k(y) - F_k(y) \right| d\mathcal{H}^{n-1}(y)$$

$$= \int_{\partial\Omega \cap C_{x,r}} \left| \frac{1}{\varepsilon} \int_0^\varepsilon [T_x f_k(y) - (f_k)^t(y)]\, dt \right| d\mathcal{H}^{n-1}(y)$$

$$\leq \frac{1}{\varepsilon} \int_0^\varepsilon \|T_x f_k - (f_k)^t\|_{L^1(\partial\Omega \cap C_{x,r}, \mathcal{H}^{n-1})}\, dt$$

$$\leq \frac{\beta}{\varepsilon} \int_0^\varepsilon \|Df_k\| [C_{x,r}(0,t)]\, dt \leq \beta \|Df_k\| [C_{x,r}(0,\varepsilon)].$$

In addition, [29, Section 3.3.3, Theorem 2] and Fubini's theorem yield

$$\int_{\partial\Omega \cap C_{x,r}} \left| F_k(y) - F_j(y) \right| d\mathcal{H}^{n-1}(y)$$

$$= \frac{1}{\varepsilon} \int_{\partial\Omega \cap C_{x,r}} \left(\int_0^\varepsilon \left| (f_k)^t(y) - (f_j)^t(y) \right| dt \right) d\mathcal{H}^{n-1}(y)$$

$$= \frac{1}{\varepsilon} \int_{U_{x,r}} \left(\int_0^\varepsilon \left| (f_k)^t[\phi(y')] - (f_j)^t[\phi(y')] \right| dt \right) J_\phi(y')\, dy'$$

$$\leq \frac{\beta}{\varepsilon} \int_{U_{x,r} \times (0,\varepsilon)} \left| f_k[\psi(y',t)] - f_j[\psi(y',t)] \right| dy'\, dt$$

$$= \frac{\beta}{\varepsilon} \int_{C_{x,r}(0,\varepsilon)} \left| f_k(y) - f_j(y) \right| dy \leq \frac{\beta}{\varepsilon} \|f_k - f_j\|_{L^1(\Omega)}.$$

Combining the previous two inequalities with

$$|T_x f_k - T f_j| \leq |T_x f_k - F_k| + |F_k - F_j| + |F_j - T_x f_j|,$$

we obtain

$$\|T_x f_k - T_x f_j\|_{L^1(\partial\Omega \cap C_{x,r}, \mathcal{H}^{n-1})}$$

$$\leq \beta \|Df_k\| [C_{x,r}(0,\varepsilon)] + \frac{\beta}{\varepsilon} \|f_k - f_j\|_{L^1(\Omega)} + \beta \|Df_j\| [C_{x,r}(0,\varepsilon)]$$

for $k, j = 1, 2, \ldots$ and $0 < \varepsilon < r$. Now choose $\eta > 0$. Since

$$\bigcap \{\Omega \cap \mathrm{cl}\, C_{x,r}(0,\varepsilon) : 0 < \varepsilon < r\} = \emptyset,$$

there is $0 < \varepsilon < r$ with $\|Df\| \left[\Omega \cap \operatorname{cl} C_{x,r}(0, \varepsilon)\right] < \eta$. Applying Proposition 5.5.4, find $p \in \mathbb{N}$ so that $\|Df_k\| \left[\Omega \cap \operatorname{cl} C_{x,r}(0, \varepsilon)\right] < \eta$ for each integer $k \geq p$. Making the integer p larger, we may assume $\|f_k - f_j\|_{L^1(\Omega)} \leq \varepsilon \eta$ whenever $k, j \geq p$. Thus for all integers $k, j \geq p$,

$$\|T_x f_k - T_x f_j\|_{L^1(\partial\Omega \cap C_{x,r}, \mathcal{H}^{n-1})} \leq 3\beta\eta.$$

As η is arbitrary, there is $T_x f \in L^1(\partial\Omega \cap C_{x,r}, \mathcal{H}^{n-1})$ such that

$$\lim \|T_x f_k - T_x f\|_{L^1(\partial\Omega \cap C_{x,r}, \mathcal{H}^{n-1})} = 0. \tag{†††}$$

Case 2b. Choose $z \in \partial\Omega \cap C_{x,r}$ so that the following holds

$$T_x f(z) = \lim T_x f_k(z), \tag{1}$$

$$\lim_{s \to 0} \frac{|\Omega \cap B_s|}{s^n} = \frac{1}{2}\alpha(n), \tag{2}$$

$$\lim_{s \to 0} \frac{\mathcal{H}^{n-1}(\partial\Omega \cap B_s)}{s^{n-1}} = \alpha(n - 1), \tag{3}$$

$$\lim_{s \to 0} \frac{\|Df\|(\Omega \cap B_s)}{s^{n-1}} = 0, \tag{4}$$

$$\lim_{s \to 0} \frac{1}{\mathcal{H}^{n-1}(B_s \cap \partial\Omega)} \int_{B_s \cap \partial\Omega} |T_x f(z) - Tf(y)| \, d\mathcal{H}^{n-1}(y) = 0. \tag{5}$$

In view of (†††), Corollary 6.4.4, and Lemma 7.5.2, equalities (1)–(4) hold for $\mathcal{H}^{n-1}$ almost all $z \in \partial\Omega \cap C_{x,r}$. The same is true for equality (5). Indeed, applying Corollary 6.2.4 to the Radon measure $\mathcal{H}^{n-1} \llcorner \partial\Omega$ and to the zero extension $\overline{T_x f}$ of $T_x f$, we infer that $\mathcal{H}^{n-1}$ almost all $z \in \partial\Omega \cap C_{x,r}$ are Lebesgue points of $T_x f$.

Define ρ, ρ', and ρ'' so that conditions (†) are satisfied. By Case 1b,

$$\int_{\Omega \cap B_{\rho''}} |T_x f_k(z) - f_k(y)| \, dy \leq \rho \int_{\partial\Omega \cap B_{\rho'}} |T_x f_k(z) - T_x f_k(y)| \, d\mathcal{H}^{n-1}(y)$$

$$+ \rho\beta \|Df_k\|(\Omega \cap B_{\rho'})$$

for $k = 1, 2, \ldots$. If $k \to \infty$, condition (1) and Proposition 5.5.4 yield

$$\int_{\Omega \cap B_{\rho''}} |T_x f(z) - f(y)| \, dy \leq \rho \int_{\partial\Omega \cap B_{\rho'}} |T_x f(z) - T_x f(y)| \, d\mathcal{H}^{n-1}(y)$$

$$+ \rho\beta \|Df\|(\Omega \cap B_{\rho'}).$$

Using this inequality and equalities (2)–(5), we calculate

$$\lim_{\rho'' \to 0} \frac{1}{|\Omega \cap B_{\rho''}|} \int_{\Omega \cap B_{\rho''}} |T_x f(z) - f(y)| \, dy = 0$$

for $\mathcal{H}^{n-1}$ almost all $z \in \partial\Omega \cap C_{x,r}$. For all such points z,

$$T_x f(z) = \lim_{s \to 0} \frac{1}{|\Omega \cap B_s|} \int_{\Omega \cap B_s} f(y) \, dy$$

by Remark 6.2.5. As the right side of the previous equality does not depend on x, neither does $T_x f(z)$. Hence we write Tf instead of $T_x f$, and the first part of the theorem is proved.

Case 2c. Assume that $f \in L^1(\Omega)$ and $\partial\Omega$ is compact. Then Tf belongs to $L^1(\partial\Omega, \mathcal{H}^{n-1})$. Select $v \in C(\operatorname{cl}\Omega; \mathbb{R}^n) \cap Adm(\operatorname{cl}\Omega; \mathbb{R}^n)$ such that $\operatorname{div} v$ belongs to $L^\infty(\Omega)$, and observe that $f \operatorname{div} v$, v, and $(Tf) v \cdot \nu_\Omega$ belong to $L^1(\Omega)$, $L^1(\Omega, Df; \mathbb{R}^n)$, and $L^1(\partial\Omega, \mathcal{H}^{n-1})$, respectively. If some $C_{x,r}$ contains $\operatorname{spt} v$, then by Case 1c,

$$\int_\Omega f_k \operatorname{div} v + \int_\Omega v \cdot Df_k = \int_{\partial\Omega} (Tf_k) v \cdot \nu_\Omega \, d\mathcal{H}^{n-1}$$

for $k = 1, 2, \dots$. In view of Theorem 2.3.9, this is still true if $\operatorname{spt} v \Subset \Omega$. We infer from Case 2a and Proposition 7.5.1 that

$$\int_\Omega f \operatorname{div} v + \int_\Omega v \cdot d(Df) = \int_{\partial\Omega} (Tf) v \cdot \nu_\Omega \, d\mathcal{H}^{n-1}$$

whenever $\operatorname{spt} v \Subset \Omega$ or $\operatorname{spt} v \subset C_{x,r}$ for some $x \in \partial\Omega$. Note that we must use Proposition 7.5.1; as $\operatorname{spt} v \subset C_{x,r}$ does not imply $v \in C_c(\Omega; \mathbb{R}^n)$, we cannot use Proposition 5.5.4.

Let $\{\varphi_j : j \in \mathbb{N}\} \subset C_c^\infty(\mathbb{R}^n)$ be a partition of unity subordinated to the family $\{\Omega\} \cup \{C_{x,r_x} : x \in \partial\Omega\}$ of open sets; see Proposition 7.4.4. Each $v_j = v(\varphi_j \restriction \operatorname{cl}\Omega)$ is admissible by Proposition 2.3.5, and either $\operatorname{spt} v_j \subset C_{x,r_x}$ for some $x \in \partial\Omega$, or $\operatorname{spt} v_j \Subset \Omega$. Thus

$$\int_\Omega f \operatorname{div} v_j + \int_\Omega v_j \cdot d(Df) = \int_{\partial\Omega} (Tf) v_j \cdot \nu_\Omega \, d\mathcal{H}^{n-1}$$

for $j = 1, 2, \dots$. Since $v(x) = \sum_{j \in \mathbb{N}} v_j(x)$ for each $x \in \operatorname{cl}\Omega$, and since $\operatorname{div} v(x) = \sum_{j \in \mathbb{N}} \operatorname{div} v_j(x)$ for each $x \in \Omega$ at which v is differentiable, equality (7.5.2) follows:

$$\int_\Omega f \operatorname{div} v + \int_\Omega v \cdot d(Df) = \sum_{j \in \mathbb{N}} \left(\int_\Omega f \operatorname{div} v_j + \int_\Omega v_j \cdot d(Df) \right)$$

$$= \sum_{j \in \mathbb{N}} \left(\int_{\partial\Omega} (Tf) v_j \cdot \nu_\Omega \, d\mathcal{H}^{n-1} \right)$$

$$= \int_\Omega (Tf) v \cdot \nu_\Omega \, d\mathcal{H}^{n-1}. \qquad \square$$

The function Tf defined in Theorem 7.5.3 is called the *trace* of f. We view Tf as the boundary value of f.

Lemma 7.5.4. *Let $\Omega \subset \mathbb{R}^n$ be an open set, $f \in BV(\Omega)$, and $\varphi \in C^1(\Omega)$. If φ and $D\varphi$ are bounded, then $f\varphi \in BV(\Omega)$ and*

$$\|D(f\varphi)\|(\Omega) \leq \|\varphi\|_{L^\infty(\Omega)}\|Df\|(\Omega) + \|f\|_{L^1(\Omega)}\|D\varphi\|_{L^\infty(\Omega)}.$$

PROOF. By Theorem 5.6.3, there is a sequence $\{f_i\}$ in $C^\infty(\Omega) \cap BV(\Omega)$ such that $\lim \|f_i - f\|_{L^1(\Omega)} = 0$ and $\lim \|Df_i\|(\Omega) = \|Df\|(\Omega)$. It follows that $\lim \|f_i\varphi - f\varphi\|_{L^1(\Omega)} = 0$, and Proposition 5.1.7 implies

$$V(f\varphi,\Omega) \leq \liminf \|D(f_i\varphi)\|(\Omega) = \liminf \int_\Omega |\varphi Df_i + f_i D\varphi|$$

$$\leq \liminf \left[\|\varphi\|_{L^\infty(\Omega)}\|Df_i\|(\Omega) + \|f_i\|_{L^1(\Omega)}\|D\varphi\|_{L^\infty(\Omega)}\right]$$

$$= \|\varphi\|_{L^\infty(\Omega)}\|Df\|(\Omega) + \|f\|_{L^1(\Omega)}\|D\varphi\|_{L^\infty(\Omega)} < \infty.$$

In particular, $f\varphi \in BV(\Omega)$. $\square$

Theorem 7.5.5. *Let $\Omega \subset \mathbb{R}^n$ be a Lipschitz domain with compact boundary. There is a constant $\gamma > 0$ such that*

$$\|Tf\|_{L^1(\partial\Omega, \mathcal{H}^{n-1})} \leq \gamma\|f\|_{BV(\Omega)}$$

for each $f \in BV(\Omega)$.

PROOF. Since $|Tf| \leq T|f|$, we may assume $f \geq 0$. Choose $x \in \partial\Omega$, and find a cylinder $C_x = C(x; r, h)$ and a Lipschitz function $g : \mathbb{R}^{n-1} \to \mathbb{R}$ such that

$$\Omega \cap C_x = \{y \in C_x : g(y') < y_n\}$$

where $y_n = y \cdot e_n$ and $y' = \pi_n(y)$. Let $\beta := \sqrt{1 + (\text{Lip } g)^2}$, and deduce from Corollary 6.4.4, (1) that $-e_n \cdot \nu_\Omega(y) \geq 1/\beta$ for $\mathcal{H}^{n-1}$ almost all $y \in \partial\Omega \cap C_x$. If spt $f \subset C_x$, then spt $Tf \subset \partial\Omega \cap C_x$ by (7.5.1). Thus applying equality (7.5.2) to f and $v = -e_n$, we obtain

$$\int_{\partial\Omega} Tf \, d\mathcal{H}^{n-1} \leq \beta \int_{\partial\Omega} (Tf)(-e_n) \cdot \nu_\Omega \, d\mathcal{H}^{n-1}$$

$$= \beta \int_\Omega (-e_n) \cdot d(Df) \leq \beta\|Df\|(\Omega).$$

If $f \in BV(\Omega)$ is arbitrary, let $\{\varphi_k : k \in \mathbb{N}\} \subset C_c^\infty(\mathbb{R}^n)$ be a partition of unity subordinated to the family $\{C_x : x \in \partial\Omega\}$. For $k = 1, 2, \ldots,$

$$\int_{\partial\Omega} T(\varphi_k f) \, d\mathcal{H}^{n-1} \leq \beta\|D(\varphi_k f)\|(\Omega)$$

$$\leq \|Df\|(\Omega) + \|f\|_{L^1(\Omega)}\|D\varphi_k\|_{L^\infty(\Omega)} \tag{$*$}$$

by Lemma 7.5.4 and the first part of the proof. Since $\partial\Omega$ is compact, and since the family $\{\text{spt } \varphi_k : k \in \mathbb{N}\}$ is locally finite, there is $p \in \mathbb{N}$ and a neighborhood U of $\partial\Omega$ such that spt $\varphi_k \cap U = \emptyset$ for all integers $k > p$. By equality (7.5.1),

$$g \mapsto Tg : BV(\Omega) \to L^1(\partial\Omega, \mathcal{H}^{n-1})$$

is a linear map. Thus by inequality $(*)$,

$$\int_{\partial\Omega} Tf \, d\mathcal{H}^{n-1} = \sum_{k=1}^{p} \int_{\partial\Omega} T(\varphi_k f) \, d\mathcal{H}^{n-1} \le \beta \sum_{i=1}^{p} \|D(\varphi_k f)\|(\Omega)$$

$$\le \beta p \|Df\|(\Omega) + \beta \|f\|_{L^1(\Omega)} \sum_{i=1}^{p} \|D\varphi_i\|_{L^\infty(\Omega)}$$

$$\le \beta \left[p + \sum_{i=1}^{p} \|D\varphi_i\|_{L^\infty(\Omega)} \right] \|f\|_{BV(\Omega)}$$

and $\gamma = \beta \left[p + \sum_{i=1}^{p} \|D\varphi_i\|_{L^\infty(\Omega)} \right]$ is the desired constant. $\square$

The following example shows that in Theorem 7.5.5, the compactness of the boundary $\partial\Omega$ cannot be relaxed to $\mathbf{P}(\Omega) < \infty$.

Example 7.5.6. Assume $n \ge 2$, and let $s = 2/(2n-3)$. Choose a sequence $\{x_k\}$ in $\mathbb{R}^n$ so that $|x_j - x_k| \ge 2$ whenever $j \ne k$, and let $U_k = U(x_k, r_k)$ where $r_k = k^{-s}$. The open balls U_k are disjoint, and $\Omega = \bigcup_{k\in\mathbb{N}} U_k$ is a Lipschitz domain such that $|\Omega| < \infty$ and $\mathbf{P}(\Omega) < \infty$. Letting

$$f = \sum_{k=1}^{\infty} k^s \chi_{U_k},$$

we calculate $\mathbf{V}(f, \Omega) = 0$ and $\|f\|_{L^1(\Omega)} < \infty$. Thus $f \in BV(\Omega)$. On the other hand, from (7.5.1) we see that for almost all $x \in \partial\Omega$,

$$Tf(x) = \sum_{k=1}^{\infty} k^s \chi_{\partial U_k}(x)$$

and consequently $\|Tf\|_{L^1(\partial\Omega, \mathcal{H}^{n-1})} = \infty$.

Proposition 7.5.7. *Let $\Omega \subset \mathbb{R}^n$ be a Lipshitz domain with compact boundary. Assume $f \in BV(\Omega)$ and $g \in BV(\mathbb{R}^n - \mathrm{cl}\,\Omega)$, and let $h : \mathbb{R}^n \to \overline{\mathbb{R}}$ be such that $h \upharpoonright \Omega = f$ and $h \upharpoonright (\mathbb{R}^n - \mathrm{cl}\,\Omega) = g$. Then $h \in BV(\mathbb{R}^n)$ and*

$$\|Dh\|(\mathbb{R}^n) = \|Df\|(\Omega) + \|Dg\|(\mathbb{R}^n - \mathrm{cl}\,\Omega) + \int_{\partial\Omega} |Tf - Tg| \, d\mathcal{H}^{n-1}.$$

PROOF. Note $\mathbb{R}^n - \mathrm{cl}\,\Omega = \mathrm{ext}_*\Omega$. Select $v \in C_c^1(\mathbb{R}^n; \mathbb{R}^n)$, and apply equality (7.5.2) to Ω and $\mathrm{ext}_*\Omega$. As $\partial(\mathrm{ext}_*\Omega) = \partial\Omega$ and $\nu_{\mathrm{ext}_*\Omega} = -\nu_\Omega$, we obtain

$$-\int_{\mathbb{R}^n} h \, \mathrm{div}\, v \, dx = -\int_{\Omega} f \, \mathrm{div}\, v \, dx - \int_{\mathrm{ext}_*\Omega} g \, \mathrm{div}\, v \, dx$$

$$= \int_{\Omega} v \cdot d(Df) - \int_{\partial\Omega} (Tf) \, v \cdot \nu_\Omega \, d\mathcal{H}^{n-1} \qquad (*)$$

$$+ \int_{\mathrm{ext}_*\Omega} v \cdot d(Dg) + \int_{\partial\Omega} (Tg) \, v \cdot \nu_\Omega \, d\mathcal{H}^{n-1}.$$

If $\|v\|_{L^\infty(\mathbb{R}^n;\mathbb{R}^n)} \leq 1$, then

$$\int_{\mathbb{R}^n} h \operatorname{div}(-v)\, dx \leq \|Df\|(\Omega) + \|Dg\|(\operatorname{ext}_*\Omega) + \int_{\partial\Omega} |Tg - Tf|\, d\mathcal{H}^{n-1}.$$

The arbitrariness of v implies $h \in BV(\mathbb{R}^n)$ and

$$\|Dh\|(\mathbb{R}^n) \leq \|Df\|(\Omega) + \|Dg\|(\operatorname{ext}_*\Omega) + \int_{\partial\Omega} |Tg - Tf|\, d\mathcal{H}^{n-1}. \qquad (**)$$

Now $h \in BV(\mathbb{R}^n)$ implies $\int_{\mathbb{R}^n} v \cdot d(Dh) = -\int_{\mathbb{R}^n} h \operatorname{div} v\, dx$. By equality $(*)$,

$$\int_{\mathbb{R}^n} v \cdot d(Dh) = \int_\Omega v \cdot d(Df) + \int_{\operatorname{ext}_*\Omega} v \cdot d(Dg)$$
$$+ \int_{\partial\Omega} (Tg - Tf)\, v \cdot \nu_\Omega\, d\mathcal{H}^{n-1} \qquad (\dagger)$$

for each $v \in C_c^1(\mathbb{R}^n; \mathbb{R}^n)$. In particular,

$$\int_\Omega v \cdot d(Dh) = \int_\Omega v \cdot d(Df)$$

when $\operatorname{spt} v \subset \Omega$. Thus viewing Df as a measure in $\mathbb{R}^n$ that lives in Ω, Proposition 5.3.11 shows that $Dh \llcorner \Omega = Df$. Similarly, viewing Dg as a measure in $\mathbb{R}^n$ that lives in $\operatorname{ext}_*\Omega$, we show that $Dh \llcorner \operatorname{ext}_*\Omega = Dg$. Denoting the zero extension of $Tg - Tf$ by $\overline{Tg - Tf}$, equality $(\dagger)$ translates to

$$\int_{\mathbb{R}^n} v \cdot d(Dh) = \int_{\mathbb{R}^n} v \cdot d(Dh \llcorner \Omega) + \int_{\mathbb{R}^n} v \cdot d(Dh \llcorner \operatorname{ext}_*\Omega)$$
$$+ \int_{\mathbb{R}^n} \left(\overline{Tg - Tf}\right) v \cdot \nu_\Omega\, d\mathcal{H}^{n-1}$$

for every $v \in C_c^1(\mathbb{R}^n; \mathbb{R}^n)$. Another application of Proposition 5.3.11 yields

$$Dh = Dh \llcorner \Omega + Dh \llcorner \operatorname{ext}_*\Omega + \mathcal{H}^{n-1} \llcorner \left(\overline{Tg - Tf}\right) \nu_\Omega.$$

We infer $Dh \llcorner \partial\Omega = \mathcal{H}^{n-1} \llcorner \left(\overline{Tg - Tf}\right) \nu_\Omega$, and consequently

$$Dh(\partial\Omega) = \int_{\partial\Omega} (Tg - Tf)\, \nu_\Omega\, d\mathcal{H}^{n-1}.$$

According to Proposition 5.3.7,

$$\|Dh\|(\partial\Omega) = \int_{\partial\Omega} |Tg - Tf|\, d\mathcal{H}^{n-1}$$

and the proposition follows from Observation 5.3.12:

$$\|Df\|(\Omega) + \|Dg\|(\operatorname{ext}_*\Omega) + \int_{\partial\Omega} |Tg - Tf|\, d\mathcal{H}^{n-1}$$
$$= \|Dh\|(\Omega) + \|Dh\|(\operatorname{ext}_*\Omega) + \|Dh\|(\partial\Omega) = \|Dh\|(\mathbb{R}^n). \qquad \square$$

Corollary 7.5.8. *Let $\Omega \subset \mathbb{R}^n$ be a Lipschitz domain with compact boundary, and let $f \in BV(\Omega)$. The zero extension $\overline{f}$ of f belongs to $BV(\mathbb{R}^n)$, and there is a constant $\kappa > 0$ independent of f such that*

$$\|\overline{f}\|_{BV(\mathbb{R}^n)} \leq \kappa \|f\|_{BV(\Omega)}.$$

PROOF. Proposition 7.5.7 and Theorem 7.5.5 imply $\overline{f} \in BV(\mathbb{R}^n)$ and

$$\|\overline{f}\|_{BV(\mathbb{R}^n)} = \|\overline{f}\|_{L^1(\mathbb{R}^n)} + \|D\overline{f}\|(\mathbb{R}^n)$$
$$= \|f\|_{L^1(\Omega)} + \|Df\|(\Omega) + \|Tf\|_{L^1(\partial\Omega, \mathcal{H}^{n-1})}$$
$$\leq \|f\|_{BV(\Omega)} + \gamma \|f\|_{BV(\Omega)} = (\gamma + 1)\|f\|_{BV(\Omega)}$$

where $\gamma > 0$ is a constant independent of f. $\square$

Remark 7.5.9. Corollary 7.5.8 generalizes Theorem 5.8.3. It follows that Corollary 5.8.4, and hence Lemma 5.9.2, holds when Ω is a bounded Lipschitz domain. The same is true about Poincaré's inequality (Theorem 5.9.11) and the relative isoperimetric inequality (Theorem 5.9.12), except that Ω_r may not be defined.

Lemma 7.5.10. *If $g \in BV(\mathbb{R}^n)$ then $\lim \|g - g\chi_{B(0,k)}\|_{BV(\mathbb{R}^n)} = 0$.*

PROOF. If $U_k = \mathbb{R}^n - B(0, k)$ then $g - g\chi_{B_k} = g\chi_{U_k}$, and we may assume that $g \geq 0$. Since for $k = 1, 2, \ldots$, the unit exterior normal

$$\nu_{U_k} : x \mapsto -x|x|^{-1} : \partial U_k \to \mathbb{R}^n$$

is Lipschitz with $\operatorname{Lip} \nu_{U_k} \leq 2$, it can be extended by Proposition 1.5.2 to a Lipschitz vector field $v_k : \mathbb{R}^n \to \mathbb{R}$ so that

$$\|v_k\|_{L^\infty(\mathbb{R}^n;\mathbb{R}^n)} \leq 1 \quad \text{and} \quad \operatorname{Lip} v_k \leq 2\sqrt{n};$$

in particular $\|\operatorname{div} v_k\|_{L^\infty(\mathbb{R}^n)} \leq 2n^{3/2}$. According to equality (7.5.2),

$$\int_{\partial U_k} T(g \upharpoonright U_k) \, d\mathcal{H}^{n-1} = \int_{\partial U_k} [T(g \upharpoonright U_k)] v_k \cdot \nu_{U_k} \, d\mathcal{H}^{n-1}$$
$$= \int_{U_k} g(x) \operatorname{div} v_k(x) \, dx + \int_{U_k} v_k \cdot d(Dg)$$
$$\leq 2n^{3/2} \int_{U_k} g(x) \, dx + \|Dg\|(U_k).$$

Since $\bigcap_{k=1}^\infty U_k = \emptyset$ and Proposition 7.5.7 implies

$$\|D(g\chi_{U_k})\|(\mathbb{R}^n) = \|Dg\|(U_k) + \int_{\partial U_k} T(g \upharpoonright U_k) \, d\mathcal{H}^{n-1},$$

the lemma follows. $\square$

The next proposition is the predicted generalization of Theorem 5.10.2, and by extrapolation, of Corollary 5.10.3.

Proposition 7.5.11. *Let $\phi : \mathbb{R}^n \to \mathbb{R}^n$ be a Lipschitz map. If g belongs to $BV(\mathbb{R}^n)$, then so does $\phi_\# g$ and*

$$\|D\phi_\# g\|(\mathbb{R}^n) \le (\operatorname{Lip} \phi)^{n-1}\|Dg\|(\mathbb{R}^n).$$

PROOF. For $k = 1, 2, \ldots$, let $B_k = B(0, k)$ and $g_k = g\chi_{B_k}$. From (5.10.1) and (5.10.2) we see that $\lim \phi_\# g_k(y) = \phi_\# g(y)$ for each $y \in \mathbb{R}^n$, and

$$\sup \|\phi_\# g_k\|_{L^1(\mathbb{R}^n)} \le (\operatorname{Lip} \phi)^n \|g\|_{L^1(\mathbb{R}^n)}.$$

Thus $\phi_\# g = \lim \phi_\# g_k$ in $L^1(\mathbb{R}^n)$, and the desired inequality

$$\mathbf{V}(\phi_\# g) \le \liminf \|D\phi_\# g_k\|(\mathbb{R}^n)$$
$$\le (\operatorname{Lip} \phi)^{n-1} \lim \|Dg_k\|(\mathbb{R}^n) = (\operatorname{Lip} \phi)^{n-1}\|Dg\|(\mathbb{R}^n)$$

follows from Proposition 5.1.7, Theorem 5.10.2, and Lemma 7.5.10. $\square$

7.6. BV vector fields

For an open set $\Omega \subset \mathbb{R}^n$ define the linear space $BV(\Omega; \mathbb{R}^n)$ in the obvious way. Select $v = (v_1, \ldots, v_n)$ in $BV(\Omega; \mathbb{R}^n)$, and note that

$$\operatorname{div} v := \sum_{i=1}^n D_i v_i$$

is a signed measure in Ω; see Section 5.5. If μ_+ and μ_- are, respectively, the positive and negative parts of $\operatorname{div} v$, let $L_{\operatorname{div} v} := L_{\mu_+} - L_{\mu_-}$ where $L_{\mu_\pm}$ are the distributions defined in Example 3.1.2, (2). Since for each test function $\varphi \in \mathcal{D}(\Omega)$, Theorem 5.5.1 implies

$$L_{\operatorname{div} v}(\varphi) = \int_\Omega \varphi \, d(\operatorname{div} v) = \sum_{i=1}^n \int_\Omega \varphi \, d(D_i v_i)$$
$$= -\sum_{i=1}^n \int_\Omega v_i \, D_i \varphi = -\int_\Omega v \cdot D\varphi$$

we see that $L_{\operatorname{div} v}$ is the distributional divergence F_v of v defined in Example 3.1.2, (3). If Ω is a Lipschitz domain, we define a vector field

$$Tv := (Tv_1, \ldots, Tv_n),$$

in $L^1(\partial\Omega, \mathcal{H}^{n-1}; \mathbb{R}^n)$, called the *trace* of v; see Theorem 7.5.3.

Theorem 7.6.1. *Let $\Omega \subset \mathbb{R}^n$ be a Lipschitz domain with compact boundary. If $w \in BV(\Omega; \mathbb{R}^n)$, then*

$$\operatorname{div} w(\Omega) = \int_{\partial\Omega} (Tw) \cdot \nu_\Omega \, d\mathcal{H}^{n-1}.$$

PROOF. Substituting $f := w_i$ and $v := e_i$ into equality (7.5.2), we obtain

$$D_i w_i(\Omega) = \int_\Omega e_i \cdot d(Dw_i) = \int_{\partial\Omega} (Tw_i)(e_i \cdot \nu_\Omega) \, d\mathcal{H}^{n-1}$$

for $i = 1, \ldots, n$. Summing up these equalities completes the proof. $\square$

Remark 7.6.2. Theorem 7.5.3 and its consequences hold for sets that are more general than Lipschitz domains [75, Definition 5.10.1]. The largest family of such sets is defined abstractly in [1, Definition 3.20]; its elements are called the *extension domains*.

Remark 7.6.3. Let $v \in L^\infty(\mathbb{R}^n; \mathbb{R}^n)$ be a vector field whose distributional divergence is a signed measure μ; explicitly, let v be such that

$$\int_{\mathbb{R}^n} \varphi \, d\mu = -\int_{\mathbb{R}^n} v \cdot D\varphi$$

for each $\varphi \in \mathcal{D}(\mathbb{R}^n)$. It has been shown that for any bounded set $A \in \boldsymbol{BV}(\mathbb{R}^n)$, normalized by the condition $A = \mathrm{cl}_* A$, the vector field v has a "trace" Tv which belongs to $L^\infty(\partial_* A, \mathcal{H}^{n-1}; \mathbb{R}^n)$ and satisfies

$$\mu(A) = \int_{\partial_* A} (Tv) \cdot \nu_A \, d\mathcal{H}^{n-1}.$$

This result is relatively recent. It is due to G.-Q. Chen, M. Torres, and W.P. Ziemer [15, 16], and independently to M. Šilhavý [74], who proved it by a different method.

Chapter 8

Unbounded vector fields

We show that the divergence theorem is still valid for admissible vector fields that are unbounded along compact sets whose upper Minkowski contents in a codimension larger than one are finite. The growth of the vector field must be controlled proportionately to the Minkowski content of the exceptional sets.

8.1. Minkowski contents

Let $E \subset \mathbb{R}^n$ and $t \geq 0$. The *t-dimensional upper Minkowski content* of E is the extended real number

$$\mathcal{M}^{*t}(E) := \limsup_{r \to 0} \frac{|B(E,r)|}{r^{n-t}}. \quad [1]$$

In general, the function $\mathcal{M}^{*t} : E \mapsto \mathcal{M}^{*t}(E)$ is not a measure in $\mathbb{R}^n$; see Example 8.1.3. Notwithstanding, upper Minkowski contents provide useful information about subsets of $\mathbb{R}^n$. We apply it only to compact sets, since $\mathcal{M}^{*t}(E) = \mathcal{M}^{*t}(\operatorname{cl} E)$ and $\mathcal{M}^{*t}(E) = \infty$ whenever E is not bounded. For a compact set $K \subset \mathbb{R}^n$, we have

$$\mathcal{M}^{*0}(K) = \alpha(n)\mathcal{H}^0(K) \quad \text{and} \quad \mathcal{M}^{*n}(K) = |K|;$$

if $t > n$ then $\mathcal{M}^{*t}(K) = 0$.

Remark 8.1.1. In order to achieve the equality $\mathcal{H}^t(E) = \mathcal{M}^{*t}(E)$ for some special sets, the definition of $\mathcal{M}^{*t}(E)$ is often presented with a normalizing constant depending on n and t; see [46, Section 5.5] and [33, Section 3.2.37]. For our purposes this is not necessary, since we will be interested only in distinguishing the cases

$$\mathcal{M}^{*t} = 0, \quad 0 < \mathcal{M}^{*t}(E) < \infty, \quad \mathcal{M}^{*t}(E) = \infty.$$

Proposition 8.1.2. *For each compact set $K \subset \mathbb{R}^n$ and each $t \geq 0$,*

$$\mathcal{H}^t(K) \leq 2^t \frac{\alpha(t)}{\alpha(n)} \mathcal{M}^{*t}(K).$$

---

[1] Replacing "lim sup" by "lim inf" defines the *t-dimensional lower Minkowski content* $\mathcal{M}^t_*(E)$ of E — a concept not used in this book.

PROOF. It suffices to assume that $K \neq \emptyset$ and $0 < t < n$. For $r > 0$, the *covering number* $N(K, r)$ is the least $p \in \mathbb{N}$ such that K can be covered by p closed balls of radius r, and the *packing number* $P(K, r)$ is the largest $q \in \mathbb{N}$ such that there are q disjoint closed balls of radius r centered at points of K.

Claim. $N(K, 2r) \leq P(K, r)$ and $P(K, r)\alpha(n)r^n \leq |B(K, r)|$.

Proof. Let $N := N(K, 2r)$ and $P := P(K, r)$. If $N > P$, there are disjoint balls $B(x_i, r)$ with $x_i \in K$, and $x \in K - \bigcup_{i=1}^{P} B(x_i, 2r)$. Thus the balls $B(x_1, r), \ldots, B(x_P, r), B(x, r)$ are disjoint, a contradiction. Since $\bigcup_{i=1}^{P} B(x_i, r) \subset B(K, r)$, the second inequality follows.

Choose $\delta > 0$, and cover K by $N(K, 2\delta)$ closed balls of radius 2δ. Using the claim, we calculate

$$\mathcal{H}_{5\delta}^{t}(K) \leq \alpha(t)N(K, 2\delta)(2\delta)^t \leq 2^t \frac{\alpha(t)}{\alpha(n)}\delta^{t-n}P(K, \delta)\alpha(n)\delta^n$$

$$\leq 2^t \frac{\alpha(t)}{\alpha(n)} \cdot \frac{|B(K, \delta)|}{\delta^{n-t}},$$

and the proposition follows by letting $\delta \to 0$. $\qquad\square$

Example 8.1.3. Assume $n = 2$ and $0 < t < 2$. Let $x_0 := (0, 0)$, and for $j \in \mathbb{N}$, let $x_j := (1/j, 0)$ and $r_j := \frac{1}{2}|x_j - x_{j+1}|$. Consider compact sets $K := \{x_0, x_1, \ldots\}$ and $K_j = [0, 1/j] \times \{0\}$. Given $0 < r \leq 1/4$, there is a unique $j \in \mathbb{N}$ with $r_{j+1} < r \leq r_j$. Since

$$\sum_{k=1}^{j} \frac{|B(x_k, r_{j+1})|}{(r_j)^{2-t}} \leq \frac{|B(K, r)|}{r^{2-t}} \leq \frac{|B(K_j, r_j)|}{(r_{j+1})^{2-t}} + \sum_{k=1}^{j} \frac{|B(x_k, r_j)|}{(r_{j+1})^{2-t}},$$

we calculate $0 < \mathcal{M}^{*1/2}(K) < \infty$, and

$$\mathcal{M}^{*t}(K) = \limsup_{r \to 0} \frac{|B(K, r)|}{r^{2-t}} = \begin{cases} 0 & \text{if } t > 1/2, \\ \infty & \text{if } t < 1/2. \end{cases}$$

Note that $\mathcal{M}^{*t}(\{x_j\}) = 0$ for $j = 0, 1, \ldots$, and that $\mathcal{H}^t(K) = 0$.

Example 8.1.4. Let $I \subset \mathbb{R}^n$ be a closed segment of unit length. As the upper Minkowski volume is invariant with respect to rotations and translations, we may assume that

$$I = \{(0, \ldots, 0, t) \in \mathbb{R}^n : 0 \leq t \leq 1\}.$$

Let I_0 and I_1 be disjoint closed subsegments of I, each of length $1/3$, such that $I - (I_0 \cup I_1)$ is an open segment, necessarily in the middle of I and of length $1/3$. For $k \in \mathbb{N}$, define recursively disjoint closed segments $I_{i_1 \cdots i_k}$, $i_j = 0, 1$, each of length 3^{-k}, so that $I_{i_1 \cdots i_{k+1}} \subset I_{i_1 \cdots i_k}$ and

$$I_{i_1 \cdots i_k} - (I_{i_1 \cdots i_k 0} \cup I_{i_1 \cdots i_k 1})$$

is an open segment, necessarily in the middle of $I_{i_1 \cdots i_k}$ and of length 3^{-k-1}. The *Cantor ternary discontinuum* [62, Section 2.44] in I is the intersection $D = \bigcap_{k=1}^{\infty} D_k$ where

$$D_k := \bigcup \{I_{i_1 \cdots i_k} : i_j = 0, 1 \text{ for } j = 1, \ldots, k\}.$$

Furthermore, denote by $I_{i_1 \cdots i_k}^*$ the closed segment of length $3 \cdot 3^{-k}$ that contains $I_{i_1 \cdots i_k}$ in the middle, and let

$$D_k^* := \bigcup \{I_{i_1 \cdots i_k}^* : i_j = 0, 1 \text{ for } j = 1, \ldots, k\}.$$

For $E \subset I$ and $r > 0$, let $C(E, r) := \pi_n[B(0, r)] \times E$. Given $0 < r < 1/3$, find $k \in \mathbb{N}$ with $3^{-k-1} \le r < 3^{-k}$. The inclusions

$$C(D_{k+1}, 3^{-k-1}) \subset B(D, r) \subset B(D_k, r) \subset C(D_k^*, 3^{-k})$$

imply

$$\frac{|C(D_{k+1}, 3^{-k-1})|}{(3^{-k})^{n-t}} \le \frac{|B(D, r)|}{r^{n-t}} \le \frac{|C(D_k^*, 3^{-k})|}{(3^{-k-1})^{n-t}}.$$

By a direct calculation,

$$|C(D_{k+1}, 3^{-k-1})| = 2^{k+1} \cdot 3^{-k-1} \cdot (3^{-k-1})^{n-1} \alpha(n-1),$$

$$|C(D_k^*, 3^{-k})| \le 2^k \cdot (3 \cdot 3^{-k}) \cdot (3^{-k})^{n-1} \alpha(n-1).$$

Now let $t := \log 2 / \log 3$. Since $3^t = 2$, we obtain

$$\frac{2}{3^n} \alpha(n-1) \le \frac{|B(D, r)|}{r^{n-t}} \le \frac{3^{n+1}}{2} \alpha(n-1).$$

As this inequality holds for all $0 < r < 1/3$,

$$\frac{2}{3^n} \alpha(n-1) \le \mathcal{M}^{*t}(D) \le \frac{3^{n+1}}{2} \alpha(n-1).$$

We note that $\mathcal{H}^t(D) = \alpha(t)/2^t$ for $t := \log 2 / \log 3$. A proof of this equality involving only the definition of $\mathcal{H}^t$ is given in [30, Theorem 1.14].

Proposition 8.1.5. *If $K \subset \mathbb{R}^n$ is a compact set, then $B(K, r)$ belongs to $\mathcal{BV}_c(\mathbb{R}^n)$ for $\mathcal{L}^1$ almost all $r > 0$, and*

$$\liminf_{r \to 0} \frac{P[B(K, r)]}{r^{n-1-t}} \le (n - t) \mathcal{M}^{*t}(K)$$

for $0 \le t \le n - 1$.

PROOF. Select $r > 0$, and observe that the function

$$f : x \mapsto \min\{r, \operatorname{dist}(x, K)\} : \mathbb{R}^n \to [0, r]$$

is Lipschitz with $\mathrm{Lip}\,(f) \leq 1$. Clearly, $\{f \leq s\} = B(K,s)$ for $0 < s < r$, and $\{f \leq s\} = \mathbb{R}^n$ for $s \geq r$. By the coarea theorem (Theorem 5.7.3),

$$P(r) := \int_0^r \mathbf{P}\big[B(K,s)\big]\,ds = \int_{B(K,r)} \big|Df(x)\big|\,dx \leq \big|B(K,r)\big|.$$

Thus $\mathbf{P}\big[B(K,s)\big] < \infty$ for $\mathcal{L}^1$ almost all $0 < s < r$, and since r is arbitrary and $\big|B(K,r)\big| < \infty$, the first claim of the proposition is satisfied. As there is nothing more to prove otherwise, we assume $\mathcal{M}^{*t}(K) < \infty$. This implies $\lim_{r \to 0} P(r) r^{t+1-n} = 0$. Given $\varepsilon > 0$, find $\delta > 0$ so that

$$\frac{\big|B(K,s)\big|}{s^{n-t}} \leq \mathcal{M}^{*t}(K) + \varepsilon$$

for all $0 < s \leq \delta$. Integrating by parts, we obtain

$$\int_0^r \frac{\mathbf{P}\big[B(K,s)\big]}{s^{n-1-t}}\,ds = P(r) r^{t+1-n} - (t+1-n) \int_0^r P(s) s^{t-n}\,ds$$

$$\leq r \frac{\big|B(K,r)\big|}{r^{n-t}} + (n-1-t) \int_0^r \frac{\big|B(K,s)\big|}{s^{n-t}}\,ds$$

$$\leq r(n-t)\big[\mathcal{M}^{*t}(K) + \varepsilon\big]$$

for each $0 < r \leq \delta$. Hence

$$\liminf_{r \to 0} \frac{\mathbf{P}\big[B(K,r)\big]}{r^{n-1-t}} \leq \liminf_{r \to 0} \frac{1}{r} \int_0^r \frac{\mathbf{P}\big[B(K,s)\big]}{s^{n-1-t}}\,ds$$

$$\leq (n-t)\big[\mathcal{M}^{*t}(K) + \varepsilon\big]$$

and the proposition follows from the arbitrariness of ε. $\qquad\square$

Let $A \in \mathbf{BV}(\mathbb{R}^n)$. For a compact set $K \subset \mathbb{R}^n$ and $t \geq 0$, the t-dimensional upper Minkowski content of K *relative* to $\partial_* A$ is the extended real number

$$\mathcal{M}^{*t}(K, \partial_* A) := \limsup_{r \to 0} \frac{\mathcal{H}^{n-1}\big[B(K,r) \cap \partial_* A\big]}{r^{n-1-t}}.$$

Since $\mathcal{H}^{n-1}(\partial_* A) < \infty$, it is clear that for $t \geq n-1$,

$$\mathcal{M}^{*t}(K, \partial_* A) = \mathcal{H}^t(K \cap \partial_* A).$$

Example 8.1.7 below shows that in general $\mathcal{M}^{*t}(K, \partial_* A)$ differs from both $\mathcal{H}^t(K \cap \partial_* A)$ and $\mathcal{M}^{*t}(K \cap \partial_* A)$, even when $K \subset \partial_* A$. This difference cannot be eliminated by introducing a normalizing constant mentioned in Remark 8.1.1.

Proposition 8.1.6. *If $A \in \mathbf{BV}(\mathbb{R}^n)$, then the following is true*

  (1) $\mathcal{M}^{*0}\big(\{x\}, \partial_* A\big) = 0$ *for $\mathcal{H}^{n-1}$ almost all $x \in \mathbb{R}^n - \partial_* A$,*

  (2) $\mathcal{M}^{*0}\big(\{x\}, \partial_* A\big) > 0$ *for each $x \in \partial_* A$,*

  (3) $\mathcal{M}^{*0}\big(\{x\}, \partial_* A\big) = \alpha(n-1)$ *for each $x \in \partial^* A$.*

PROOF. Since $\partial_* A = \partial_*(\mathbb{R}^n - A)$ and

$$\mathcal{M}^{*0}(\{x\}, \partial_* A) = \limsup_{r \to 0} \frac{\mathcal{H}^{n-1}[B(x,r) \cap \partial_* A]}{r^{n-1}},$$

we see that $\mathcal{M}^{*0}(\{x\}, \partial_* A) = 0$ if and only if $x \in \text{int}_c A \cup \text{int}_c(\mathbb{R}^n - A)$. By Proposition 7.3.1, the sets $\text{int}_c A \cup \text{int}_c(\mathbb{R}^n - A)$ and $\mathbb{R}^n - \partial_* A$ differ by an $\mathcal{H}^{n-1}$ negligible set. This establishes claim (1). In view of Theorem 6.5.2, claims (2) and (3) follow, respectively, from Lemma 5.9.14 and Corollary 6.4.4, (3). □

Example 8.1.7. Assume $n = 2$. For $k = 1, 2, \ldots$, let

$$A_k := \left\{ x \in \mathbb{R}^2 : (k + \varepsilon_k)^{-2} \le |x| \le k^{-2} \right\}$$

where $0 < \varepsilon_k < 1$ and $\lim \varepsilon_k = \varepsilon$. Then $A = \bigcup_{k=1}^\infty A_k$ is a BV set in $\mathbb{R}^2$, and after some calculation we obtain $\Theta(A, 0) = \varepsilon$. It follows that

$$\mathcal{M}^{*t}(\{0\} \cap \partial_* A) = \mathcal{H}^{*t}(\{0\} \cap \partial_* A) = \begin{cases} 1 & \text{if } 0 < \varepsilon < 1 \text{ and } t = 0, \\ 0 & \text{otherwise.} \end{cases}$$

On the other hand, for any value of ε we calculate

$$\mathcal{M}^{*t}(\{0\}, \partial_* A) = \begin{cases} \infty & \text{if } 0 \le t < 1/2, \\ 4\pi & \text{if } t = 1/2, \\ 0 & \text{if } t > 1/2. \end{cases}$$

8.2. Controlled vector fields

We consider unbounded vector fields v whose growth to infinity is controlled by an appropriate power of the distance from a compact set $K \subset \mathbb{R}^n$. The controlling power is determined by $\mathcal{M}^{*t}(K)$. In essence, we extrapolate from the fact that the function $x \mapsto |x|^t$ is locally integrable in $\mathbb{R}^n$ when $t > -n$; cf. Example 8.2.9.

Let $A \subset E \subset \mathbb{R}^n$. Given $f, g : E \to \mathbb{R}_+$, we abbreviate

$$\lim \frac{f(x)}{g(x)} < \infty \quad \text{and} \quad \lim \frac{f(x)}{g(x)} = 0 \quad \text{as} \quad \text{dist}(x, A) \to 0,$$

by writing, respectively,

$$f(x) = O(1)g(x) \quad \text{and} \quad f(x) = o(1)g(x) \quad \text{as} \quad x \to A.$$

This is a common notation that will simplify our exposition.

Let $E \subset \mathbb{R}^n$. A map $\phi : E \to \mathbb{R}^m$ is called *bounded at* $x \in \text{cl}\, E$ if there is a neighborhood U of x such that

$$\sup \left\{ |\phi(y)| : y \in E \cap U \right\} < \infty.$$

The set of all $x \in \text{cl}\, E$ at which ϕ is *not bounded* is denoted by E_∞.

Clearly E_∞ is a closed set. A map $\phi : E \to \mathbb{R}^m$ that is bounded or locally bounded is bounded at each $x \in E$, but not necessarily at $x \in \operatorname{cl} E$.

Definition 8.2.1. Let $E \subset \mathbb{R}^n$. A map $\phi : E \to \mathbb{R}^m$ is called *controlled* if its restriction to $E - E_\infty$ is admissible, and if there are numbers $0 \le t_j < n - 1$ and disjoint nonempty compact sets K_j such that $E_\infty = \bigcup_j K_j$ and for each $j \in \mathbb{N}$, the following conditions are satisfied:

(1) $\mathcal{M}^{*t_j}(K_j) < \infty$;

(2) if $\mathcal{M}^{*t_j}(K_j) = 0$, then
$$\left|\phi(x)\right| = O(1)\operatorname{dist}(x, K_j)^{t_j+1-n} \text{ as } x \to K_j;$$

(3) if $\mathcal{M}^{*t_j}(K_j) > 0$, then
$$\left|\phi(x)\right| = o(1)\operatorname{dist}(x, K_j)^{t_j+1-n} \text{ as } x \to K_j.$$

Remark 8.2.2. In Definition 8.2.1, each K_j is called a *singular set* of ϕ, and t_j is called the *size* of K_j. Note that the size of a singular set K is not determined by K alone. The *upper Minkowski dimension* of K
$$\overline{\dim} K := \inf\{s \ge 0 : \mathcal{M}^{*s}(K) < \infty\},$$

which is the optimal candidate for the size of K, can be used only when $\mathcal{M}^{*\overline{\dim} K}(K) < \infty$; see [46, Sections 4.8, 5.3 and 5.5].

Let $\phi : E \to \mathbb{R}^m$ be a controlled map. Since $\mathcal{H}^{n-1}(E_\infty) = 0$ by Proposition 8.1.2, it follows from Remark 2.3.3 that ϕ is $\mathcal{H}^{n-1}$ measurable whenever E is $\mathcal{H}^{n-1}$ measurable. Moreover, the restriction $\phi \upharpoonright (E - E_\infty)$ is Lipschitz at almost all $x \in E - E_\infty$. As E_∞ is a closed negligible set, the map ϕ is Lipschitz at almost all $x \in E$, and hence relatively differentiable at almost all $x \in E$ by Theorem 7.2.3 and Corollary 4.4.4. Clearly, ϕ is admissible if and only if $E \cap E_\infty = \emptyset$, or equivalently, if and only if ϕ is locally bounded.

Proposition 8.2.3. *Let $E \subset \mathbb{R}^n$, and let $\phi : E \to \mathbb{R}^m$ be a controlled map. The family $\{K_j\}$ of all singular sets of ϕ is locally finite. In particular, $\{K_j\}$ is finite whenever E is bounded.*

PROOF. By Definition 8.2.1, (2) and (3), each singular set K_j has a neighborhood U such that ϕ is bounded at each $x \in \operatorname{cl} E \cap (U - K_j)$. Thus $U \cap K_i = \emptyset$ whenever $i \ne j$. Choose $x \in \operatorname{cl} E$ and $B = B(x, r)$. For each K_j that meets B select $x_j \in K_j \cap B$. Since $\{x_j : K_j \cap B \ne \emptyset\}$ is a discrete subset of a compact set B, it is finite. $\square$

Lemma 8.2.4. *Let $A \subset \mathbb{R}^n$ be an $\mathcal{H}^{n-1}$ measurable set, and let K be a singular set of a controlled map $\phi : A \to \mathbb{R}^m$. There is a sequence $\{r_i\}$ in $\mathbb{R}_+$ such that $\lim r_i = 0$ and*
$$\lim \int_{\partial B(K,r_i) \cap A} |\phi| \, d\mathcal{H}^{n-1} = 0.$$

PROOF. Let $0 \leq t < n - 1$ be the size of K. By Proposition 8.1.5, there is a decreasing sequence $\{r_i\}$ with

$$\lim r_i = 0 \quad \text{and} \quad \lim \frac{\mathbf{P}\big[B(K, r_i)\big]}{r_i^{n-1-t}} \leq n\mathcal{M}^{*t}(K).$$

Choose $\varepsilon > 0$, and let $\delta(x) = \text{dist}\,(x, K)$ for $x \in \mathbb{R}^n$. If $\mathcal{M}^{*t}(K) = 0$, there are $a > 0$ and $j_a \in \mathbb{N}$ such that

$$\big|\phi(x)\big| \leq a\delta(x)^{t+1-n} \quad \text{and} \quad \mathbf{P}\big[B(K, r_i)\big] \leq \frac{\varepsilon}{a}r_i^{n-1-t}$$

whenever $x \in A$, $0 < \delta(x) \leq r_{j_a}$, and $i \geq j_a$. Thus for each $i \geq j_a$,

$$\int_{\partial_* B(K, r_i) \cap A} |\phi|\, d\mathcal{H}^{n-1} \leq a \int_{\partial_* B(K, r_i)} \delta(x)^{t+1-n}\, d\mathcal{H}^{n-1}(x)$$
$$= ar_i^{t+1-n}\mathbf{P}\big[B(K, r_i)\big] \leq \varepsilon.$$

If $\mathcal{M}^{*t}(K) > 0$, select $b \geq n\mathcal{M}^{*t}(K)$ and find an integer $j_b \in \mathbb{N}$ so that

$$\big|\phi(x)\big| \leq \frac{\varepsilon}{b}\delta(x)^{t+1-n} \quad \text{and} \quad \mathbf{P}\big[B(K, r_i)\big] \leq br_i^{n-1-t}$$

whenever $x \in A$, $0 < \delta(x) \leq r_{j_b}$, and $i \geq j_b$. Calculating as before,

$$\int_{\partial_* B(K, r_i) \cap \text{cl}_* A} |\phi|\, d\mathcal{H}^{n-1} \leq \varepsilon$$

for each $i \geq j_b$. The lemma follows from the arbitrariness of ε. $\qquad\square$

A *control function* is a decreasing function $\beta : \mathbb{R}_+ \to \mathbb{R}_+$ such that

$$\int_0^1 \beta(s)\, ds < \infty.$$

A control function need not be bounded, but $\beta(r) = o(1)r^{-1}$ as $r \to 0$. Indeed since β decreases, $r\beta(r) \leq \int_0^r \beta(s)\, ds \to 0$ as $r \to 0$.

Definition 8.2.5. Let $A \in \mathcal{BV}(\mathbb{R}^n)$. A controlled map $\phi : \text{cl}_* A \to \mathbb{R}^m$ with singular sets K_j of size t_j is called *fully controlled* if the following additional conditions are satisfied for $j = 1, 2, \ldots$:

(i) $\mathcal{M}^{*t_j}(K_j, \partial_* A) < \infty$;

(ii) there is a control function β_j such that for $\psi := \phi \restriction \partial_* A$,

$$\big|\psi(x)\big| = O(1)\beta_j\big[\text{dist}\,(x, K_j)\big]\text{dist}\,(x, K_j)^{t+2-n} \quad \text{as } x \to K_j.$$

Lemma 8.2.6. *Let $A \in \mathcal{BV}(\mathbb{R}^n)$, and let K be a singular set of a fully controlled map $\phi : \text{cl}_* A \to \mathbb{R}^m$. There is a sequence $\{r_i\}$ in $\mathbb{R}_+$ such that $\lim r_i = 0$ and*

$$\lim \int_{\partial_* [A \cap B(K, r_i)]} |\phi|\, d\mathcal{H}^{n-1} = 0.$$

PROOF. Let $0 \leq t < n - 1$ be the size of K, and let $\delta(x) = \text{dist}\,(x, K)$ for $x \in \mathbb{R}^n$. Choose $c > M^{*t}(K, \partial_* A)$ and $0 < r \leq 1$ so that

$$\mathcal{H}^{n-1}\big[B(K, s) \cap \partial_* A\big] \leq cs^{n-1-t} \quad \text{and} \quad |\phi(x)| \leq c\beta\big[\delta(x)\big]\delta(x)^{t+2-n}$$

for each $0 < s \leq r$ and each $x \in B(K, r) \cap (\partial_* A - K)$. The sets

$$D_i = B(K, r2^{-i}) - B(K, r2^{-i-1})$$

are disjoint and $B(K, r) = K \cup \bigcup_{i=0}^{\infty} D_i$. Since $\mathcal{H}^{n-1}(K) = 0$,

$$\int_{B(K,r)\cap\partial_* A} |\phi|\, d\mathcal{H}^{n-1} \leq c\sum_{i=0}^{\infty} \int_{D_i \cap \partial^* A} \beta\big[\delta(x)\big]\delta(x)^{t+2-n}\, d\mathcal{H}^{n-1}(x)$$

$$\leq c^2 \sum_{i=0}^{\infty} \beta(r2^{-i})(r2^{-i-1})^{t+2-n}(r2^{-i})^{n-1-t}$$

$$= c^2 2^{n-1-t} \sum_{i=0}^{\infty} \beta(r2^{-i})r2^{-i-1}$$

$$\leq c^2 2^n \sum_{i=0}^{\infty} \int_{r2^{-i-1}}^{r2^{-i}} \beta(s)\, ds = c^2 2^n \int_0^r \beta(s)\, ds.$$

From the definition of a control function, we obtain

$$\lim_{r\to 0} \int_{B(K,r)\cap\partial_* A} |\phi|\, \mathcal{H}^{n-1} = 0. \tag{$*$}$$

According to Corollary 4.2.5,

$$\partial_* \big[A \cap B(K, r)\big] \subset \big[B(K, r) \cap \partial_* A\big] \cup \big[\partial B(K, r) \cap \text{cl}_* A\big],$$

and the lemma follows from $(*)$ and Lemma 8.2.4. $\qquad\square$

Lemma 8.2.7. *If* $\sum_{j=1}^{\infty} \mathbf{P}(B_j) < \infty$ *and* $B = \bigcup_{j=1}^{\infty} B_j$, *then*

$$\mathbf{P}(B) \leq \sum_{j=1}^{\infty} \mathbf{P}(B_j) \quad \text{and} \quad \mathcal{H}^{n-1}\bigg(\partial_* B - \bigcup_{j=1}^{\infty} \partial_* B_j\bigg) = 0.$$

PROOF. For $k = 0, 1, \ldots$, Observation 4.2.1 yields

$$\partial_* B - \bigcup_{j=1}^{k} \partial_* B_j \subset \partial_*\bigg(\bigcup_{j=k+1}^{\infty} B_j\bigg).$$

From Theorem 6.5.2 and Propositions 4.5.2 and 5.1.7, we obtain

$$\mathcal{H}^{n-1}\bigg[\partial_*\bigg(\bigcup_{j=k+1}^{\infty} B_j\bigg)\bigg] = \mathbf{P}\bigg(\bigcup_{j=k+1}^{\infty} B_j\bigg) \leq \sum_{j=k+1}^{\infty} \mathbf{P}(B_j).$$

The lemma follows by letting $k = 0$ and $k \to \infty$. $\qquad\square$

Theorem 8.2.8. *Let $A \in \mathbf{BV}(\mathbb{R}^n)$, and let $v : \mathrm{cl}_* A \to \mathbb{R}^n$ be a fully controlled vector field. If $\mathrm{div}_{\mathrm{cl}_* A} v$ belongs to $L^1(A)$, then*

$$\int_A \mathrm{div}_{\mathrm{cl}_* A} v(x) \, dx = \int_{\partial_* A} v \cdot \nu_A \, d\mathcal{H}^{n-1}.$$

PROOF. Let K_j be the singular sets of v. Choose $\varepsilon > 0$ and $\eta > 0$. As $\mathcal{M}^{*n-1}(K_j) = 0$, Propositions 8.1.5 and 6.6.6 show that there are BV sets $B_j := A \cap B(K_j, r_j)$ such that

$$|B_j| \le \eta 2^{-j} \quad \text{and} \quad \mathbf{P}(B_j) \le \eta 2^{-j}. \tag{$*$}$$

Making r_j smaller, if necessary, Lemma 8.2.6 implies

$$\int_{\partial_* B_j} |v| \, d\mathcal{H}^{n-1} \le \varepsilon 2^{-j}.$$

By Lemma 8.2.7 and $(*)$, the union $B := \bigcup_j B_j$ is a BV set, and

$$\int_{\partial_* B} |v| \, d\mathcal{H}^{n-1} \le \int_{\bigcup_j \partial_* B_j} |v| \, d\mathcal{H}^{n-1} \le \sum_j \int_{\partial_* B_j} |v| \, d\mathcal{H}^{n-1} \le \varepsilon.$$

Since $\mathrm{div}_{\mathrm{cl}_* A} v \in L^1(A)$, and since $|B| \le \eta$ by $(*)$, we obtain

$$\int_B \left| \mathrm{div}_{\mathrm{cl}_* A} v(x) \right| dx \le \varepsilon$$

by making η sufficiently small. The restriction $v \restriction \left[\mathrm{cl}_*(A - B) \right]$ is admissible and bounded, and $\mathrm{div}_{\mathrm{cl}_* A} v(x) = \mathrm{div}_{\mathrm{cl}_* A - B} v(x)$ for almost all $x \in A - B$ in accordance with (7.2.1). Thus

$$\int_{A-B} \mathrm{div}_{\mathrm{cl}_* A} v(x) \, dx = \int_{\partial_*(A-B)} v \cdot \nu_{A-B} \, d\mathcal{H}^{n-1}$$

by Proposition 7.4.3. In view of this equality and Theorem 6.6.1,

$$\left| \int_{\partial_* A} v \cdot \nu_A \, d\mathcal{H}^{n-1} - \int_A \mathrm{div}_{\mathrm{cl}_* A} v(x) \, dx \right|$$

$$= \left| \int_{\partial_* B} v \cdot \nu_B \, d\mathcal{H}^{n-1} - \int_B \mathrm{div}_{\mathrm{cl}_* A} v(x) \, dx \right|$$

$$\le \int_{\partial_* B} |v| \, d\mathcal{H}^{n-1} + \int_B \left| \mathrm{div}_{\mathrm{cl}_* A} v(x) \right| dx < 2\varepsilon.$$

The theorem follows from the arbitrariness of ε. $\qquad\qquad\square$

Example 8.2.9. Assume $n = 2$ and $0 \le s \le 1$. If

$$v(x) := \begin{cases} x|x|^{-1-s} & \text{if } x \in \mathbb{R}^2 - \{0\}, \\ 0 & \text{if } x = 0, \end{cases}$$

then $\operatorname{div} v(x) = (1 - s)|x|^{-1-s}$ for each $x \in \mathbb{R}^2 - \{0\}$. It is clear that $\{0\}$ is the only singular set of v, and that $\mathcal{M}^{*0}(\{0\}) = 1$. For $s < 1$, the vector field v is fully controlled in $B(0, 1)$, since

$$\big|v(x)\big| = |x|^{-s} = o(1)|x|^{-1} \text{ as } x \to 0.$$

Thus Theorem 8.2.8 applies for $s < 1$. When $s = 1$, then $\operatorname{div} v \equiv 0$ in $\mathbb{R}^2 - \{0\}$ and $\int_{\partial B} v \cdot \nu_B \, d\mathcal{H}^1 = 2\pi$.

Now let $A \subset B(0, 1)$ be the BV set defined in Example 8.1.7, and recall that $\mathcal{M}^{*t}(\{0\}, \partial_* A) < \infty$ if and only if $t \geq 1/2$. If $s < 1/2$, then $w := v \restriction \operatorname{cl}_* A$ is fully controlled and Theorem 8.2.8 applies. Indeed, for $t = 1/2$ and the control function $\beta : r \mapsto r^{-\frac{1}{2}-s}$, we have

$$\big|v(x)\big| = \beta\big(|x|\big)|x|^t$$

for each $x \in B(0, 1)$. On the other hand, if $1/2 \leq s \leq 1$ then

$$\int_A \operatorname{div} w(x) \, dx < \infty \quad \text{and} \quad \int_{\partial_* A} w \cdot \nu_A \, d\mathcal{H}^1 = \infty.$$

8.3. Integration by parts

Lemma 8.3.1. *Let $A \subset \mathbb{R}^n$ be a measurable set, and let K be a singular set of a controlled map $\phi : A \to \mathbb{R}^m$. Then*

$$\lim_{r \to 0} \int_{A \cap B(K, r)} \big|\phi(x)\big| \, dx = 0.$$

PROOF. The proof is similar to that of Lemma 8.2.6. Let $0 \leq t < n - 1$ be the size of K, and let $\delta(x) = \operatorname{dist}(x, K)$ for each $x \in \mathbb{R}^n$. There are $c > \mathcal{M}^{*t}(K)$ and $r > 0$ such that

$$\big|B(K, s)\big| \leq cs^{n-t} \quad \text{and} \quad \big|\phi(x)\big| \leq c\delta(x)^{t+1-n}$$

for each $0 < s \leq r$ and each $x \in B(K, r) - K$. The sets

$$D_i := B(K, r2^{-i}) - B(K, r2^{-i-1})$$

are disjoint, and $B(K, r) = K \cup \bigcup_{i=1}^{\infty} B_i$. Since $|K| = 0$, we obtain

$$\int_{A \cap B(K, r)} \big|\phi(x)\big| \, dx \leq c \sum_{i=1}^{\infty} \int_{D_i} \delta(x)^{t+1-n} \, dx$$

$$\leq c \sum_{i=1}^{\infty} (r2^{-i-1})^{t+1-n}\big|B(K, r2^{-i})\big|$$

$$\leq c^2 \sum_{i=1}^{\infty} (r2^{-i-1})^{t+1-n}(r2^{-i})^{n-t} = c^2 2^{n-t} r. \qquad \square$$

Corollary 8.3.2. *Let $\Omega \subset \mathbb{R}^n$ be an open set. If $\phi : \Omega \to \mathbb{R}^m$ is a controlled map, then $\phi \in L^1_{\mathrm{loc}}(\Omega; \mathbb{R}^m)$.*

PROOF. By Lemma 8.3.1, each $x \in \Omega \cap \Omega_\infty$ has a neighborhood $U \subset \Omega$ with $\phi \in L^1(U; \mathbb{R}^m)$. As $\phi \upharpoonright (\Omega - \Omega_\infty)$ is locally bounded, the same is true for every $x \in \Omega - \Omega_\infty$. □

Theorem 8.3.3 (Integration by parts). *Let $\Omega \subset \mathbb{R}^n$ be an open set, let $g : \Omega \to \mathbb{R}$ be an admissible function, and let $v : \Omega \to \mathbb{R}^n$ be a controlled vector field. If $\operatorname{div} v \in L^1_{\mathrm{loc}}(\Omega)$ and $Dg \in L^\infty_{\mathrm{loc}}(\Omega; \mathbb{R}^n)$, then*

$$\int_\Omega g(x) \operatorname{div} v(x)\, dx = - \int_\Omega Dg(x) \cdot v(x)\, dx$$

whenever $\operatorname{spt}(gv) \Subset \Omega$.

PROOF. Although the assumptions differ, the proof is similar to that of Theorem 2.3.9. Find a figure $A \subset \Omega$ whose interior contains the compact set $C := \operatorname{spt}(gv)$, and let $w := gv \upharpoonright A$. Since $g \upharpoonright A$ is bounded by Remark 2.3.3, the vector field w is controlled, and if K_j are the singular sets of v, then $C \cap K_j$ are the singular sets of w. Moreover, w is fully controlled, since $C \Subset A$. By our assumptions and Corollary 8.3.2,

$$\operatorname{div} w = g \operatorname{div} v + Dg \cdot v$$

belongs to $L^1(A)$. As $w \equiv 0$ on ∂A, Theorem 8.2.8 implies

$$0 = \int_A \operatorname{div} w(x)\, dx = \int_A g(x) \operatorname{div} v(x)\, dx + \int_A Dg(x) \cdot v(x)\, dx. \qquad (*)$$

Choose $x \in \Omega - A$ so that g and v are differentiable at x; this is true for almost all $x \in \Omega - A$. Since $gv \equiv 0$ in the open set $\Omega - A$,

$$0 = \operatorname{div}(gv)(x) = g(x) \operatorname{div} v(x) + Dg(x) \cdot v(x).$$

Hence $g(x) \operatorname{div} v(x) = -Dg(x) \cdot v(x)$. As either $g(x) = 0$ or $v(x) = 0$, the theorem follows from $(*)$. □

Remark 8.3.4. In view of Theorem 8.3.3, the techniques introduced in Sections 3.3–3.4 can be applied to controlled vector fields. For illustration, consider an open set $\Omega \subset \mathbb{C}$ containing the Cantor set D of Example 8.1.4, and a holomorphic function $f : \Omega - D \to \mathbb{C}$ such that

$$\left| f(z) \right| \le \operatorname{dist}(z, D)^{t-1}$$

for $\log 2 / \log 3 < t < 1$ and each $z \in U - D$ where $U \Subset \Omega$ is an open set containing D. Following the proof of Theorem 3.3.1, we can show that f has a unique extension to a holomorphic function in Ω.

Chapter 9

Mean divergence

The mean divergence of a vector field is the density of its flux. For controlled vector fields the mean and ordinary divergence coincide almost everywhere. In general, the mean divergence may exist even if the ordinary divergence does not. Viewing the flux of a vector field as an additive function of dyadic figures, we give a sufficient condition under which the mean divergence exists and determines the flux. This is accomplished by replacing the classical variation of an additive function by a suitable Borel measure — the idea originally introduced by B.S. Thomson [73] in the real line. Throughout this chapter, Ω is a fixed open subset of $\mathbb{R}^n$.

9.1. The derivative

By $\mathbf{DC}(\Omega)$ and $\mathbf{DF}(\Omega)$ we denote, respectively, the family of all dyadic cubes and dyadic figures contained in Ω. For $F : \mathbf{DF}(\Omega) \to \mathbb{R}$ and $x \in \Omega$, let

$$\underline{D}F(x) = \sup_{\eta>0} \inf_C \frac{F(C)}{|C|} \quad \text{and} \quad \overline{D}F(x) = \inf_{\eta>0} \sup_C \frac{F(C)}{|C|}$$

where $C \subset \Omega$ is a dyadic cube with $x \in C$ and $d(C) < \eta$. When

$$\underline{D}F(x) = \overline{D}F(x) \neq \pm\infty,$$

we call this common value the *derivative* of F at x, denoted by $F'(x)$.

Example 9.1.1. Let $v \in L^\infty_{\text{loc}}(\Omega, \mathcal{H}^{n-1}; \mathbb{R}^n)$, and let

$$F : C \mapsto \int_{\partial C} v \cdot \nu_C \, d\mathcal{H}^{n-1} : \mathbf{DF}(\Omega) \to \mathbb{R}.$$

If v is differentiable at $x \in \Omega$, then Corollary 2.1.3 implies

$$F'(x) = \operatorname{div} v(x).$$

Observation 9.1.2. *Let $f \in L^1_{\text{loc}}(\Omega)$. If*

$$F : C \mapsto \int_C f(x) \, dx : \mathbf{DF}(\Omega) \to \mathbb{R},$$

then $F'(x) = f(x)$ for almost all $x \in \Omega$.

The proof of the previous observation is almost identical to that of Theorem 4.3.4. Properties of dyadic cubes are used in lieu of Vitali's theorem. For illustration, we present it in the small print.

PROOF OF OBSERVATION 9.1.2. Denote by N the set of all $x \in \Omega$ at which either $F'(x)$ does not exist or $F'(x)$ differs from $f(x)$. Given $x \in N$, find $\gamma_x > 0$ so that for each $\eta > 0$, there is $C \in \mathbf{DC}(\Omega)$ satisfying $x \in C$, $d(C) < \eta$, and $\left| F(C) - f(x)|C| \right| \geq \gamma_x |C|$. Select open sets $\Omega_k \Subset \Omega$ with $\Omega = \bigcup_{k=1}^{\infty} \Omega_k$, and let $N_k := \{ x \in N \cap \Omega_k : \gamma_x \geq 1/k \}$. Choose $\varepsilon > 0$, and use Henstock's lemma to find $\delta : \Omega_k \to \mathbb{R}_+$ so that

$$\sum_{i=1}^{q} \left| f(x_i)|Q_i| - F(Q_i) \right| < \frac{\varepsilon}{k}$$

for each δ-fine dyadic partition $\{(Q_1, x_1), \ldots, (Q_q, x_q)\}$ in Ω_k. The set N_k is covered by a family $\mathcal{C}$ of all $C \in \mathbf{DC}(\Omega_k)$ such that $d(C) < \delta(x_C)$ for some $x_C \in C \cap N$. There are nonoverlapping cubes $C_i \in \mathcal{C}$ with $\bigcup_i C_i = \bigcup \mathcal{C}$. Since for $p = 1, 2, \ldots$, the collections $\{(C_1, x_{C_1}), \ldots, (C_p, x_{C_p})\}$ are δ-fine dyadic partitions in Ω_k,

$$|N_k| \leq \sum_i |C_i| \leq k \sum_i \left| f(x_i)|C_i| - F(C_i) \right| \leq \varepsilon.$$

As ε is arbitrary, N_k is a negligible set and so is $N = \bigcup_{k=1}^{\infty} N_k$. $\square$

Proposition 9.1.3. *For each $F : \mathbf{DF}(\Omega) \to \mathbb{R}$, the functions*

$$\underline{D}F : x \mapsto \underline{D}F(x) : \Omega \to \overline{\mathbb{R}} \quad and \quad \overline{D}F : x \mapsto \overline{D}F(x) : \Omega \to \overline{\mathbb{R}}$$

are Borel measurable.

PROOF. Choose $c \in \mathbb{R}$, and observe that $E := \{ x \in \Omega : \underline{D}F(x) < c \}$ consists of all $x \in \Omega$ which satisfy the following condition: there is $j \in \mathbb{N}$ such that for each $k \in \mathbb{N}$ we can find $C \in \mathbf{DC}(\Omega)$ with

$$x \in C, \quad d(C) < 1/k, \quad and \quad \frac{F(C)}{|C|} < c - \frac{1}{j}. \tag{$*$}$$

The family $\mathcal{C}_{j,k}$ of all dyadic cubes $C \subset \Omega$ satisfying $(*)$ contains a countable subfamily with the same union as $\mathcal{C}_{j,k}$. Thus $E_{j,k} = \bigcup \mathcal{C}_{j,k}$ is a Borel set and so is $E = \bigcup_{j=1}^{\infty} \bigcap_{k=1}^{\infty} E_{j,k}$. As $\overline{D}F = -\underline{D}(-F)$, the proposition follows. $\square$

Each k-cube C is contained in a unique $(k-1)$-cube C^*, called the *mother* of C. Clearly, C^* is simultaneously the mother of 2^n nonoverlapping k-cubes.

Lemma 9.1.4. *Let $G : \mathbf{DF}(\Omega) \to \mathbb{R}$, let $X \subset \Omega$ have positive measure, and let $\underline{D}G(x) > 0$ for each $x \in X$. There are $\delta > 0$ and a set $E \subset X$ of positive measure such that $G(Q) > 0$ for each $Q \in \mathbf{DC}(\Omega)$ satisfying*

$$d(Q) < \delta \quad and \quad Q^* \cap E \neq \emptyset.$$

PROOF. For each $x \in X$ there is a $\delta_x > 0$ such that $G(Q) > 0$ for every $Q \in \mathbf{DC}(\Omega)$ with $x \in Q$ and $d(Q) < \delta_x$. Since

$$X = \bigcup_{i=1}^{\infty} \left\{ x \in X : \delta_x > \tfrac{1}{i} \right\}$$

there is an integer $j \geq 1$ such that $Y := \{x \in X : \delta_x > 1/j\}$ has positive measure, and the following holds:

$$G(Q) > 0 \text{ for each } Q \in \mathbf{DC}(\Omega) \text{ with } Y \cap Q \neq \emptyset \text{ and } d(Q) < \tfrac{1}{j}. \qquad (*)$$

Let $x \in \mathrm{int}_* Y$. By Lemma 4.2.3 there is $\eta_x > 0$ such that $|Y \cap Q|/|Q| > 1/2$ for each $Q \in \mathbf{DC}(\Omega)$ with $x \in Q$ and $d(Q) < \eta_x$. As

$$\mathrm{int}_* Y = \bigcup_{j=1}^{\infty} \left\{ x \in \mathrm{int}_* Y : \eta_x > \tfrac{1}{j} \right\},$$

there is an integer $k \geq 1$ such that the set $E = \{\mathrm{int}_* Y : \eta_x > 1/k\}$ has positive measure, and the following is true:

$$|Y \cap Q|/|Q| > \tfrac{1}{2} \text{ for each } Q \in \mathbf{DC}(\Omega) \text{ with } E \cap Q \neq \emptyset \text{ and } d(Q) < \tfrac{1}{k}. \quad (**)$$

Now let $\delta := \tfrac{1}{2} \min\{1/j, 1/k\}$, and choose $Q \in \mathbf{DC}(\Omega)$ with $d(Q) < \delta$ and $Q^* \cap E \neq \emptyset$. As $d(Q^*) = 2d(Q) < 2\delta \leq 1/k$, the inequality

$$|Y \cap Q^*| > \tfrac{1}{2}|Q^*| = 2^{n-1}|Q| \qquad (\dagger)$$

follows from $(**)$. On the other hand, $|Q^*| = 2^n|Q|$. Thus if $Y \cap Q = \emptyset$, then $|Y \cap Q^*| \leq 2^{n-1}|Q|$ contrary to $(\dagger)$. This shows that $Y \cap Q \neq \emptyset$, and since $d(Q) < \delta < 1/j$, the lemma follows from $(*)$. $\qquad \square$

A function $F : \mathbf{DF}(\Omega) \to \mathbb{R}$ is called *additive* if

$$F(A \cup B) = F(A) + F(B)$$

for each pair $A, B \in \mathbf{DF}(\Omega)$ with $|A \cap B| = 0$. For additive functions of dyadic figures, a version of the classical *Ward theorem* [65, Chapter 4, Theorem 11.15] is not difficult to prove.

Theorem 9.1.5 (Ward). *If $F : \mathbf{DF}(\Omega) \to \mathbb{R}$ is an additive function, then $F'(x)$ exists at almost all $x \in \Omega$ for which $\overline{D}|F|(x) < \infty$.*

PROOF. Let $N = \{x \in \Omega : \overline{D}|F|(x) < \infty\}$, and observe that $\underline{D}F(x)$ and $\overline{D}F(x)$ are real numbers for every $x \in N$. Since

$$\{x \in N : \underline{D}F(x) < \overline{D}F(x)\} = \bigcup_{i=1}^{\infty} \left\{ x \in N : \overline{D}F(x) - \underline{D}F(x) > \tfrac{1}{i} \right\},$$

we proceed to a contradiction by assuming that for some $0 < \eta < 1$, the set

$$N_\eta := \{x \in N : \overline{D}F(x) - \underline{D}F(x) > \eta\}$$

has positive measure. Choose $0 < \varepsilon < \eta/4$ and for $k \in \mathbb{Z}$, let

$$X_k := \{x \in \Omega_1 : \varepsilon k < \underline{D}F(x) \leq \varepsilon(k+1)\}.$$

As $N_\eta = \bigcup_{k \in \mathbb{Z}} X_k$, there is an integer s such that the set $X := X_s$ has positive measure. Letting $G := F - \varepsilon p \mathcal{L}^n$, we calculate that for each $x \in X$,

$$0 < \underline{D}G(x) \le \varepsilon \quad \text{and} \quad \overline{D}G(x) > \eta . \tag{$*$}$$

The remainder of the proof relies only on inequalities $(*)$.

By Lemma 9.1.4, there are $\delta > 0$ and a set $E \subset X$ of positive measure such that $G(Q) > 0$ for each $Q \in \mathcal{DC}(\Omega)$ with $d(Q) < \delta$ and $Q^* \cap E \ne \emptyset$. Choose $z \in E \cap \operatorname{int}_* E$, and using $(*)$ and Lemma 4.2.3, find $Q \in \mathcal{DC}(\Omega)$ so that $z \in Q$, $d(Q) < \delta$, and the following inequalities are satisfied:

$$G(Q) < 2\varepsilon|Q| \quad \text{and} \quad |E \cap Q| > (1 - \varepsilon)|Q| > 0. \tag{$**$}$$

The second inequality of $(*)$ shows that for each $x \in E \cap \operatorname{int} Q$, there is a dyadic cube $C_x \subset Q$ with $x \in C_x$ and $G(C_x) > \eta|C_x|$. Since

$$E \cap \operatorname{int} Q \subset \bigcup \{C_x : x \in E \cap \operatorname{int} Q\},$$

there are nonoverlapping C_{x_i} such that $E \cap \operatorname{int} Q \subset \bigcup_i C_{x_i} \subset Q$. Thus

$$\sum_i |C_i| \ge |E \cap Q| > (1 - \varepsilon)|Q|,$$

and there is an integer $p \ge 1$ such that

$$\sum_{i=1}^{p} G(C_{x_i}) > \eta \sum_{i=1}^{p} |C_{x_i}| > \eta(1 - \varepsilon)|Q|. \tag{$\dagger$}$$

If $\bigcup_{i=1}^{p} C_{x_i} \ne Q$, let $A = Q - \operatorname{int} \left(\bigcup_{i=1}^{p} C_{x_i} \right)$. Given $y \in A$, denote by Q_y the *largest* dyadic cube such that $y \in Q_y$ and Q_y does not overlap $\bigcup_{i=1}^{p} C_{x_i}$. As $Q_y \subsetneq Q$, we have $Q_y^* \subset Q$. It follows from the maximality of Q_y that Q_y^* overlaps, and thus contains, some C_{x_i} for $1 \le i \le p$. In particular, $x_i \in Q_y^* \cap E$, and if $d = \min\{d(C_i) : i = 1, \dots, p\}$, then $d(Q_x^*) \ge d$. Moreover $d(Q_y) \le d(Q) < \delta$, and the choice of δ and E implies $G(Q_y) > 0$. Since $\bigcup_{y \in A} Q_y = A$ and $d(Q_y) \ge d/2 > 0$, the set A is the union of nonoverlapping dyadic cubes $Q_{y_1}, \dots, Q_{y_q}$ where $q \in \mathbb{N}$. If $\bigcup_{i=1}^{p} C_{x_i} = Q$, let $q = 0$. In either case, Q is the union of nonoverlapping dyadic cubes $C_{x_1}, \dots, C_{x_p}, Q_{y_1}, \dots, Q_{y_q}$ where $q \ge 0$ is an integer. Consequently,

$$2\varepsilon|Q| > G(Q) = \sum_{i=1}^{p} G(C_{x_i}) + \sum_{i=1}^{q} G(Q_{y_i}) \ge \sum_{i=1}^{p} G(C_{x_i}) > \eta(1 - \varepsilon)|Q|$$

follows from $(**)$ and $(\dagger)$. We infer $2\varepsilon > \eta(1 - \varepsilon) > \eta/2$, contrary to our choice of $\varepsilon < \eta/4$. $\qquad \square$

9.2. The critical variation

Let $F : \mathbf{DF}(\Omega) \to \mathbb{R}$. Given $E \subset \Omega$ and $\delta : E \to \mathbb{R}_+$, we let

$$V_\delta F(E) := \sup_P \sum_{i=1}^p \big|F(C_i)\big|$$

where $P = \big\{(C_1, x_1), \ldots, (C_p, x_p)\big\}$ is a δ-fine dyadic partition in Ω; as usual, we define $\sum_{i=1}^p \big|F(C_i)\big| := 0$ when $P = \emptyset$. We further let

$$VF(E) := \inf_\delta V_\delta F(E)$$

where $\delta : E \to \mathbb{R}_+$. The function $VF : E \mapsto VF(E)$, defined for every $E \subset \Omega$, is called the *critical variation* of F.

Proposition 9.2.1. *Let $F : \mathbf{DF}(\Omega) \to \mathbb{R}$ be an additive function. Then*

$$\big|F(A)\big| \leq VF(A)$$

for each $A \in \mathbf{DF}(\Omega)$, and $VF(\operatorname{int} A) \leq F(A)$ if F is nonnegative.

PROOF. Choose $A \in \mathbf{DF}(\Omega)$ and $\delta : A \to \mathbb{R}_+$. By Proposition 2.2.2 there is a δ-fine dyadic partition $P = \big\{(A_1, x_1), \ldots, (A_p, x_p)\big\}$ with $[P] = A$. Then

$$\big|F(A)\big| = \bigg|\sum_{i=1}^p F(C_i)\bigg| \leq \sum_{i=1}^p \big|F(C_i)\big| \leq V_\delta F(A),$$

and $\big|F(A)\big| \leq VF(A)$ by the arbitrariness of δ. Now assume F is nonnegative, and select $\sigma : \operatorname{int} A \to \mathbb{R}_+$ so that $\sigma(x) \leq \operatorname{dist}(x, \partial A)$ for each $x \in \operatorname{int} A$. If $Q = \big\{(B_1, y_1), \ldots, (B_s, y_s)\big\}$ is a σ-fine dyadic partition, then $[Q] \subset A$. Thus $\sum_{i=1}^s F(B_i) = F\big([Q]\big) \leq F(A)$. As Q is arbitrary,

$$VF(\operatorname{int} A) \leq V_\sigma F(\operatorname{int} A) \leq F(A). \qquad \square$$

Proposition 9.2.2. *If $F : \mathbf{DF}(\Omega) \to \mathbb{R}$, then VF is a Borel measure in Ω.*

PROOF. Clearly $VF(\emptyset) = 0$. If $A \subset B \subset \Omega$ and $\delta : B \to \mathbb{R}_+$, then

$$VF(A) \leq V_{\delta \restriction A} F(A) \leq V_\delta F(B)$$

and $VF(A) \leq VF(B)$ by the arbitrariness of δ. Let E be the union of a sequence $\{E_k\}$ of disjoint subsets of Ω, and let $\delta_k : E_k \to \mathbb{R}_+$. Define a function $\delta : E \to \mathbb{R}_+$ by letting $\delta(x) := \delta_k(x)$ when $x \in E_k$. If P is a δ-fine dyadic partition in Ω, then $P_k := \big\{(C, x) \in P : x \in E_k\big\}$ are δ_k-fine disjoint dyadic partitions in Ω and $P = \bigcup_k P_k$. Thus

$$VF(E) \leq V_\delta F(E) \leq \sum_{(C,x) \in P} \big|F(C)\big| = \sum_k \sum_{(C,x) \in P_k} \big|F(C)\big| \leq \sum_k V_{\delta_k} F(E_k),$$

and the arbitrariness of δ_k implies $VF(E) \leq \sum_k VF(E_k)$. If E_k are any sets, then $A_k := E_k - \bigcup_{j=1}^{k-1} E_k$ are disjoint sets and $E = \bigcup_k A_k$. Hence

$$VF(E) \leq \sum_k VF(A_k) \leq \sum_k VF(E_k).$$

Finally, given sets $A, B \subset \Omega$ such that $d := \text{dist}\,(A, B) > 0$, choose a function $\delta : A \cup B \to (0, d/3)$ and observe that

$$VF(A) + VF(B) \leq V_{\delta \restriction A} F(A) + V_{\delta \restriction B} F(B) = V_\delta F(A \cup B).$$

As δ is arbitrary, $VF(A) + VF(B) \leq VF(A \cup B)$, and VF is a Borel measure by Theorem 1.3.1. $\qquad\square$

Proposition 9.2.3. *Let* $F : \mathbf{DF}(\Omega) \to \mathbb{R}$. *If the measure* VF *is absolutely continuous, it is Borel regular.*

PROOF. The set $B := \bigcup\{\partial Q : Q \in \mathbf{DC}(\Omega)\}$ is negligible and Borel. Thus to verify the Borel regularity of an absolutely continuous Borel measure, it suffices to consider a set $E \subset \Omega - B$. As there is nothing to prove otherwise, assume that $VF(E) < \infty$. Fix an integer $k \geq 1$, and find $\delta_k : E \to \mathbb{R}_+$ so that $V_{\delta_k} F(E) < VF(E) + 1/k$. For $j \in \mathbb{N}$ define

$$E_j = \{x \in E : \delta_k(x) > 1/j\}.$$

Let $\sigma \equiv 1/j$ be a constant function on $\text{cl}\, E_j - B$, and choose a σ-fine dyadic partition $\{(Q_1, y_1), \ldots, (Q_p, y_p)\}$ in Ω. Since $y_j \in \text{cl}\, E_j \cap \text{int}\, Q_j$, the intersection $E_j \cap \text{int}\, Q_j$ is not empty. Choosing $x_j \in E_j \cap \text{int}\, Q_j$, the dyadic partition $\{(Q_1, x_1), \ldots, (Q_p, x_p)\}$ is δ_k-fine. We infer

$$VF(\text{cl}\, E_j - B) \leq V_\sigma F(\text{cl}\, E_j - B) \leq V_{\delta_k} F(E) < VF(E) + 1/k.$$

Now $C_k = \bigcup_{j \in \mathbb{N}}(\text{cl}\, E_j - B)$ is a Borel set containing E, and

$$VF(C_k) = \lim VF(\text{cl}\, E_j - B) \leq VF(E) + 1/k.$$

Hence $C = \bigcap_{k \in \mathbb{N}} C_k$ is a Borel set containing E, and the proposition follows, since $VF(C) \leq VF(E)$. $\qquad\square$

Lemma 9.2.4. *Let* $F : \mathbf{DF}(\Omega) \to \mathbb{R}$. *If* VF *is absolutely continuous, then* $\overline{D}|F|(x) < \infty$ *for almost all* $x \in \Omega$.

PROOF. By Proposition 9.1.3, the set $E := \{x \in \Omega : \overline{D}|F|(x) = \infty\}$ is Borel. Seeking a contradiction, suppose $|E| > 0$ and find a compact set $K \subset E$ with $|K| > 0$. Replacing K by $\text{spt}\,(\mathcal{L}^n \llcorner K)$ if necessary, we may assume that $|K \cap V| > 0$ for each open set $V \subset \mathbb{R}^n$ which meets K. Select a countable dense subset C of K, and find open sets $U_k \subset \Omega$ so that $C \subset U_k$ and $|U_k| < 1/k$. If $U = \bigcap_{k=1}^\infty U_k$, then $D = K \cap U$ is a negligible G_δ set, still dense in K by the *Baire category theorem* [61, Chapter 7, Section 7]. According to our assumptions $VF(D) = 0$.

Now choose $\sigma : D \to \mathbb{R}_+$, and using Baire's theorem again, find $t > 0$ and an open set $V \subset \Omega$ such that $D \cap V \neq \emptyset$ and the set

$$D_t := \{x \in D \cap V : \sigma(x) > t\}$$

is dense in $D \cap V$, and hence in $K \cap V$. Denote by $\mathfrak{Q}$ the family of all cubes $Q \in \mathbf{DC}(\Omega)$ satisfying

$$d(Q) < t \quad \text{and} \quad |F(Q)| > \frac{|Q|}{|K \cap V|}.$$

Observe $E \subset \bigcup \mathfrak{Q}$. Passing to a subfamily if necessary, we may assume that $\mathfrak{Q}$ consists of nonoverlapping cubes. Since the boundary of each cube is a negligible set, the subfamily $\{Q \in \mathfrak{Q} : K \cap V \cap \operatorname{int} Q \neq \emptyset\}$, enumerated as $Q_1, Q_2, \ldots$, covers $K \cap V$ almost entirely. Thus

$$\sum_i |F(Q_i)| > \frac{1}{|K \cap V|} \sum_i |Q_i| \geq 1,$$

and consequently $\sum_{i=1}^p |F(Q_i)| > 1$ for some $p \in \mathbb{N}$. Each Q_i contains a point $x_i \in D_t$, and the collection $\{(Q_1, x_1), \ldots, (Q_p, x_p)\}$ is a σ-fine dyadic partition in Ω. Hence $V_\sigma F(D) > 1$, and $VF(D) \geq 1$ by the arbitrariness of σ — a contradiction. $\qquad\square$

Example 9.2.5. Assume $n = 1$ and $\Omega = \mathbb{R}$. Let D be the Cantor ternary discontinuum in the interval $I = [0, 1]$, and let $\mathcal{U}$ be the family of all connected components of $I - D$; see Example 8.1.4. A variant of the *Cantor-Vitali function* ("devil's staircase") is an increasing continuous function $v : \mathbb{R} \to \mathbb{R}$ such that

$$v(x) := \begin{cases} 0 & \text{if } x < 0, \\ (a+b)/2 & \text{if } a < x < b \text{ and } (a,b) \in \mathcal{U}, \\ 1 & \text{if } x > 1; \end{cases}$$

cf. [26, Section 4.2] and [54]. If F is the flux of v, then

$$F([a,b]) = \int_{\partial([a,b])} v \cdot \nu_{[a,b]} \, d\mathcal{H}^0 = v(b) - v(a) \geq 0$$

for every interval $[a, b] \subset \mathbb{R}$. If $P = \{(A_1, x_1), \ldots, (A_p, x_p)\}$ is any dyadic partition, then $\sum_{i=1}^p F(A_i) \leq 1$. Hence by Proposition 9.2.1,

$$1 = F(I) \leq VF(I) \leq VF(\mathbb{R}) \leq 1.$$

Now define $\sigma : x \mapsto \operatorname{dist}(x, D) : \mathbb{R} - D \to \mathbb{R}_+$, and observe that

$$VF(\mathbb{R} - D) \leq V_\sigma F(\mathbb{R} - D) = 0.$$

Thus $VF(D) = 1$, and as $|D| = 0$, the measure VF is not absolutely continuous. Still $F'(x) = v'(x) = 0$ for each x in $\mathbb{R} - D$.

Lemma 9.2.6. *Let $F : \mathbf{DF}(\Omega) \to \mathbb{R}$ be additive. If $\underline{D}F(x) \geq 0$ for VF almost all $x \in \Omega$, then F is nonnegative.*

PROOF. Let $E = \{x \in \Omega : \underline{D}F(x) < 0\}$, and choose $A \in \mathbf{DF}(\Omega)$ and $\varepsilon > 0$. There is $\sigma : E \to \mathbb{R}_+$ such that $V_\sigma F(E) < \varepsilon$. Given $x \in A - E$, find $\eta_x > 0$ so that $F(Q) > -\varepsilon|Q|$ for every cube $Q \in \mathbf{DC}(\Omega)$ with $x \in Q$ and $d(Q) < \eta_x$. Define a function $\delta : A \to \mathbb{R}_+$ by the formula

$$\delta(x) := \begin{cases} \sigma(x) & \text{if } x \in E, \\ \eta_x & \text{it } x \in A - E. \end{cases}$$

There is a δ-fine dyadic partition $P = \{(C_1, x_1), \ldots, (C_p, x_p)\}$ with $[P] = A$; see Proposition 2.2.2. By additivity,

$$F(A) = \sum_{i=1}^{p} F(C_i) \geq -\varepsilon \sum_{x_i \in A - E} |C_i| - \sum_{x_i \in E} |F(C_i)|$$
$$> -\varepsilon|A| - V_\sigma F(E) > -\varepsilon(|A| + 1)$$

and $F(A) \geq 0$ by the arbitrariness of ε. $\qquad\qquad\square$

We denote by $AC_*(\Omega)$ the linear space of all additive functions

$$F : \mathbf{DF}(\Omega) \to \mathbb{R}$$

whose critical variation VF is *absolutely continuous*.

Theorem 9.2.7. *Let $F \in AC_*(\Omega)$. The derivative $F'(x)$ exists at almost all $x \in \Omega$, and the linear map*

$$F \mapsto F' : AC_*(\Omega) \to L^0(\Omega)$$

is injective. If $F' \in L^1_{\mathrm{loc}}(\Omega)$ then for each $A \in \mathbf{DF}(\Omega)$,

$$F(A) = \int_A F'(x) \, dx.$$

PROOF. Theorem 9.1.5 and Lemmas 9.2.4 and 9.2.6 imply the first two assertions. Suppose $F' \in L^1_{\mathrm{loc}}(\Omega)$, and for each $A \in \mathbf{DF}(\Omega)$, let

$$G(A) := \int_A F'(x) \, dx.$$

Choose a negligible set $N \subset \Omega$ and $\varepsilon > 0$. Making N larger, we may assume that $F'(x)$ exists for each $x \in \Omega - N$. There are open sets $\Omega_k \Subset \Omega$ such that $\Omega = \bigcup_{k \in \mathbb{N}} \Omega_k$. Fix $k \in \mathbb{N}$, let $N_k = N \cap \Omega_k$, and define $g \in L^1(\Omega_k)$ by

$$g(x) := \begin{cases} F'(x) & \text{if } x \in \Omega_k - N_k, \\ 0 & \text{if } x \in N_k. \end{cases}$$

Find $\delta : \Omega_k \to \mathbb{R}_+$ that corresponds to g, Ω_k, and ε according to Henstock's lemma. We may assume that $\delta(x) < \mathrm{dist}\,(x, \partial\,\Omega_k)$ for each $x \in N_k$. Thus if

$P = \{(C_1, x_1), \ldots, (C_p, x_p)\}$ is a $(\delta \upharpoonright N_k)$-fine dyadic partition, then $[P] \subset \Omega_k$. By Henstock's lemma,

$$\sum_{i=1}^{p} |G(C_i)| = \sum_{i=1}^{p} \Big| g(x_i)|C_i| - G(C_i) \Big| < \varepsilon$$

and hence $VG(N_k) \leq V_{\delta \upharpoonright N_k}(N_k) \leq \varepsilon$. The arbitrariness of ε shows that N_k is VG negligible, and so is $N = \bigcup_{k \in \mathbb{N}} N_k$. It follows that $G \in AC_*(\Omega)$. Since $G'(x) = F'(x)$ for almost all $x \in \Omega$ by Observation 9.1.2, the first part of the proof implies $G = F$. $\qquad\square$

The flux of $v \in L^\infty_{\mathrm{loc}}(\Omega, \mathcal{H}^{n-1}; \mathbb{R}^n)$ is the additive function

$$F : A \mapsto \int_{\partial A} v \cdot \nu_A \, d\mathcal{H}^{n-1} : \boldsymbol{DF}(\Omega) \to \mathbb{R}.$$

Let $x \in \Omega$. If the derivative $F'(x)$ exists, we call it the *mean divergence* of v at the point x, and denote it by $\mathfrak{div}\, v(x)$. The following corollary is a mere reformulation of Theorem 9.2.7.

Corollary 9.2.8. *Let the flux F of $v \in L^\infty_{\mathrm{loc}}(\Omega, \mathcal{H}^{n-1}; \mathbb{R}^n)$ belong to $AC_*(\Omega)$. Then F is uniquely determined by the mean divergence $\mathfrak{div}\, v$, which is defined almost everywhere in Ω. If $\mathfrak{div}\, v \in L^1_{\mathrm{loc}}(\Omega)$, then for each $A \in \boldsymbol{DF}(\Omega)$,*

$$\int_A \mathfrak{div}\, v(x) \, dx = \int_{\partial A} v \cdot \nu_A \, d\mathcal{H}^{n-1}. \tag{9.2.1}$$

Remark 9.2.9. If $v \in C(\Omega; \mathbb{R}^n)$, then it follows from Sections 2.4 and 6.7 that under the assumptions of Corollary 9.2.8, equality (9.2.1) holds for each $A \Subset \Omega$ such that $A \in \boldsymbol{BV}(\mathbb{R}^n)$. This fact will be made more explicit in the next chapter.

Remark 9.2.10. Linear spaces akin to $AC_*(\Omega)$ provide descriptive definitions of various forms of multidimensional conditionally convergent integrals [55, 5, 12, 6, 13, 7, 50]. It is easy to verify that in dimension one, the functions in $AC_*(\mathbb{R})$ coincide with *Denjoy-Perron primitives*, called ACG$_*$ *functions* in [65, Chapter 7, Section 8]; cf. [50]. However, the classical definition of ACG$_*$ functions depends in an essential way on the order structure of the real line, and its extension to higher dimensions is far from obvious [60]. Thus it is significant that the space $AC_*(\Omega)$ is defined directly in *any dimension*. A typical element of $AC_*(\Omega)$ is the flux of a pointwise Lipschitz vector field $v : \Omega \to \mathbb{R}^n$ whose divergence is not locally integrable; cf. Remark 2.3.8, (2). Although we do not use such vector fields in this book, we believe that they occur naturally and deserve attention. Under less restrictive assumptions, deeper properties of the space $AC_*(\Omega)$ are established in [51].

The following two examples are due to Z. Buczolich.

Example 9.2.11. Let $n = 2$ and $\Omega = \mathbb{R}^2$. For $r > 0$ and $x \in \mathbb{R}^n$, let

$$\varphi_r(x) := \begin{cases} \exp\left(\frac{|x|^2}{|x|^2 - \varepsilon^2}\right) & \text{if } |x| < r, \\ 0 & \text{if } |x| \geq r, \end{cases}$$

and $u_r(x) := (\varphi_r(x), 0)$. Clearly $u_r \in C^\infty(\mathbb{R}^2; \mathbb{R}^2)$, and we have

$$\|u_r\|_{L^\infty(\mathbb{R}^2; \mathbb{R}^2)} = 1 \quad \text{and} \quad \int_{\mathbb{R}^2} |\mathrm{div}\, u_r| \leq 2\pi r.$$

Enumerating a dense countable subset of $\mathbb{R}^2$, define recursively sequences $\{z_k\}$ in $\mathbb{R}^2$ and $\{r_k\}$ in $\mathbb{R}_+$ so that

(1) $r_1 \leq 1/2$ and $r_{k+1} \leq r_k/2$ for $k = 1, 2, \ldots$,
(2) the family $\{B(z_k, r_k) : k = 1, 2, \ldots\}$ is disjoint,
(3) the open set $U = \bigcup_{k=1}^\infty U(z_k, r_k)$ is a dense in $\mathbb{R}^2$.

Note that $0 < |U| < 2\pi r_1^2$ and $\mathbf{P}(U) \leq 4\pi r_1$.

Claim. If $Z = \{z_k : k = 1, 2, \ldots\}$ then $\mathbb{R}^2 - U \subset \mathrm{cl}\, Z$.

Proof. Suppose there are $x \in \mathbb{R}^2 - U$ and $\varepsilon > 0$ such that $U(x, \varepsilon)$ does not meet Z. As $\lim r_k = 0$, there exists an integer $p \geq 1$ which satisfies $U(x, \varepsilon/2) \cap B(z_k, r_k) = \emptyset$ for $k > p$. Since $x \notin U$, and since the compact sets $B(z_1, r_1), \ldots, B(z_p, r_p)$ are disjoint, we see at once that the open set $U(x, \varepsilon/2) - \bigcup_{k=1}^p B(z_k, \varepsilon_k)$ is not empty and does not meet U. As U is dense in $\mathbb{R}^2$, this is a contradiction.

Let $v_k(x) := u_{r_k}(x - z_k)$ for each $x \in \mathbb{R}^2$, and $v := \sum_{k=1}^\infty v_k$. Clearly v is bounded, Borel measurable, and C^∞ in U. By the Claim, v is discontinuous at every $x \in \mathbb{R}^2 - U$. Let $A \in \mathcal{BV}(\mathbb{R}^2)$. For $k = 1, 2, \ldots$,

$$\int_A \mathrm{div}\, v_k(x)\, dx = \int_{\partial_* A} v_k \cdot \nu_A\, d\mathcal{H}^1 \tag{$*$}$$

by Theorem 6.5.4, and

$$\sum_{k=1}^\infty \int_{\partial_* A} |v_k \cdot \nu_A|\, d\mathcal{H}^1 \leq \sum_{k=1}^\infty \mathcal{H}^1[\partial_* A \cap B(z_k, r_k)] \leq \mathbf{P}(A)$$

by (2). The function $g = \sum_{k=1}^\infty \mathrm{div}\, v$ belongs to $L^1(\mathbb{R}^2)$, since

$$\sum_{k=1}^\infty \int_{\mathbb{R}^2} |\mathrm{div}\, v_k| \leq 2\pi \sum_{k=1}^\infty r_k \leq 2\pi.$$

In view of the previous two inequalities and $(*)$,

$$\int_A g(x)\, dx = \int_{\partial_* A} v \cdot \nu_A\, d\mathcal{H}^1.$$

It follows from Observation 9.1.2 that $g(x) = \mathfrak{div}\, v(x)$ for almost all $x \in \mathbb{R}^2$.

Example 9.2.12. Adhering to the notation of Example 9.2.11, we modify $v : \mathbb{R}^2 \to \mathbb{R}^2$ so that it is continuous but not Lipschitz at any $x \in \mathbb{R}^2 - U$.

Claim. Let $A \subset \mathbb{R}^2$ be a bounded set. Given $\varepsilon > 0$ and $p \in \mathbb{N}$, there is an integer $q \geq p$ such that dist $(x, \{z_p, \ldots, z_q\}) < \varepsilon$ for each point $x \in A - U$.

Proof. As $A - U$ is bounded, it contains points $x_1, \ldots, x_r$ such that $A - U \subset \bigcup_{i=1}^{r} U(x_i, \varepsilon/2)$. By the claim of Example 9.2.11, each $U(x_i, \varepsilon/2)$ contains z_{k_i} with $k_i \geq p$. Thus $q := \max\{k_1, \ldots, k_r\}$ is the desired integer.

Let $p(1) = 0$, and define recursively a strictly increasing sequence $\{p(i)\}$ of integers so that

$$\text{dist}\,(x, \{z_{p(i)+1}, \ldots, z_{p(i+1)}\}) < 2^{-i}$$

for every x in $U(0, i) - U$, $i = 1, 2, \ldots$. Given an integer $k \geq 1$, there is a unique integer $i_k \geq 1$ with $p(i_k) < k \leq p(i_k + 1)$, and we let $c_k = 1/i_k$. Observe that $i_k \leq i_{k+1}$ for $k = 1, 2, \ldots$ and $\lim i_k = \infty$.

If $w := \sum_{k=1}^{\infty} c_k v_k$, then calculating as in Example 9.2.11, we obtain

$$\text{div}\, v(x) = \begin{cases} \text{div}\, w(x) & \text{if } x \in U, \\ 0 & \text{if } x \in \mathbb{R}^2 - U. \end{cases}$$

Clearly w is C^∞ in U. Select x in $\mathbb{R}^2 - U$ and $\varepsilon > 0$. There are an integer $k \geq 1$ with $c_k < \varepsilon$, and $\eta > 0$ such that $U(x, \eta) \cap U(z_j, r_j) = \emptyset$ for $j = 1, \ldots, k - 1$. Each $y \in U(x, \eta)$ is either in $\mathbb{R}^2 - V$, in which case $w(y) = w(x) = 0$, or in $U(z_j, r_j)$ where $j \geq k$, in which case

$$|w(y) - w(x)| = c_j v_j(y) \leq c_j \leq c_k < \varepsilon\,.$$

Consequently w is continuous at x. On the other hand, if $x \in \mathbb{R}^2 - U$ then $x \in U(0, i_k) - U$ for all sufficiently large $k \in \mathbb{N}$. Since $p(i_k) < k \leq p(i_k + 1)$, we obtain $|z_k - x| < 2^{-i_k}$. Hence

$$\frac{|w(z_k) - w(x)|}{|z_k - x|} \geq 2^{i_k} c_k v_k(z_k) \leq \frac{2^{i_k}}{i_k}$$

for all sufficiently large $k \in \mathbb{N}$. It follows that w is not Lipschitz at x.

We note that V. Shapiro [67] studied the mean divergence of vector fields by a completely different method.

Chapter 10

Charges

A charge is a distribution whose continuity properties resemble those of the distributional divergence of a continuous vector field. We define a locally convex topology $\mathcal{T}$ in the space $\mathcal{D}$ of all test functions so that charges are $\mathcal{T}$ continuous, and that the space BV_c of all BV functions with compact support is the sequential completion of $(\mathcal{D}, \mathcal{T})$. In the sense of Mackey-Arens theorem, the space of all charges is in duality with BV_c — a fact we employ in Chapter 11. Some properties of locally convex spaces are stated without proofs. In such cases, we provide references to standard texts.

10.1. Continuous vector fields

Throughout this section, $\Omega \subset \mathbb{R}^n$ is a fixed open set and

$$\mathcal{D}(\Omega) := \{\varphi \in \mathcal{D}(\mathbb{R}^n) : \operatorname{spt} \varphi \subset \Omega\}.$$

Thus when U is an open subset of Ω, we write $\mathcal{D}(U) \subset \mathcal{D}(\Omega)$; see (1.1.2).

The *distributional divergence* of $v \in L^1_{\text{loc}}(\Omega; \mathbb{R}^n)$ is the distribution

$$F_v : \varphi \mapsto - \int_\Omega v(x) \cdot D\varphi(x) \, dx : \mathcal{D}(\Omega) \to \mathbb{R}$$

introduced in Example 3.2.1, (3). For $v \in C(\Omega; \mathbb{R}^n)$ we state the essential continuity property of F_v.

Observation 10.1.1. *Let F_v be the distributional divergence of a vector field $v \in C(\Omega; \mathbb{R}^n)$. Given an open set $U \Subset \Omega$ and $\varepsilon > 0$, there is $\theta > 0$ such that*

$$F_v(\varphi) \leq \theta \|\varphi\|_{L^1(U)} + \varepsilon \|D\varphi\|_{L^1(U;\mathbb{R}^n)}$$

for each test function $\varphi \in \mathcal{D}(U)$.

PROOF. Choose an open set $U \Subset \Omega$ and $\varepsilon > 0$. There is $w \in C^1(\Omega; \mathbb{R}^n)$ such that $\|w - v\|_{L^\infty(U;\mathbb{R}^n)} < \varepsilon$. Now select $\theta > \|\operatorname{div} w\|_{L^\infty(U)}$ and a test function $\varphi \in \mathcal{D}(U)$. The following calculation, which involves integration by parts, completes the argument:

$$F_v(\varphi) = -\int_U w \cdot D\varphi + \int_U (w - v) \cdot D\varphi$$

$$\leq \int_U \varphi \operatorname{div} w + \varepsilon \|D\varphi\|_{L^1(U;\mathbb{R}^n)}$$

$$\leq \theta \|\varphi\|_{L^1(U)} + \varepsilon \|D\varphi\|_{L^1(U;\mathbb{R}^n)}. \qquad \square$$

Formulating the continuity property of Observation 10.1.1 without reference to a particular vector field leads to the following definition.

Definition 10.1.2. A distribution $F \in \mathcal{D}'(\Omega)$ is called a *charge*[1] in Ω if given an open set $U \Subset \Omega$ and $\varepsilon > 0$, there is $\theta > 0$ such that

$$F(\varphi) \leq \theta \|\varphi\|_{L^1(U)} + \varepsilon \|D\varphi\|_{L^1(U;\mathbb{R}^n)}$$

for each test function $\varphi \in \mathcal{D}(U)$.

The linear space of all charges in Ω is denoted by $CH(\Omega)$. We define a locally convex topology in $CH(\Omega)$ by seminorms

$$\|F\|_U := \sup\{F(\varphi) : \varphi \in \mathcal{D}(U) \text{ and } \|D\varphi\|_{L^1(U;\mathbb{R}^n)} \leq 1\} \qquad (10.1.1)$$

where $F \in CH(\Omega)$ and $U \Subset \Omega$ is an open set. Arguing as in Example 1.2.2, we see that $CH(\Omega)$ is a Fréchet space. When the open set $\Omega \subset \mathbb{R}^n$ is understood, we usually say only "charge" instead of "charge in Ω".

Observation 10.1.1 asserts that the distributional divergence F_v of a continuous vector field $v : \Omega \to \mathbb{R}^n$ is a charge. If $U \Subset \Omega$ is an open set, then

$$F_v(\varphi) = -\int_U v \cdot D\varphi \leq \|D\varphi\|_{L^1(U;\mathbb{R}^n)} \|v\|_{L^\infty(U;\mathbb{R}^n)}$$

for each $\varphi \in \mathcal{D}(U)$, and hence

$$\|F_v\|_U \leq \|v\|_{L^\infty(U;\mathbb{R}^n)}. \qquad (10.1.2)$$

The following example and proposition show that charges may arise from other sources than continuous vector fields. However, see Theorem 11.3.8 below.

Example 10.1.3. Assume $n = 2$. For $x = (\xi, \eta)$ in $\mathbb{R}^2$, define

$$v(x) := \begin{cases} |x|^{-3/2}(-\eta, \xi) & \text{if } |x| > 0, \\ 0 & \text{if } |x| = 0. \end{cases}$$

---

[1]The word "charge" has been used in the literature to describe various concepts. Some differ completely from our definition [58], while others are similar [51, 52, 22]. Our notion of "charge", called "strong charge" in [23], was introduced without name in [51, Proposition 2.1.7]. It evolved from the concept of "continuous additive function" defined in [48].

Clearly v is discontinuous at $x = 0$, controlled, and $\operatorname{div} v(x) = 0$ for each $x \neq 0$; cf. Example 8.2.9. As Theorems 8.3.3 show that $F_v(\varphi) = 0$ for each $\varphi \in \mathcal{D}(\mathbb{R}^2)$, we see that F_v is a charge in $\mathbb{R}^2$.

Proposition 10.1.4. *If $f \in L_{\mathrm{loc}}^n(\Omega)$, then*

$$L_f : \varphi \mapsto \int_\Omega f\varphi : \mathcal{D}(\Omega) \to \mathbb{R}$$

is a charge. In addition, there is a constant $\gamma = \gamma(n) > 0$ such that

$$\|L_f\|_U \leq \gamma \|f\|_{L^n(U)}$$

for each open set $U \Subset \Omega$.

PROOF. Choose an open set $U \Subset \Omega$ and a test function $\varphi \in \mathcal{D}(\Omega)$. For $c > 0$, let $U_c := \{x \in U : |f(x)|^n \geq c\}$. By the Hölder and Sobolev inequalities,

$$L_f(\varphi) \leq c \int_{U-U_c} |\varphi| + \|f\|_{L^n(U_c)} \|\varphi\|_{L^{\frac{n}{n-1}}(U_c)} \tag{$*$}$$

$$\leq c\|\varphi\|_{L^1(U)} + \gamma \|f\|_{L^n(U_c)} \|D\varphi\|_{L^1(U;\mathbb{R}^n)}$$

where $\gamma = \gamma(n) > 0$. Observe $\lim_{c\to\infty} |U_c| = 0$, since $\chi_{U_c} \leq |f|^n/c$. Thus $\int_{U_\theta} |f|^n \leq (\varepsilon/\gamma)^n$ for some $\theta > 0$, and we obtain

$$L_f(\varphi) \leq \theta \|g\|_{L^1(U)} + \varepsilon \|D\varphi\|_{L^1(U;\mathbb{R}^n)}.$$

This shows that L_f is a charge. Since $U_c = U$ when $c = 0$,

$$L_f(\varphi) \leq \gamma \|f\|_{L^n(U)} \|D\varphi\|_{L^1(U;\mathbb{R}^n)}$$

by inequality $(*)$. The proposition follows. □

Our next goal is to define a topology $\mathcal{T}$ in $\mathcal{D}(\Omega)$ so that $CH(\Omega)$ is the dual space of $(\mathcal{D}(\Omega), \mathcal{T})$. This problem is best approached in the abstract setting of locally convex spaces. The definition of $\mathcal{T}$ resembles that of the internal injective limit topology discussed in Section 3.6. We follow ideas presented previously in [21, Section 6], [51, Section 1.2], and [24, Section 3].

10.2. Localized topology

For a family $\mathcal{A}$ of sets and a set E, we defined in Section 3.6 the family

$$\mathcal{A} \sqcap E := \{A \cap E : A \in \mathcal{A}\}.$$

A topology $\mathcal{U}$ in a set X is called *regular* if each neighborhood of every $x \in X$ contains a closed neighborhood of x. Equivalently, $\mathcal{U}$ is regular if each closed set $C \subset X$ and every point $x \in X - C$ have disjoint neighborhoods. Each locally convex topology in a linear space is regular [64, Theorem 1.10].

Proposition 10.2.1. *Let $(X, \mathcal{U})$ be a topological space, and let X be the union of a family $\mathcal{C}$ of its subsets. There is a unique topology $\mathcal{U}_{\mathcal{C}}$ in X such that*

(i) *$\mathcal{U}_{\mathcal{C}} \llcorner C \subset \mathcal{U} \llcorner C$ for each $C \in \mathcal{C}$,*

(ii) *given a topological space Y, a map $\phi : (X, \mathcal{U}_{\mathcal{C}}) \to Y$ is continuous whenever the restrictions $\phi \restriction C : (C, \mathcal{U} \llcorner C) \to Y$ are continuous for each $C \in \mathcal{C}$.*

The topology $\mathcal{U}_{\mathcal{C}}$ has the following properties:

(1) *$\mathcal{U}_{\mathcal{C}} := \{E \subset X : E \cap C \in \mathcal{U} \llcorner C \text{ for each } C \in \mathcal{C}\}$; in particular $\mathcal{U} \subset \mathcal{U}_{\mathcal{C}}$ and $\mathcal{U}_{\mathcal{C}} \llcorner C = \mathcal{U} \llcorner C$ for every $C \in \mathcal{C}$.*

(2) *$E \subset X$ is $\mathcal{U}_{\mathcal{C}}$ closed if and only if the intersection $E \cap C$ is $\mathcal{U} \llcorner C$ closed for every $C \in \mathcal{C}$.*

(3) *If $Y \subset X$, then $\mathcal{U}_{\mathcal{C}} \llcorner Y \subset (\mathcal{U} \llcorner Y)_{\mathcal{C} \llcorner Y}$ and the equality occurs when Y is $\mathcal{U}$ open or $\mathcal{U}$ closed.*

PROOF. If $\mathcal{S}$ and $\mathcal{S}'$ are topologies in X satisfying conditions (i) and (ii), then

$$x \mapsto x : (X, \mathcal{S}) \to (X, \mathcal{S}')$$

is a homeomorphism. Thus $\mathcal{S} = \mathcal{S}'$, and the uniqueness of $\mathcal{U}_{\mathcal{C}}$ is proved. A direct verification validates property (1), which establishes the existence of the topology $\mathcal{U}_{\mathcal{C}}$.

Let $E \subset X$ and $C \in \mathcal{C}$. Clearly $E \cap C$ is $\mathcal{U} \llcorner C$ closed if and only if

$$(X - E) \cap C = C - E \cap C$$

belongs to $\mathcal{U} \llcorner C$. It follows from property (1) that $E \cap C$ is closed for every $C \in \mathcal{C}$ if and only if $X - E$ belongs to $\mathcal{U}_{\mathcal{C}}$, or alternatively if and only if E is $\mathcal{U}_{\mathcal{C}}$ closed.

Given $Y \subset X$, choose an arbitrary but fixed $C \in \mathcal{C}$. Select $A \in \mathcal{U}_{\mathcal{C}} \llcorner Y$, and find $U \in \mathcal{U}_{\mathcal{C}}$ with $U \cap Y = A$. By property (1), there is $V \in \mathcal{U}$ such that

$$A \cap (C \cap Y) = (U \cap Y) \cap C = (U \cap C) \cap Y = (V \cap C) \cap Y.$$

As $V \cap C \in \mathcal{U} \llcorner C$, the intersection $A \cap (C \cap Y)$ belongs to

$$(\mathcal{U} \llcorner C) \llcorner Y = (\mathcal{U} \llcorner Y) \llcorner (C \cap Y).$$

The arbitrariness of C and property (1) yield $A \in (\mathcal{U} \llcorner Y)_{\mathcal{C} \llcorner Y}$.

If $Y \in \mathcal{U}$ then $\mathcal{U} \llcorner Y \subset \mathcal{U}$. Select $B \in (\mathcal{U} \llcorner Y)_{\mathcal{C} \llcorner Y}$ and using property (1), observe that $B \cap (C \cap Y) = B \cap C$ belongs to $(\mathcal{U} \llcorner Y) \llcorner (C \cap Y)$. In other words, there is W in $\mathcal{U} \llcorner Y$, and hence in $\mathcal{U}$, such that

$$B \cap C = W \cap (C \cap Y) = W \cap C.$$

Since C is arbitrary, $B \in \mathcal{U}_{\mathcal{C}} \llcorner Y$ by property (1). In view of property (2), the case when Y is $\mathcal{U}$ closed is proved similarly. $\qquad\square$

Throughout we adhere to the notation of Proposition 10.2.1, and call $\mathcal{U}_{\mathcal{C}}$ the *localization* of the original topology $\mathcal{U}$ by $\mathcal{C}$, or simply the *localized topology* when the original topology $\mathcal{U}$ and localizing family $\mathcal{C}$ are understood from the context.

Proposition 10.2.2. *Let $(X, \mathcal{U})$ be a regular space, and let X be the union of an increasing sequence $\mathcal{C} = \{C_k\}$ of **closed** subsets of $(X, \mathcal{U})$. The localized topology $\mathcal{U}_{\mathcal{C}}$ has the following properties:*

(1) *A sequence $\{x_i\}$ in X converges in $(X, \mathcal{U}_{\mathcal{C}})$ if and only if $\{x_i\}$ is a sequence in some C_k and $\{x_i\}$ converges in $(X, \mathcal{U})$.*

(2) *If each $\mathcal{U} \llcorner C_k$ is sequential, then so is $\mathcal{U}_{\mathcal{C}}$.*

PROOF. (1) Let $\{x_i\}$ be a sequence in X converging to $x \in X$ in $(X, \mathcal{U}_{\mathcal{C}})$. As $\mathcal{U} \subset \mathcal{U}_{\mathcal{C}}$, the sequence $\{x_i\}$ converges to x in $(X, \mathcal{U})$. If $\{x_i\}$ is not a sequence in any C_k, construct recursively a subsequence $\{y_k\}$ of $\{x_i\}$ so that $y_k \notin C_k$ for $k = 1, 2, \ldots$. By regularity, there are $U_k \in \mathcal{U}$ such that $C_k \subset U_k$ and $y_k \notin U_k$. If $x \in C_p$, let $U := \bigcap_{j=p}^{\infty} U_j$. Since $\{C_k\}$ is increasing,

$$C_k \cap U = \begin{cases} C_k & \text{if } k \leq p, \\ C_k \cap \bigcap\{U_j : p \leq j < k\} & \text{if } k > p \end{cases}$$

for $k = 1, 2, \ldots$. Thus U is a neighborhood of x in $(X, \mathcal{U}_{\mathcal{C}})$ that contains no y_k with $k \geq p$ — a contradiction. Since each C_k is $\mathcal{U}$ closed, the converse follows from Proposition 10.2.1, (i).

(2) Suppose that each $\mathcal{U} \llcorner C_k$ is sequential, and that $E \subset X$ is sequentially closed in $(X, \mathcal{U}_{\mathcal{C}})$. Let $\{x_i\}$ be a sequence in $E \cap C_j$ converging in $(X, \mathcal{U})$. Property (1) shows that $\{x_i\}$ converges to $x \in C_j$ in $(X, \mathcal{U}_{\mathcal{C}})$. By our assumption, $x \in E \cap C_j$. Thus $E \cap C_j$ is sequentially closed, and hence $\mathcal{U} \llcorner C_j$ closed. Since this is true for any C_j, the set E is closed in $(X, \mathcal{U}_{\mathcal{C}})$ according to Proposition 10.2.1, (2). □

Remark 10.2.3. Sequences $\{C_j\}$ and $\{D_k\}$ of sets are called *interlacing* if each C_j is contained in some D_k, and each D_k is contained in some C_j. Let $(X, \mathcal{U})$ be a topological space, and let $\mathcal{C}$ and $\mathcal{D}$ be interlacing increasing sequences of subsets of X. If $X = \bigcup \mathcal{C}$, then $X = \bigcup \mathcal{D}$ and $\mathcal{U}_{\mathcal{C}} = \mathcal{U}_{\mathcal{D}}$ by Proposition 10.2.1, (1); cf. Proposition 3.6.2.

10.3. Locally convex spaces

The following lemma is a convenient tool for applying the results of Section 10.2 to locally convex spaces.

Lemma 10.3.1. *Let X be a linear space, and let $\mathcal{U}$ be any topology in X that has a neighborhood base at zero consisting of convex sets. If the maps*

$$z \mapsto x + z : X \to X \quad and \quad z \mapsto tz : X \to X$$

are continuous for every $x \in X$ and every $t \in \mathbb{R}$, then $(X, \mathcal{U})$ is a locally convex space.

PROOF. According to our assumptions, for every $t \in \mathbb{R} - \{0\}$, the map

$$z \mapsto tz : (X, \mathcal{U}) \to (X, \mathcal{U})$$

is a surjective homeomorphism. Thus if U is a convex neighborhood of zero, then so is tU for each $t \neq 0$. In particular, $U \cap (-U)$ is a neighborhood of zero, and $\mathcal{U}$ has a neighborhood base at zero consisting of convex symmetric sets. Our assumptions also show that for $u, x \in X$ and $t \in \mathbb{R} - \{0\}$, the maps

$$(v, z) \mapsto (u + v, x + z) \quad and \quad (t, z) \mapsto (t, x + tz)$$

are surjective homeomorphisms of the spaces $(X, \mathcal{U}) \times (X, \mathcal{U})$ and $\mathbb{R} \times (X, \mathcal{U})$, respectively. Thus it suffices to show that the addition $(x, y) \mapsto x + y$ is continuous at $(0, 0)$, and the scalar multiplication $(t, x) \mapsto tx$ is continuous at $(t, 0)$ for each $t \in \mathbb{R}$. Choose a convex symmetric neighborhood $U \subset X$ of zero. Since $\frac{1}{2}U + \frac{1}{2}U \subset U$, the continuity of addition at $(0, 0)$ follows. Next choose $t \in \mathbb{R}$. If $|t| < 1/2$, then $sx \in U$ for $x \in U$ and $|s - t| < 1/2$, since $|s| < 1$. If $|t| \geq 1$, then $sx \in U$ for $x \in \frac{1}{2|t|}U$ and $|s - t| < |t|$, since $|s| < 2|t|$. This proves the continuity of scalar multiplication at $(t, 0)$. $\square$

Definition 10.3.2. A *localizing sequence*[2] in a locally convex space X is an increasing sequence $\{C_k\}$ of **compact** convex sets such that $0 \in C_1$, and for each $k \in \mathbb{N}$, $t \in \mathbb{R}$, and $x \in X$ there is $k(t, x) \in \mathbb{N}$ with

$$x + tC_k \subset C_{k(t,x)}.$$

Note that $X = \bigcup_{k=1}^{\infty} C_k$, since $x \in C_{1(0,x)}$ for each $x \in X$, and that $(X, \mathcal{U})$ is a regular space [64, Theorem 1.11]. According to Proposition 10.2.1, each localizing sequence $\mathcal{C}$ in a locally convex space $(X, \mathcal{U})$ defines a localized topology $\mathcal{U}_\mathcal{C}$.

Theorem 10.3.3. *If $\mathcal{C} = \{C_k\}$ is a localizing sequence in a locally convex space $(X, \mathcal{U})$, then*

    (1) *$\mathcal{U}_\mathcal{C}$ is locally convex and sequentially complete,*
    (2) *$E \subset X$ is $\mathcal{U}_\mathcal{C}$ bounded if and only if E is contained in some C_k.*

---

[2]The terms "localized topology" and "localizing sequence" were introduced in [24] by T. De Pauw.

PROOF. Let $x \in X$ and $t \in \mathbb{R}$. By Proposition 10.2.1, (ii),

$$\tau_x : z \mapsto x + z \quad \text{and} \quad \mu_t : z \mapsto tz$$

are continuous maps of $(X, \mathcal{U}_\mathcal{C})$, since $\tau_x(C_k) \subset C_{k(1,x)}$ and $\mu_t(C_k) \subset C_{k(t,0)}$ for $k = 1, 2, \ldots$. In view of Lemma 10.3.1, we prove that $\mathcal{U}_\mathcal{C}$ is a locally convex topology by showing that it has a neighborhood base at zero consisting of convex sets. To this end, choose $U \in \mathcal{U}_\mathcal{C}$ containing zero. Using Proposition 10.2.1, (1), find $U_k \in \mathcal{U}$ so that $C_k \cap U = C_k \cap U_k$ for $k = 1, 2, \ldots$, and observe that each U_k contains zero. There is a convex set $V_0 \in \mathcal{U}$ such that $0 \in V_0$ and $\operatorname{cl}_\mathcal{U} V_0 \subset U_1$. As $C_1 \cap \operatorname{cl}_\mathcal{U} V_0$ is a compact convex set and

$$C_1 \cap \operatorname{cl}_\mathcal{U} V_0 \subset C_1 \cap U_1 = C_1 \cap U \subset C_2 \cap U = C_2 \cap U_2 \subset U_2,$$

there is a convex set $V_1 \in \mathcal{U}$ such that $C_1 \cap \operatorname{cl}_\mathcal{U} V_0 \subset V_1$ and $\operatorname{cl}_\mathcal{U} V_1 \subset U_2$; see [64, Theorem 1.10]. Recursively, we define convex sets $V_k \in \mathcal{U}$ so that $C_k \cap \operatorname{cl}_\mathcal{U} V_{k-1} \subset V_k$ and $\operatorname{cl}_\mathcal{U} V_{k-1} \subset U_k$ for each $k \in \mathbb{N}$. In particular, we have $C_k \cap V_{k-1} \subset V_k \subset U_{k+1}$ and hence

$$C_k \cap V_{k-1} \subset C_{k+1} \cap V_k \subset C_{k+1} \cap U_{k+1} = C_{k+1} \cap U \subset U.$$

As $0 \in C_1 \cap V_0$, the union $V = \bigcup_{k=1}^{\infty}(C_k \cap V_{k-1})$ is a convex subset of U containing zero. In addition $V \in \mathcal{U}_\mathcal{C}$, because $C_j \cap V = C_j \cap \bigcup_{k=j}^{\infty} V_k$ for $j = 1, 2, \ldots$.

Let $E \subset X$ be $\mathcal{U}_\mathcal{C}$ bounded. Proceeding to a contradiction, assume that E is contained in no C_k. It follows from Definition 10.3.2 that there is an increasing sequence $\{j_k\}$ in $\mathbb{N}$ such that $kC_k \subset C_{j_k}$ for each $k \in \mathbb{N}$. By our assumption, E contains a sequence $\{x_k\}$ such that $x_k \notin C_{j_k}$, or alternatively $x_k/k \notin C_k$, for all $k \in \mathbb{N}$. As E is $\mathcal{U}_\mathcal{C}$ bounded, the sequence $\{x_k/k\}$ converges to zero in $(X, \mathcal{U}_\mathcal{C})$, contrary to Proposition 10.2.2, (1). Since each C_k is compact, the converse is clear.

As each Cauchy sequence in a topological linear space is bounded, it follows from the previous paragraph that a Cauchy sequence $\{x_j\}$ in $(X, \mathcal{U}_\mathcal{C})$ is a sequence in some C_k. By compactness, $\{x_j\}$ has a $\mathcal{U}$ convergent subsequence, and being Cauchy, it $\mathcal{U}$ converges itself. Thus $\{x_j\}$ converges in $(X, \mathcal{U}_\mathcal{C})$ by Proposition 10.2.2, (1). $\qquad\square$

Example 10.3.4. Denote by $\mathbb{R}^\infty$ the linear space of all sequences $\{x_i\}$ in $\mathbb{R}$ such that $x_i = 0$ for all but finitely many i. Given $x = \{x_i\}$ and $y = \{y_i\}$ in $\mathbb{R}^\infty$, let $x \cdot y = \sum_{i=1}^{\infty} x_i y_i$ and give $\mathbb{R}^\infty$ the locally convex topology $\mathcal{U}$ induced by the norm $|x| := \sqrt{x \cdot x}$. Via the embedding

$$(x_1, \ldots, x_k) \mapsto (x_1, \ldots, x_k, 0, 0, \ldots) : \mathbb{R}^k \to \mathbb{R}^\infty,$$

we identify $\mathbb{R}^k$ with a subspace of $\mathbb{R}^\infty$, still denoted by $\mathbb{R}^k$. It is clear that the sets $C_k := \{x \in \mathbb{R}^k : |x| \leq k\}$, $k = 1, 2, \ldots$, form a localizing sequence $\mathcal{C} := \{C_k\}$ in $(\mathbb{R}^\infty, \mathcal{U})$. According to Proposition 10.2.2, (2), the space $(X, \mathcal{U}_\mathcal{C})$

is sequential. We show next that it is not first countable. For $k, j = 1, 2, \ldots$, let $x_k := (1/k, 0, 0, \ldots)$ and

$$x_{k,j} := (1/k, \underbrace{1/j, \ldots, 1/j}_{k\text{-times}}, 0, 0, \ldots).$$

In $(\mathbb{R}^\infty, \mathcal{U}_e)$, we have $\lim x_k = 0$ and $\lim_{j \to \infty} x_{k,j} = x_k$ for $k = 1, 2, \ldots$. On the other hand, it follows from Proposition 10.2.2, (1) that for no sequence $\{j_k\}$ of positive integers, the sequence $\{x_{k,j_k}\}$ converges to zero in $(\mathbb{R}^\infty, \mathcal{U}_e)$. Thus $(\mathbb{R}^\infty, \mathcal{U}_e)$ is not first countable; in particular it is not metrizable. Using Section 3.6, observe that

$$(\mathbb{R}^\infty, \mathcal{U}_e) = \operatorname*{inj\,lim}_{k \to \infty} (\mathbb{R}^k, \mathcal{U} \llcorner \mathbb{R}^k).$$

10.4. Duality

Let $X^* = (X, \mathcal{U})^*$ be the dual space of a locally convex space $(X, \mathcal{U})$. Each $x \in X$ defines a linear functional

$$f_x : x^* \mapsto \langle x^*, x \rangle : X^* \to \mathbb{R}.$$

It is clear that the weak* topology $\mathcal{W}^*$, defined at the end of Section 1.2, is the smallest locally convex topology in X^* such that the functional

$$f_x : (X^*, \mathcal{W}^*) \to \mathbb{R}$$

is continuous for each $x \in X$. It follows from [64, Theorem 3.10 and Section 3.14] that each continuous linear functional $f : (X^*, \mathcal{W}^*) \to \mathbb{R}$ has the form $f = f_x$ for some $x \in X$. Thus

$$\mathfrak{E} : x \mapsto f_x : X \to (X^*, \mathcal{W}^*)^*$$

is a surjective linear map, called the *evaluation map*. The Hahn-Banach theorem shows that $\mathfrak{E}$ is also injective [64, Theorem 3.4, (b)].

If $\mathcal{V}$ is a locally convex topology in X^* that is larger than $\mathcal{W}^*$, then $(X^*, \mathcal{W}^*)^* \subset (X^*, \mathcal{V})^*$. In this case, the evaluation map

$$\mathfrak{E} : X \to (X^*, \mathcal{V})^*$$

is still defined and injective, but it may not be surjective. The Mackey-Arens theorem [27, Theorem 8.3.2], stated below without proof, describes all topologies $\mathcal{V}$ in X^* such that $(X^*, \mathcal{V})^* = (X^*, \mathcal{W}^*)^*$, or equivalently, such that the evaluation map $\mathfrak{E} : X \to (X^*, \mathcal{V})^*$ is bijective.

The *smallest* locally convex topology in X for which all linear functionals in $(X, \mathcal{U})^*$ are still continuous is called the *weak topology* in X, denoted by $\mathcal{W}$. In other words, $\mathcal{W}$ is the smallest locally convex topology in X such that $(X, \mathcal{W})^* = (X, \mathcal{U})^*$. A set $E \subset X$ is called *weakly compact* if it is compact in $(X, \mathcal{W})$. For emphasis, compact subsets of $(X, \mathcal{U})$ are sometimes called *originally compact*.

Theorem 10.4.1 (Mackey-Arens). *Let $(X, \mathcal{U})$ be a locally convex space, let $\mathcal{W}$ be the weak topology in X, and let $\mathcal{V}$ be a locally convex topology in the dual space $X^* = (X, \mathcal{U})^*$. The evaluation map*

$$\mathfrak{E} : X \to (X^*, \mathcal{V})^*$$

is bijective if and only if there is a family $\mathcal{C}$ consisting of weakly compact subsets of X such that $\mathcal{V}$ is defined by seminorms

$$\|x^*\|_C := \sup\Big\{|\langle x^*, x\rangle| : x \in C\Big\}$$

where $x^* \in X^*$ and $C \in \mathcal{C}$.

For a locally convex space $(X, \mathcal{U})$, the strong topology $\mathcal{S}^*$ in $(X, \mathcal{U})^*$ is defined at the end of Section 1.2. It can be described as the topology of uniform convergence on each bounded subset of $(X, \mathcal{U})$.

Corollary 10.4.2. *Let $\mathcal{C} = \{C_k\}$ be a localizing sequence in a locally convex space $(X, \mathcal{U})$, and let $\mathcal{U}_\mathcal{C}$ be the localization of $\mathcal{U}$ by $\mathcal{C}$. If $X^* = (X, \mathcal{U}_\mathcal{C})^*$, then the evaluation map $\mathfrak{E} : X \to (X^*, \mathcal{S}^*)^*$ is bijective.*

PROOF. By Theorem 10.3.3, (2), each bounded subset of $(X, \mathcal{U}_\mathcal{C})$ is contained in some C_k. Thus the strong topology $\mathcal{S}^*$ in X^* is defined by seminorms

$$\|x^*\|_{C_k} = \sup\Big\{|\langle x^*, x\rangle| : x \in C_k\Big\}$$

where $x^* \in X^*$ and $k = 1, 2, \ldots$. As each C_k is originally compact, it is weakly compact. The corollary follows from Mackey-Arens theorem. In fact, the strong topology $\mathcal{S}^*$ is the *Arens topology* in X^*; see [27, Section 8.3.3]. $\square$

In Bourbaki's terminology, Corollary 10.4.2 states that the space $(X, \mathcal{U}_\mathcal{C})$ is *semireflexive* [9, page 87].

10.5. The space $BV_c(\Omega)$

We assume that $\Omega \subset \mathbb{R}^n$ is a fixed open set, and let

$$BV_c(\Omega) := \big\{f \in BV(\mathbb{R}^n) : \mathrm{spt}\, f \Subset \Omega\big\}.$$

Remark 10.5.1. Since the elements of $BV(\mathbb{R}^n)$ are equivalence classes, the definition of $BV_c(\Omega)$ is meaningful only in light of Convention 1.3.2, according to which we assume that for each $f \in BV(\Omega)$,

$$\mathrm{spt}\, f = \mathrm{ess\,spt}\, f.$$

We define the original topology $\mathcal{U}$ in $BV_c(\Omega)$ as the metric topology given by the L^1 norm $\|\cdot\|_{L^1(\Omega)}$. Choose open sets $\Omega_k \Subset \Omega_{k+1}$ so that $\Omega = \bigcup_{k=1}^{\infty} \Omega_k$. Using Theorem 5.5.12, observe that

$$BV_k := \left\{ g \in BV_c(\Omega) : \operatorname{spt} g \subset \operatorname{cl}\Omega_k \text{ and } \|Dg\|(\Omega) \leq k \right\}$$

are compact subsets of a normed space $\left(BV_c(\Omega), \|\cdot\|_{L^1(\Omega)}\right)$. It follows that $\{BV_k\}$ is a localizing sequence. Throughout the remainder of this book, we denote by $\mathcal{T}$ the localized topology $\mathcal{U}_{\{BV_k\}}$. If $X \subset BV_c(\Omega)$ and no confusion is possible, we write $(X, \mathcal{T})$ instead of the cumbersome $(X, \mathcal{T} \llcorner X)$.

Letting $\mathcal{D}_k = BV_k \cap \mathcal{D}(\Omega)$, we have two topologies in $\mathcal{D}(\Omega)$:

$$\mathcal{T} := \mathcal{T} \llcorner \mathcal{D}(\Omega) \quad \text{and} \quad \overline{\mathcal{T}} := \big[\mathcal{U} \llcorner \mathcal{D}(\Omega) \big]_{\{\mathcal{D}_k\}}.$$

The inclusion $\mathcal{T} \llcorner \mathcal{D}(\Omega) \subset \overline{\mathcal{T}}$ may be proper, and the topology $\overline{\mathcal{T}}$ may not be locally convex. By Proposition 10.2.2 (2), the necessary condition for the equality $\mathcal{T} \llcorner \mathcal{D}(\Omega) = \overline{\mathcal{T}}$ is that $\mathcal{T} \llcorner \mathcal{D}(\Omega)$ is a sequential topology. It is easy to see that this condition is also sufficient. Whether it holds is unclear.

Proposition 10.5.2. *A linear functional $F : \mathcal{D}(\Omega) \to \mathbb{R}$ is a charge if and only if it is $\mathcal{T}$ continuous.*

PROOF. Assume F is a charge. Fix $k \in \mathbb{N}$ and choose $\varepsilon > 0$. By Definition 10.1.2, there is $\theta > 0$ such that $F(\varphi) \leq \theta \|\varphi\|_{L^1(\Omega_{2k})} + 2\varepsilon k$ for each $\varphi \in \mathcal{D}_{2k}$. If $\varphi, \psi \in \mathcal{D}_k$ and $\|\varphi - \psi\|_{L^1(\Omega_k)} < \varepsilon$, then $\varphi - \psi \in \mathcal{D}_{2k}$ and

$$F(\varphi) - F(\psi) = F(\varphi - \psi) \leq (\theta + 2k)\varepsilon.$$

Thus $F \upharpoonright \mathcal{D}_k$ is uniformly $\mathcal{U}$ continuous. By Lemma 5.5.6, the space $(BV_k, \mathcal{U})$ is the completion of $(\mathcal{D}_k, \mathcal{U})$. We infer that F has a unique $\mathcal{U}$ continuous extension to BV_k. As this is true for each $k \in \mathbb{N}$, the functional F has an extension to $BV_c(\Omega)$, which is $\mathcal{T}$ continuous by Proposition 10.2.1, (ii). In particular, F is $\mathcal{T}$ continuous.

Conversely, assume F is $\mathcal{T}$ continuous, and hence $\overline{\mathcal{T}}$ continuous according to Proposition 10.2.1, (3). Choose $\varepsilon > 0$ and an open set $U \Subset \Omega$. Find $k \in \mathbb{N}$ with $U \subset \Omega_k$. Since F is $\mathcal{U}$ continuous in $\mathcal{D}_k$, there is $\eta > 0$ such that $F(\psi) \leq \varepsilon$ for every $\psi \in \mathcal{D}(U)$ satisfying $\|D\psi\|_{L^1(U;\mathbb{R}^n)} \leq 1$ and $\|\psi\|_{L^1(U)} \leq \eta$. Now let $\theta := \varepsilon/\eta$, and select $\varphi \in \mathcal{D}(U)$ so that $\|D\varphi\|_{L^1(U;\mathbb{R}^n)} = 1$. Letting $c := \|\varphi\|_{L^1(U)}$, we distinguish two cases.

(i) If $c \leq \eta$, then $F(\varphi) \leq \varepsilon = \varepsilon \|D\varphi\|_{L^1(U;\mathbb{R}^n)}$.

(ii) If $c > \eta$, let $\psi := (\eta/c)\varphi$. As $\|\psi\|_{L^1(U)} = \eta$ and

$$\|D\psi\|_{L^1(U;\mathbb{R}^n)} = (\eta/c)\|D\varphi\|_{L^1(U;\mathbb{R}^n)} = \eta/c < 1,$$

we obtain $F(\psi) \leq \varepsilon = \eta\theta$, and consequently

$$F(\varphi) = (c/\eta)F(\psi) \leq c\theta = \theta\|\varphi\|_{L^1(U)}.$$

In either case the inequality

$$F(\varphi) \leq \theta\|\varphi\|_{L^1(U)} + \varepsilon\|D\varphi\|_{L^1(U;\mathbb{R}^n)} \tag{$*$}$$

holds, and it remains to remove the assumption $\|D\varphi\|_{L^1(U;\mathbb{R}^n)} = 1$. Assuming $b = \|D\varphi\|_{L^1(U;\mathbb{R}^n)} > 0$, inequality $(*)$ is valid for φ/b, and hence for φ. If $\|D\varphi\|_{L^1(U;\mathbb{R}^n)} = 0$, then $\varphi = 0$ and $(*)$ is satisfied, since $\operatorname{spt}\varphi \Subset U$ implies $V - \operatorname{spt}\varphi \neq \emptyset$ for each connected component V of U. $\qquad\square$

Proposition 10.5.3. *The space* $(BV_c(\Omega), \mathcal{T})$ *is the sequential completion of the space* $(\mathcal{D}(\Omega), \mathcal{T})$.

PROOF. Lemma 5.5.6 shows that $(BV_c(\Omega), \mathcal{T})$ is the sequential closure of $\mathcal{D}(\Omega)$. Since $(BV_c(\Omega), \mathcal{T})$ is a sequentially complete space by Theorem 10.3.3, the proposition is proved. $\qquad\square$

Proposition 10.5.4. *Each* $F \in CH(\Omega)$ *has a unique* $\mathcal{T}$ *continuous extension*

$$\widetilde{F} : BV_c(\Omega) \to \mathbb{R},$$

which is linear. Given an open set $U \Subset \Omega$ *and* $\varepsilon > 0$, *there is* $\theta > 0$ *such that*

$$\widetilde{F}(g) \leq \theta\|g\|_{L^1(U)} + \varepsilon\|Dg\|(U) \tag{10.5.1}$$

for each $g \in BV_c(U)$. *Moreover, for every open set* $U \Subset \Omega$,

$$\|F\|_U = \sup\{\widetilde{F}(g) : g \in BV_c(U) \text{ and } \|Dg\|(U) \leq 1\}. \tag{10.5.2}$$

If F_v *is the distributional divergence of* $v \in C(\Omega; \mathbb{R}^n)$, *then*

$$\widetilde{F_v}(g) = -\int_\Omega v \cdot d(Dg) \tag{10.5.3}$$

for each $g \in BV_c(\Omega)$.

PROOF. Let $F \in CH(\Omega)$. The first claim is an immediate consequence of Propositions 10.5.3 and 10.5.2. The linearity of $\widetilde{F}$ follows from that of F and the continuity of $\widetilde{F}$. Next fix an open set $U \Subset \Omega$ and $g \in BV_c(U)$. By Lemma 5.5.6, there is $\{\varphi_j\}$ in $\mathcal{D}(U)$ such that

$$\lim \|\varphi_j - g\|_{L^1(U)} = 0 \quad \text{and} \quad \sup\|D\varphi_j\|_{L^1(U;\mathbb{R}^n)} \leq \|Dg\|(U).$$

Choose $\varepsilon > 0$ and find $\theta > 0$ so that for $j = 1, 2, \ldots$,

$$F(\varphi_j) \leq \theta\|\varphi_j\|_{L^1(U)} + \varepsilon\|D\varphi_j\|_{L^1(U;\mathbb{R}^n)}.$$

Since $\{\varphi_j\}$ converges to g in $(BV_c(\Omega), \mathcal{T})$, we obtain

$$\widetilde{F}(g) = \lim F(\varphi_k) \leq \theta\|g\|_{L^1(U)} + \varepsilon\|Dg\|(U), \tag{$*$}$$

$$\widetilde{F}(g) = \lim F(\varphi_k) \leq \|F\|_U \|D\varphi_k\|_{L^1(U;\mathbb{R}^n)} \leq \|F\|_U \|Dg\|(U). \tag{$**$}$$

Now $(*)$ is the inequality (10.5.1), and $(**)$ implies

$$\sup\{\widetilde{F}(g) : g \in BV_c(U) \text{ and } \|Dg\|(U) \leq 1\} \leq \|F\|_U.$$

Since the reverse inequality is obvious, equality (10.5.2) is proved. Finally, Proposition 5.5.4 yields w-lim $D\varphi_k = Dg$ in Ω. As there is $w \in C_c(\Omega; \mathbb{R}^n)$ such that $w \restriction U = v \restriction U$, we conclude

$$\widetilde{F_v}(g) = \lim F_v(\varphi_k) = -\lim \int_\Omega v \cdot D\varphi_k = -\int_\Omega v \cdot d(Dg). \qquad \square$$

In view of Proposition 10.5.4, throughout we assume that all charges are continuous linear functionals defined on $(BV_c(\Omega), \mathcal{T})$. Accordingly we identify F with $\widetilde{F}$, and denote both by F.

Theorem 10.5.5. *There are equalities*

$$CH(\Omega) = (\mathcal{D}(\Omega), \mathcal{T})^* = (BV_c(\Omega), \mathcal{T})^*,$$

and the evaluation map $\mathfrak{E} : BV_c(\Omega) \to CH(\Omega)^*$, defined by

$$\langle \mathfrak{E}(g), F \rangle = \langle F, g \rangle \qquad (10.5.4)$$

for $g \in BV_c(\Omega)$ and $F \in CH(\Omega)$, is linear and bijective.

PROOF. The equalities follow from Propositions 10.5.2 and 10.5.4. Observe that for each $F \in CH(\Omega)$ and $k = 2, 3, \dots$,

$$\tfrac{1}{k-1} \sup\{F(g) : g \in BV_{k-1}\} \le \|F\|_{\Omega_k} \le \tfrac{1}{k} \sup\{F(g) : g \in BV_k\}.$$

This and Theorem 10.3.3, (2) show that the topology in $CH(\Omega)$ defined by seminorms (10.1.1) is the strong topology $\mathcal{S}^*$; see Section 1.2. The theorem follows from Corollary 10.4.2. $\qquad \square$

10.6. Streams

Fix an open set $\Omega \subset \mathbb{R}^n$, and let

$$\mathcal{BV}_c(\Omega) := \{E \subset \mathbb{R}^n : \chi_E \in BV_c(\Omega)\}.$$

Note that for $\Omega = \mathbb{R}^n$, the family $\mathcal{BV}_c(\Omega)$ coincides with the family $\mathcal{BV}_c(\mathbb{R}^n)$ defined in Section 6.7. Via the injection $E \mapsto \chi_E : \mathcal{BV}_c(\Omega) \to BV_c(\Omega)$, we identify $\mathcal{BV}_c(\Omega)$ with a closed subset of $(BV_c(\Omega), \mathcal{T})$ where $\mathcal{T}$ is the localized topology defined in Section 10.5. It follows from Proposition 6.7.3 that $(\mathcal{BV}_c(\Omega), \mathcal{T})$ is the sequential completion of $(\mathcal{DF}(\Omega), \mathcal{T})$.

Let $v \in C(\Omega; \mathbb{R}^n)$. The flux of v is the additive function

$$\phi_v : A \mapsto \int_A v \cdot \nu_E \, d\mathcal{H}^{n-1} : \mathcal{DF}(\Omega) \to \mathbb{R}$$

introduced in Section 2.1. If F_v is the distributional divergence of v, then

$$\phi_v(A) = F_v(\chi_A)$$

for each $A \in \mathcal{DF}(\Omega)$ according to Proposition 10.5.4. Thus given an open set $U \Subset \Omega$ and $\varepsilon > 0$, there is $\theta > 0$ such that for each $A \in \mathcal{DF}(U)$,

$$|\phi_v(A)| \le \theta|A| + \varepsilon \mathbf{P}(A).$$

This motivates a definition analogous to that of a charge (Definition 10.1.2).

Definition 10.6.1. An additive function $\phi : \mathcal{DF}(\Omega) \to \mathbb{R}$ is called a *stream* in Ω if given an open set $U \Subset \Omega$ and $\varepsilon > 0$, there is $\theta > 0$ such that for each $A \in \mathcal{DF}(U)$,

$$|\phi(A)| \le \theta |A| + \varepsilon \mathbf{P}(A).$$

Proposition 10.6.2. *Each stream* $\phi : \mathcal{DF}(\Omega) \to \mathbb{R}$ *is* $\mathcal{T}$ *continuous and has a unique continuous extension* $\widetilde{\phi} : (\mathcal{BV}_c(\Omega), \mathcal{T}) \to \mathbb{R}$. *The extension is additive with respect to nonoverlapping sets. Moreover, given an open set* $U \Subset \Omega$ *and* $\varepsilon > 0$, *there is* $\theta > 0$ *such that for each* $E \in \mathcal{BV}_c(U)$,

$$|\widetilde{\phi}(E)| \le \theta |E| + \varepsilon \mathbf{P}(E). \tag{10.6.1}$$

PROOF. Proving the $\mathcal{T}$ continuity of ϕ is similar to proving the $\mathcal{T}$ continuity of a charge; see the proof of Proposition 10.5.2. As $(\mathcal{BV}_c(\Omega), \mathcal{T})$ is a sequential completion of $(\mathcal{DF}(\Omega), \mathcal{T})$, the existence and uniqueness of the continuous extension $\widetilde{\phi} : (\mathcal{BV}_c(\Omega), \mathcal{T}) \to \mathbb{R}$ is obvious. The proof of additivity of $\widetilde{\phi}$ is the same as that of Proposition 2.4.3. Choose an open set $U \Subset \Omega$ and $\varepsilon > 0$. Given $E \in \mathcal{BV}_c(U)$, Proposition 6.7.3 implies that there is a constant $\gamma = \gamma(n) > 0$ and a sequence $\{A_k\}$ in $\mathcal{DF}(U)$ such that

$$\sup \mathbf{P}(A_k) \le \gamma \mathbf{P}(E) \quad \text{and} \quad \lim |E \bigtriangleup A_k| = 0.$$

Find $\theta > 0$ corresponding to U and ε/γ according to Definition 10.6.1. Then

$$|\widetilde{\phi}(E)| = \lim |F(A_k)| \le \theta \lim |A_k| + \tfrac{\varepsilon}{\gamma} \limsup \mathbf{P}(A_k) \le \theta |E| + \varepsilon \mathbf{P}(E). \qquad \square$$

In view of Proposition 10.6.2, we assume that all streams are defined on $\mathcal{BV}_c(\Omega)$ and identify ϕ with $\widetilde{\phi}$. It is clear that each charge F defines a stream $\phi = F \restriction \mathcal{BV}_c(\Omega)$. The converse is also true: in a slightly different context, the following theorem is proved in [51, Section 4.1].

Theorem 10.6.3. *Let* $\phi : \mathcal{BV}_c(\Omega) \to \mathbb{R}$ *be a stream. Then* $F : BV_c(\Omega) \to \mathbb{R}$ *defined by*

$$F(g) := \int_0^\infty \phi(\{g > t\}) \, dt - \int_0^\infty \phi(\{-g > t\}) \, dt$$

for each $g \in BV_c(\Omega)$ *is a charge and* $F \restriction \mathcal{BV}_c(\Omega) = \phi$. *Moreover, if* $G : BV_c(\Omega) \to \mathbb{R}$ *is a charge such that* $G \restriction \mathcal{BV}_c(\Omega) = \phi$, *then* $G = F$.

In accordance with our agreement, each stream is a $\mathcal{T}$ continuous additive function defined on $\mathcal{BV}_c(\Omega)$. The next example shows that the converse is false.

Example 10.6.4. Let $v(x) = x/|x|$ for $x \in \mathbb{R}^2 - \{0\}$, and $v(0) = 0$. Since v is admissible and $\operatorname{div} v \in L^1_{\text{loc}}(\mathbb{R}^2)$, Proposition 7.4.3 implies that

$$\int_A \operatorname{div} v(x) \, dx = \int_{\partial_* A} v \cdot \nu_A \, d\mathcal{H}^1$$

for each $A \in \mathcal{BV}_c(\mathbb{R}^2)$. As the Lebesgue integral is absolutely continuous,

$$\phi : A \mapsto \int_A \operatorname{div} v(x) \, dx : (\mathcal{BV}_c(\mathbb{R}^2), \mathcal{T}) \to \mathbb{R}$$

is an additive $\mathcal{T}$ continuous function. On the other hand, for $B_k = B(0, 1/k)$ we calculate

$$\phi(B_k) = \int_{\partial B_k} v \cdot \nu_{B_k} \, d\mathcal{H}^1 = \mathbf{P}(B_k).$$

In particular $\lim [\phi(B_k)/\mathbf{P}(B_k)] = 1$. Proceeding toward a contradiction, assume that ϕ is a stream. Choose $0 < \varepsilon < 1$, and let $U = U(0, 2)$. There is $\theta > 0$ such that

$$|\phi(B)| \le \theta |B| + \varepsilon \mathbf{P}(B)$$

for each $B \in \mathcal{BV}_c(U)$. Since $\{B_k\}$ is a sequence in $\mathcal{BV}_c(U)$, we obtain a contradiction:

$$1 = \lim \frac{\phi(B_k)}{\mathbf{P}(B_k)} \leq \theta \lim \frac{|B_k|}{\mathbf{P}(B_k)} + \varepsilon = \varepsilon.$$

The following result, proved in [23, Section 6], describes the structure of $\mathcal{T}$ continuous additive functions defined on $\mathcal{BV}_c(\Omega)$.

Theorem 10.6.5. *Each continuous additive function* $F : (\mathcal{BV}_c(\Omega), \mathcal{T}) \to \mathbb{R}$ *is the sum of a stream and an absolutely continuous signed measure.*

Chapter 11

The divergence equation

For bounded vector fields, and then for continuous vector fields, we characterize distributions which are their distributional divergences. These results, obtained jointly by T. De Pauw and the author [23], extend parts of an earlier work of J. Bourgain and H. Brezis [10]. As the proofs involve functional analysis in an essential way, some familiarity with basic properties of locally convex spaces is assumed. More specialized facts are presented with precise references, but often without proofs.

11.1. Background

Throughout this chapter, $n \geq 2$ and $\Omega \subset \mathbb{R}^n$ is a fixed open set. Consistently we denote by p and q a *conjugate pair* of extended real numbers, i.e., we assume that $1 \leq p \leq \infty$, $1 \leq q \leq \infty$, and defining $1/\infty := 0$, we also assume

$$\frac{1}{p} + \frac{1}{q} = 1.$$

The *Sobolev conjugate* of $1 \leq p \leq n$ is the extended real number

$$p^* := \begin{cases} \frac{np}{n-p} & \text{if } p < n, \\ \infty & \text{if } p = n. \end{cases}$$

Let $L \in \mathcal{D}'(\Omega)$ be any distribution. We want to find a vector field v in $L^1_{\text{loc}}(\Omega; \mathbb{R}^n)$ that is a *weak solution* of the *divergence equation*

$$\text{div}\, v = L.$$

In other words, we wish to find $v \in L^1_{\text{loc}}(\Omega; \mathbb{R}^n)$ whose distributional divergence F_v equals L; see Example 3.1.2, (3). Depending on L we are interested in finding v that satisfies some regularity conditions, such as boundedness, or continuity, or both.

An important special case is when $L = L_f$ for a function f in $L^p_{\text{loc}}(\Omega)$; see Example 3.1.2, (1). As usual, instead of $\text{div}\, v = L_f$, we write

$$\text{div}\, v = f. \tag{11.1.1}$$

The naive approach of solving equation (11.1.1) in $\Omega = \mathbb{R}^n$ by letting

$$v_1(\xi_1,\ldots,\xi_n) = \int_0^{\xi_1} f(t,\xi_2,\ldots,\xi_n)\,dt$$

and $v = (v_1,0,\ldots,0)$ is not satisfactory. It provides regular solutions only to the extent to which f is regular in the variables $\xi_2,\ldots,\xi_n$. Utilizing *Poisson's equation* is more promising. Since $\triangle = \mathrm{div} \circ D$, we find a weak solution $u : \Omega \to \mathbb{R}$ of Poisson's equation

$$\triangle u = f$$

and let $v := Du$. Assuming that $f \in L^p(\Omega)$ has compact support, the regularity of Du depends on the relationship between p and the dimension n. Out of many successful applications, we quote two that emphasize the critical nature of the exponent $p = n$.

(i) If $1 < p < n$, then $Du \in L^{p^*}(\Omega;\mathbb{R}^n)$; see [71, Chapter 8, Section 4.2].

(ii) If $p > n$, then Du is continuous [45, Theorem 10.2].

If $p = n$, regular solutions of the divergence equation (11.1.1) cannot be obtained by solving Poisson's equation. The following example of L. Nirenberg shows that Du need not be locally bounded when $p = n$.

Example 11.1.1. For each $x = (\xi_1,\ldots,\xi_n)$ in $\mathbb{R}^n$, let

$$u(x) := \begin{cases} \xi_1 \big|\log|x|\big|^s \varphi(x) & \text{if } x \neq 0, \\ 0 & \text{if } x = 0, \end{cases}$$

where $0 < s < (n-1)/n$ and $\varphi \in C_c^\infty(\mathbb{R}^n)$ equals one in a neighborhood of 0. A direct calculation reveals that Du is not bounded about 0. Since $\triangle u$ has compact support, and since there is $c > 0$ such that

$$\big|\triangle u(x)\big| \leq \frac{c}{|x|}\big|\log|x|\big|^{s-1}$$

for each $x \in \mathbb{R}^n$, we see that $f := \triangle u$ belongs to $L^n(\mathbb{R}^n)$.

Notwithstanding Nirenberg's example, Bourgain and Brezis have shown that a locally bounded, in fact continuous, solution of equation (11.1.1) exists [10, Proposition 1]. We elaborate on their ideas.

In broad terms, our approach to solving the divergence equation can be described as follows. We consider a space X equal to $L^p(\Omega;\mathbb{R}^n)$, or to $C(\Omega;\mathbb{R}^n)$, and identify a linear space $Y \subset \mathcal{D}'(\Omega)$ such that

$$v \mapsto F_v : L^1_{\mathrm{loc}}(\Omega) \to \mathcal{D}'(\Omega)$$

maps X into Y. Then, using functional analysis, we show that the map

$$v \mapsto F_v : X \to Y$$

is surjective. With no reference to Poisson's equation, we prove

(i) if $f \in L^p(\Omega)$ and $1 < p \leq n$, then equation (11.1.1) has a weak solution in $L^{p^*}(\Omega; \mathbb{R}^n)$; see Theorem 11.2.3 below.

(ii) if $f \in L^n_{\text{loc}}(\Omega)$, then equation (11.1.1) has a continuous weak solution in $L^\infty(\Omega; \mathbb{R}^n)$; see Theorem 11.3.9 below.

11.2. Solutions in $L^p(\Omega; \mathbb{R}^n)$

If $v \in L^q(\Omega; \mathbb{R}^n)$ then by Hölder's inequality,

$$F_v(\varphi) = -\int_\Omega v \cdot D\varphi \leq \|v\|_{L^q(\Omega; \mathbb{R}^n)} \|D\varphi\|_{L^p(\Omega; \mathbb{R}^n)} \tag{11.2.1}$$

for each $\varphi \in \mathcal{D}(\Omega)$. This suggests to define

$$\|L\|_p := \sup\{L(\varphi) : \varphi \in \mathcal{D}(\Omega) \text{ and } \|D\varphi\|_{L^p(\Omega; \mathbb{R}^n)} \leq 1\}$$

for each $L \in \mathcal{D}'(\Omega)$. Observe that $\|\cdot\|_p$ is a norm in the linear space

$$\mathcal{D}'_p(\Omega) := \{L \in \mathcal{D}'(\Omega) : \|L\|_p < \infty\}.$$

Since inequality (11.2.1) implies

$$\|F_v\|_p \leq \|v\|_{L^q(\Omega; \mathbb{R}^n)} \tag{11.2.2}$$

for every $v \in L^q(\Omega; \mathbb{R}^n)$, there is a continuous linear map

$$v \mapsto F_v : \left(L^q(\Omega; \mathbb{R}^n), \|\cdot\|_{L^q(\Omega; \mathbb{R}^n)}\right) \to \left(\mathcal{D}'_p(\Omega), \|\cdot\|_p\right).$$

Proposition 11.2.1. *Let $p < \infty$. The map*

$$v \mapsto F_v : L^q(\Omega; \mathbb{R}^n) \to \mathcal{D}'_p(\Omega)$$

is surjective. Moreover, given $L \in \mathcal{D}'_p(\Omega)$, there is $v \in L^q(\Omega; \mathbb{R}^n)$ such that $F_v = L$ and $\|v\|_{L^q(\Omega; \mathbb{R}^n)} = \|L\|_p$.

PROOF. Let $X = \{D\varphi : \varphi \in \mathcal{D}(\Omega)\}$. As the support of $\varphi \in \mathcal{D}(\Omega)$ differs from each connected component of Ω, the map

$$D : \varphi \mapsto D\varphi : \mathcal{D}(\Omega) \to X$$

is a bijection. Choose $L \in \mathcal{D}'_p(\Omega)$, and define a linear functional

$$G := L \circ D^{-1} : X \to \mathbb{R}.$$

If $w \in X$, let $\varphi = D^{-1}(w)$ and observe

$$G(w) = L(\varphi) \leq \|L\|_p \|D\varphi\|_{L^p(\Omega; \mathbb{R}^n)} = \|L\|_p \|w\|_{L^p(\Omega; \mathbb{R}^n)}.$$

By the Hahn-Banach theorem, the functional G extends to a linear functional $\widetilde{G} : L^p(\Omega; \mathbb{R}^n) \to \mathbb{R}$ so that

$$\widetilde{G}(w) \leq \|L\|_p \|w\|_{L^p(\Omega; \mathbb{R}^n)} \tag{$*$}$$

for each $w \in L^p(\Omega; \mathbb{R}^n)$. The following diagram commutes:

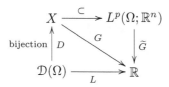

Since $p < \infty$, the standard duality properties of L^p spaces [63, Theorem 6.16] show that there is $v \in L^q(\Omega; \mathbb{R}^n)$ such that

$$\widetilde{G}(w) = -\int_\Omega v \cdot w$$

for each $w \in L^p(\Omega; \mathbb{R}^n)$, and that

$$\|v\|_{L^q(\Omega;\mathbb{R}^n)} = \sup\{\widetilde{G}(w) : w \in L^p(\Omega; \mathbb{R}^n) \text{ and } \|w\|_{L^p(\Omega;\mathbb{R}^n)} \le 1\};$$

in particular, $\|v\|_{L^q(\Omega;\mathbb{R}^n)} \le \|L\|_p$ by (*). If $\varphi \in \mathcal{D}(\Omega)$ then

$$L(\varphi) = G(D\varphi) = \widetilde{G}(D\varphi) = -\int_\Omega v \cdot D\varphi = F_v(\varphi).$$

We conclude that $L = F_v$, and hence $\|v\|_{L^q(\Omega;\mathbb{R}^n)} \le \|F_v\|_p$. The proposition follows from inequality (11.2.2). $\square$

Corollary 11.2.2. *Let $q > 1$. The equation* $\operatorname{div} v = L$ *has a weak solution* $v \in L^q(\Omega; \mathbb{R}^n)$ *if and only if $L \in \mathcal{D}'_p(\Omega)$. Moreover, the solution v can be selected so that* $\|v\|_{L^q(\Omega;\mathbb{R}^n)} = \|L\|_p$.

In order to apply Corollary 11.2.2 to the equation $\operatorname{div} v = f$ for a function $f \in L^p(\Omega)$, we need the *Gagliardo-Nirenberg-Sobolev inequality*.

Theorem 11.2.3. *Let $1 \le r < n$. There is $\kappa = \kappa(n, r) > 0$ such that*

$$\|\varphi\|_{L^{r^*}(\Omega)} \le \kappa \|D\varphi\|_{L^r(\Omega;\mathbb{R}^n)} \qquad \text{(GNS)}$$

for each $\varphi \in \mathcal{D}(\Omega)$.

PROOF. If $r = 1$, inequality (GNS) is the Sobolev's inequality of Theorem 5.9.8. Thus assume $1 < r < n$ and find $\beta = \beta(n, r) > 0$ so that

$$\frac{r}{r-1} = \beta\left(\frac{r}{r-1} - \frac{n}{n-1}\right).$$

Applying the Sobolev and Hölder inequalities to $|\varphi|^\beta$, we obtain

$$\left(\int_\Omega \left(|\varphi|^\beta\right)^{\frac{n}{n-1}}\right)^{\frac{n-1}{n}} \le \gamma \int_\Omega \left|D\left(|\varphi|^\beta\right)\right| = \beta\gamma \int_\Omega |\varphi|^{\beta-1}|D\varphi|$$

$$\le \beta\gamma \left(\int_\Omega \left(|\varphi|^{\beta-1}\right)^{\frac{r}{r-1}}\right)^{\frac{r-1}{r}} \left(\int_\Omega |D\varphi|^r\right)^{\frac{1}{r}}.$$

A direct calculation reveals $\beta = r(n-1)/(n-r)$, and hence

$$(\beta - 1)\frac{r}{r-1} = \beta\frac{n}{n-1} = r^*.$$

Thus letting $\kappa = \beta\gamma$, the previous inequality translates to

$$\left(\int_\Omega |\varphi|^{r^*}\right)^{\frac{n-1}{n}} \le \kappa\left(\int_\Omega |\varphi|^{r^*}\right)^{\frac{r-1}{r}}\left(\int_\Omega |D\varphi|^r\right)^{\frac{1}{r}}.$$

The proposition follows, since

$$\frac{n-1}{n} - \frac{r-1}{r} = \frac{1}{r^*}.\qquad\square$$

Theorem 11.2.4. *There is $\kappa = \kappa(n,p) > 0$ such that for each $f \in L^p(\Omega)$, $1 < p \le n$, the equation $\operatorname{div} v = f$ has a weak solution $v \in L^{p^*}(\Omega; \mathbb{R}^n)$ with*

$$\|v\|_{L^{p^*}(\Omega;\mathbb{R}^n)} \le \kappa\|f\|_{L^p(\Omega)}.$$

PROOF. If $p = n$ then $p^* = \infty$. By the Hölder and (GNS) inequalities,

$$L_f(\varphi) = \int_\Omega f\varphi \le \|f\|_{L^n(\Omega)}\|\varphi\|_{L^{\frac{n}{n-1}}(\Omega)} \le \kappa\|f\|_{L^n(\Omega)}\|D\varphi\|_{L^1(\Omega;\mathbb{R}^n)}$$

for every $\varphi \in \mathcal{D}(\Omega)$; here $\kappa = \kappa(n,1)$. It follows $\|L_f\|_1 \le \kappa\|f\|_{L^n(\Omega)}$, and consequently $L_f \in \mathcal{D}'_1(\Omega)$. According to Corollary 11.2.2, the equation $\operatorname{div} v = L_f$ has a weak solution $v \in L^\infty(\Omega; \mathbb{R}^n)$ such that

$$\|v\|_{L^\infty(\Omega;\mathbb{R}^n)} = \|L_f\|_1 \le \kappa\|f\|_{L^n(\Omega)}.$$

If $1 < p < n$, then $\frac{n}{n-1} < p^* < \infty$ and there is $1 < r < n$ such that $\frac{1}{r} + \frac{1}{p^*} = 1$. A direct calculation shows that $\frac{1}{r^*} + \frac{1}{p} = 1$. Given $\varphi \in \mathcal{D}(\Omega)$, the Hölder and (GNS) inequalities yield

$$L_f(\varphi) \le \|f\|_{L^p(\Omega)}\|\varphi\|_{L^{r^*}(\Omega)} \le \kappa\|f\|_{L^p(\Omega)}\|D\varphi\|_{L^r(\Omega,\mathbb{R}^n)}$$

where $\kappa = \kappa(n,r)$. We infer $\|L_f\|_r \le \kappa\|f\|_{L^p(\Omega)}$, and consequently L_f belongs to $\mathcal{D}'_r(\Omega)$. Corollary 11.2.2 implies that the equation $\operatorname{div} v = L_f$ has a weak solution $v \in L^{p^*}(\Omega; \mathbb{R}^n)$ such that

$$\|v\|_{L^{p^*}(\Omega;\mathbb{R}^n)} = \|L_f\|_r \le \kappa\|f\|_{L^p(\Omega)}.\qquad\square$$

Unlike in Corollary 11.2.2, the sufficient condition of Theorem 11.2.4 is by no means necessary.

Example 11.2.5. Assume $n = 2$, and let $\Omega_a := (0,a)^2$ where $0 < a \le 1$. For $(\xi,\eta) \in \Omega_a$, define $f(\xi,\eta) = \xi^{-\eta}$. If $1 \le p \le 1/a$, we calculate

$$\int_{\Omega_a} f(\xi,\eta)^p \, d\xi d\eta = \frac{1}{p}\int_{1-ap}^1 t^{-1}a^t \, dt.$$

Thus $f \in L^p(\Omega_a)$ if and only if $1 \leq p < 1/a$. On the other hand,

$$v : (\xi, \eta) \mapsto \left(\frac{\xi^{1-\eta}}{1-\eta}, 0 \right) : \Omega_1 \to \mathbb{R}^2$$

belongs to $C^\infty(\Omega_1; \mathbb{R}^2)$ and $\operatorname{div} v(x) = f(x)$ for each $x \in \Omega_1$. In particular, $\operatorname{div} v = f$ has a weak solution in $L^\infty(\Omega_{1/2}, \mathbb{R}^2)$ for a function f that does not belong to $L^2(\Omega_{1/2})$ — a fact unobtainable from Theorem 11.2.4.

11.3. Continuous solutions

Throughout, we topologize $C(\Omega; \mathbb{R}^n)$ as a subspace of $L^\infty_{\mathrm{loc}}(\Omega; \mathbb{R}^n)$. In view of Example 1.2.2, with this topology $C(\Omega; \mathbb{R}^n)$ is a Fréchet space. The distributional divergence of $v \in C(\Omega; \mathbb{R}^n)$ is denoted by F_v; see Example 3.1.2, (3). It follows from (10.1.2) that there is a continuous linear map

$$\Gamma : v \mapsto F_v : C(\Omega; \mathbb{R}^n) \to CH(\Omega) \tag{11.3.1}$$

and we show that Γ is surjective. To this end, we establish first that the *range* of Γ, i.e., the space $\Gamma\left[C(\Omega; \mathbb{R}^n)\right]$, is dense in $CH(\Omega)$. Then we prove that the range of the *adjoint map* $\Gamma^* : CH(\Omega)^* \to C(\Omega; \mathbb{R}^n)^*$ is closed in the strong topology $\mathbf{S}^*$ of $C(\Omega; \mathbb{R}^n)^*$, and infer the surjectivity of Γ form the *close range theorem* (Theorem 11.3.5 below).

Lemma 11.3.1. *The linear space $\Gamma\left[C(\Omega; \mathbb{R}^n)\right]$ is dense in $CH(\Omega)$.*

PROOF. Choose $\alpha \in CH(\Omega)^*$ so that $\alpha(F_v) = 0$ for each $v \in C(\Omega; \mathbb{R}^n)$. Using Theorem 10.5.5, find $g \in BV_c(\Omega)$ with $\mathfrak{E}(g) = \alpha$. By equality (10.5.4),

$$0 = \langle \alpha, F_v \rangle = \langle \mathfrak{E}(g), F_v \rangle = \langle F_v, g \rangle = - \int_\Omega v \cdot d(Dg)$$

for every $v \in C(\Omega; \mathbb{R}^n)$. It follows from Proposition 5.3.11 that $Dg \equiv 0$. Since $|U - \operatorname{spt} g| > 0$ for every connected component U of Ω, Proposition 5.5.9 shows that $g \equiv 0$; see Convention 1.3.2. Thus $\alpha = \mathfrak{E}(g) = 0$, and the lemma is a consequence of the Hahn-Banach theorem [64, Theorem 3.5]. □

Let $U \subset \mathbb{R}^n$ be an open set. An *amiable subset* of U is a compact set $K \subset U$ such that each connected component V of $U - K$ satisfies either $d(V) = \infty$ or $\partial V \cap \partial U \neq \emptyset$. Observe that a compact set K is an amiable subset of $\mathbb{R}^n$ if and only if $\mathbb{R}^n - K$ is connected.

Lemma 11.3.2. *Let $U \subset \mathbb{R}^n$ be an open set. Each compact set $C \subset U$ is contained in an amiable subset K of U.*

PROOF. Denote by $\mathcal{V}$ the collection of all bounded connected components V of $U - C$ such that $\partial V \subset C$, and by $\mathcal{W}$ the collection of all other connected components of $U - C$. For $V \in \mathcal{V}$,

$$\operatorname{dist}(V, \partial U) = \operatorname{dist}(\partial V, \partial U) \geq \operatorname{dist}(C, \partial U)$$

and since V is bounded, also $d(V) = d(\partial V) \leq d(C)$. Consequently

$$\operatorname{dist}\left(\bigcup \mathcal{V}, \partial U\right) \geq \operatorname{dist}(C, \partial U) \quad \text{and} \quad d\left(\bigcup \mathcal{V}\right) \leq 3d(C).$$

Thus $\bigcup \mathcal{V}$ is a relatively closed subset of U whose closure is contained in U. It follows that $\bigcup \mathcal{V}$ is a compact set, and so is $K = C \cup \bigcup \mathcal{V}$. If $W \in \mathcal{W}$ is bounded, then ∂W is a subset of $\partial(U - C) = \partial U \cup \partial C$, but not a subset of C. Thus each $W \in \mathcal{W}$ is either unbounded or $\partial W \cap \partial U \neq \emptyset$. Since $\mathcal{W}$ consists of all connected components of $U - K$, we conclude that K is an amiable subset of U. $\square$

Lemma 11.3.3. *Let $g \in BV(\Omega)$. If the support of Dg is contained in an amiable subset of Ω, then so is the support of g.*

PROOF. Assume $\operatorname{spt} Dg$ is contained in an amiable subset K of Ω. Choose a connected component V of $\Omega - K$. If $V \subset S$ then $d(V) < \infty$, and the definition of K leads to a contradiction:

$$\emptyset \neq \partial V \cap \partial \Omega \subset \operatorname{spt} Dg \subset \Omega.$$

Thus $V - \operatorname{spt} Dg$ is a nonempty open set, and hence $|V - \operatorname{spt} Dg| > 0$. We infer $g(x) = 0$ for almost all $x \in V$, since g is constant almost everywhere in V by Proposition 5.5.9. As $\Omega - K$ has only countably many connected components, $g(x) = 0$ for almost all $x \in \Omega - K$, and the lemma follows from Convention 1.3.2; cf. Remark 10.5.1. $\square$

Lemma 11.3.4. *Let $\{g_i\}$ be a sequence in $BV_c(\Omega)$ such that*

$$\gamma(B) := \sup\left\{\left|\int_\Omega v \cdot d(Dg_i)\right| : v \in B \text{ and } i = 1, 2, \dots\right\} < \infty$$

for every bounded set $B \subset C(\Omega; \mathbb{R}^n)$. Then $\sup \|Dg_i\|(\Omega) < \infty$, and there is a compact set $K \subset \Omega$ containing the support of each g_i.

PROOF. As the set $C = \{v \in C(\Omega; \mathbb{R}^n) : \|v\|_{L^\infty(\Omega;\mathbb{R}^n)} \leq 1\}$ is bounded in $C(\Omega; \mathbb{R}^n)$, Proposition 5.2.3 and Theorem 5.5.1 show that for $i = 1, 2, \dots$,

$$\|Dg_i\|(\Omega) = \sup\left\{\int_\Omega g_i \operatorname{div} v : v \in C \cap C_c^\infty(\Omega; \mathbb{R}^n)\right\}$$

$$= \sup\left\{-\int_\Omega v \cdot d(Dg_i) : v \in C \cap C_c^\infty(\Omega; \mathbb{R}^n)\right\}$$

$$\leq \sup\left\{\left|\int_\Omega v \cdot d(Dg_i)\right| : v \in C\right\} \leq \gamma(C) < \infty.$$

Next find open sets Ω_i such that $\Omega_i \Subset \Omega_{i+1}$ and $\bigcup_{i=1}^{\infty} \Omega_i = \Omega$. In view of Lemmas 11.3.2 and 11.3.3, it suffices to show that $\operatorname{spt} Dg_i \subset \Omega_j$ for some $j \in \mathbb{N}$ and every $i \in \mathbb{N}$. Suppose this is not true, and construct recursively subsequences of $\{g_i\}$ and $\{\Omega_i\}$, still denoted by $\{g_i\}$ and $\{\Omega_i\}$, so that $(\Omega_{i+1} - \operatorname{cl}\Omega_i) \cap \operatorname{spt} Dg_i \neq \emptyset$. There are $v_i \in C(\Omega; \mathbb{R}^n)$ such that $\|v_i\|_{L^{\infty}(\Omega;\mathbb{R}^n)} \leq 1$, $\operatorname{spt} v_i \subset \Omega_{i+1} - \operatorname{cl}\Omega_i$, and $a_i = \left|\int_{\Omega} v_i \cdot Dg_i\right| > 0$. Let $b_i := \max\{a_1^{-1}, \dots, a_i^{-1}\}$, and observe that

$$B := \left\{ v \in C(\Omega; \mathbb{R}^n) : \|v\|_{L^{\infty}(\Omega_{i+1};\mathbb{R}^n)} \leq ib_i \text{ for } i = 1, 2, \dots \right\}$$

is a bounded subset of $C(\Omega; \mathbb{R}^n)$ containing all $w_i = (ib_i)v_i$; see Section 1.2. On the other hand,

$$\gamma(B) \geq \left|\int_{\Omega} w_i \cdot d(Dg_i)\right| = ib_i \left|\int_{\Omega} v_i \cdot d(Dg_i)\right| \geq i$$

for $i = 1, 2, \dots$, contrary to our assumption. $\square$

Let X^* and Y^* be the duals of locally convex spaces X and Y, respectively. If $\Phi \colon X \to Y$ is a continuous linear map, then the formula

$$\langle \Phi^*(y^*), x \rangle := \langle y^*, \Phi(x) \rangle, \tag{11.3.2}$$

where $x \in X$ and $y^* \in Y^*$, defines a linear map $\Phi^* : Y^* \to X^*$, called the *adjoint map* of Φ. Recall that the weak* topology $\mathcal{W}^*$ and the strong topology $\mathcal{S}^*$, both in X^*, are defined in Section 1.2, and that some of their properties were discussed in Seciton 10.4. The next important result is called the *close range theorem.*

Theorem 11.3.5. *Let X and Y be Fréchet spaces. If $\Phi : X \to Y$ is a continuous linear map, then the following conditions are equivalent:*

(1) $\Phi(X)$ *is closed in Y;*
(2) $\Phi^*(Y^*)$ *is closed in $(X^*, \mathcal{W}^*)$;*
(3) $\Phi^*(Y^*)$ *is closed in $(X^*, \mathcal{S}^*)$;*
(4) $\Phi^*(Y^*)$ *is sequentially closed in $(X^*, \mathcal{S}^*)$.*

PROOF. The equivalence of conditions (1), (2), and (3) is proved in [27, Theorem 8.6.13]. Analyzing the proof of implication (3) $\Rightarrow$ (2) ibid., it is not difficult to show that (3) can be relaxed to (4). For the details we refer the reader to [24, Proposition 6.8]. $\square$

Lemma 11.3.6. $\Gamma\big[C(\Omega; \mathbb{R}^n)\big]$ *is a closed subset of $CH(\Omega)$.*

PROOF. Following Theorem 11.3.5, we consider the adjoint map

$$\Gamma^* : CH(\Omega)^* \to C(\Omega; \mathbb{R}^n)^*,$$

and show that $\Gamma^*\big[CH(\Omega)^*\big]$ is sequentially closed in $\big(C(\Omega;\mathbb{R}^n)^*,\mathcal{S}^*\big)$. To this end, denote by $M_c(\Omega;\mathbb{R}^n)$ the linear space of all compactly supported $\mathbb{R}^n$-valued measures in Ω. For each $\mu \in M_c(\Omega;\mathbb{R}^n)$,

$$L_\mu : v \mapsto -\int_\Omega v \cdot d\mu : C(\Omega;\mathbb{R}^n) \to \mathbb{R}$$

is a continuous linear functional, and there is a linear map

$$\Lambda : \mu \mapsto L_\mu : M_c(\Omega;\mathbb{R}^n) \to C(U;\mathbb{R}^n)^*.$$

According to Theorem 10.5.5, the evaluation map

$$\mathfrak{E} : BV_c(\Omega) \to CH(\Omega)^*$$

is a linear bijection. The maps Γ^*, Λ, $\mathfrak{E}$, and

$$D : g \mapsto Dg : BV_c(\Omega) \to M_c(\Omega;\mathbb{R}^n)$$

are linked by the following commutative diagram:

$$
\begin{array}{ccc}
BV_c(\Omega) & \xrightarrow{\ D\ } & M_c(\Omega;\mathbb{R}^n) \\[4pt]
{\scriptstyle\text{bijection}}\Big\downarrow{\mathfrak{E}} & & \Big\downarrow{\Lambda} \\[4pt]
CH(\Omega)^* & \xrightarrow{\ \Gamma^*\ } & C(\Omega;\mathbb{R}^n)^*
\end{array}
\qquad (11.3.3)
$$

Indeed for $g \in BV_c(\Omega)$ and $v \in C(\Omega;\mathbb{R}^n)$, the defining equalities (10.5.3), (10.5.4), and (11.3.2) yield

$$\big\langle (\Lambda \circ D)(g), v \big\rangle = \langle L_{Dg}, v \rangle = -\int_\Omega v \cdot d(Dg) = \langle F_v, g \rangle$$

$$= \big\langle \mathfrak{E}(g), F_v \big\rangle = \big\langle \mathfrak{E}(g), \Gamma(v) \big\rangle = \big\langle (\Gamma^* \circ \mathfrak{E})(g), v \big\rangle.$$

For a bounded set $B \subset C(\Omega;\mathbb{R}^n)$ and $S \in C(\Omega;\mathbb{R}^n)^*$, let

$$|S|_B := \sup\big\{ |S(v)| : v \in B \big\}$$

and note that $|\cdot|_B$ are seminorms which define the strong topology $\mathcal{S}^*$ in $C(\Omega;\mathbb{R}^n)^*$; see Section 1.2.

Proceeding to the actual proof, select $\{\alpha_i\}$ in $CH(\Omega)^*$ such that $\{\Gamma^*(\alpha_i)\}$ converges in $\big(C(\Omega;\mathbb{R}^n)^*,\mathcal{S}^*\big)$ to some $T \in C(\Omega;\mathbb{R}^n)^*$. Define $g_i := \mathfrak{E}^{-1}(\alpha_i)$, and infer from diagram (11.3.3) that

$$\big\langle \Gamma^*(\alpha_i), v \big\rangle = -\int_\Omega v \cdot d(Dg_i)$$

for each $v \in C(\Omega; \mathbb{R}^n)$. Being convergent, the sequence $\{\Gamma^*(\alpha_i)\}$ is bounded in $\left(C(\Omega; \mathbb{R}^n)^*, \mathcal{S}^*\right)$. Thus for each bounded set $B \subset C(\Omega; \mathbb{R}^n)$,

$$\sup\left\{\left|\int_\Omega v \cdot d(Dg_i)\right| : v \in B \text{ and } i = 1, 2, \ldots\right\}$$

$$= \sup\left\{\left|\langle\Gamma^*(\alpha_i), v\rangle\right| : v \in B \text{ and } i = 1, 2, \ldots\right\}$$

$$= \sup\left\{\left|\Gamma^*(\alpha_i)\right|_B : i = 1, 2, \ldots\right\} < \infty.$$

Lemma 11.3.4 implies that $\sup \|Dg_i\|(\Omega) < \infty$, and that there is a compact set $K \subset \Omega$ containing the support of each g_i. By Theorem 5.5.12, the sequence $\{g_i\}$ has a subsequence, still denoted by $\{g_i\}$, such that $\lim \|g - g_i\|_{L^1(\Omega)} = 0$ for some $g \in BV_c(\Omega)$. Moreover $Dg = \text{w-}\lim Dg_i$ by Proposition 5.5.4. Let $\alpha := \mathfrak{E}(g)$ and choose $v \in C(\Omega; \mathbb{R}^n)$. Since the supports of g and of all g_i are contained in K, we can apply the weak convergence to v:

$$\langle T, v\rangle = \lim\langle\Gamma^*(\alpha_i), v\rangle = -\lim\int_\Omega v \cdot d(Dg_i) =$$

$$= -\int_\Omega v \cdot d(Dg) = \langle\Gamma^*(\alpha), v\rangle.$$

From the arbitrariness of v, we conclude $T = \Gamma^*(\alpha)$.   □

Remark 11.3.7. Although we did not need this for the proof of Lemma 11.3.6, using the Riesz theorem (Theorem 5.4.4), one can show that the map

$$\Lambda : \mu \mapsto L_\mu : M_c(\Omega; \mathbb{R}^n) \to C(\Omega; \mathbb{R}^n)^*$$

is bijective [27, Theorem 4.10.1]. Loosely speaking, it means that

$$D : BV_c(\Omega) \to M_c(\Omega; \mathbb{R}^n) \quad \text{and} \quad \Gamma^* : CH(\Omega) \to C(\Omega; \mathbb{R}^n)^*$$

are two representations of the same map. A geometric rationale supports our vague assertion. Identifying vectors in $C(\Omega; \mathbb{R}^n)$ with the corresponding $(n-1)$-dimensional *differential forms*, the weak divergence operator Γ becomes the weak *exterior derivative*. In this context, interpreting elements of $BV_c(\Omega)$ as n-dimensional *normal currents* [33, Section 4.5.1], it is clear that the adjoint map Γ * is the *boundary operator* D. This reasoning suggests that diagram (11.3.3) may commute. The actual verification of it is routine.

The following theorem is an easy corollary of Lemmas 11.3.1 and 11.3.6.

Theorem 11.3.8. *Let $L \in \mathcal{D}'(\Omega)$. The equation $\text{div } v = L$ has a weak solution $v \in C(\Omega; \mathbb{R}^n)$ if and only if L is a charge in Ω.*

Let $f \in L^n_{\text{loc}}(\Omega)$. Combining Theorem 11.3.8 and Proposition 10.1.4, we see at once that the equation $\text{div } v = f$ has a weak solution $v \in C(\Omega; \mathbb{R}^n)$. In view of this and Theorem 11.2.4, the equation $\text{div } v = f$ has bounded weak solutions, as well as continuous weak solutions. The next theorem shows that it has a solution v which is bounded and continuous simultaneously. The idea of the proof is due to T. De Pauw.

Theorem 11.3.9. *Let $f \in L^n(\Omega)$. There is $\kappa = \kappa(n) > 0$ such that the equation $\operatorname{div} v = f$ has a weak solution $v \in C(\Omega; \mathbb{R}^n)$ with*

$$\|v\|_{L^\infty(U;\mathbb{R}^n)} \leq \kappa \|f\|_{L^n(\Omega)}.$$

PROOF. Avoiding a triviality assume $\|f\|_{L^n(\Omega)} > 0$, and let

$$\|L_f\| := \sup\{L_f(g) : g \in BV_c(\Omega) \text{ and } \|Dg\|(\Omega) \leq 1\}.$$

By Hölder and Sobolev inequalities there is $\beta = \beta(n) > 0$ such that

$$L_f(g) = \int_\Omega fg \leq \|f\|_{L^n(\Omega)} \|g\|_{L^{\frac{n}{n-1}}(\Omega)} \leq \beta \|f\|_{L^n(\Omega)} \|Dg\|(\Omega)$$

for each $g \in BV_c(\Omega)$. Thus $\|L_f\| \leq \beta \|f\|_{L^n(\Omega)}$, and we claim $\|L_f\| > 0$. Indeed, $\|L_f\| = 0$ implies $\int_\Omega f\varphi = 0$ for each $\varphi \in \mathcal{D}(\Omega)$, and consequently $f(x) = 0$ for almost all $x \in \Omega$ contrary to our assumption.

Letting $\kappa = 3\beta$, it suffices to show that the nonempty convex sets

$$A := \{v \in C(\Omega; \mathbb{R}^n) : \|v\|_{L^\infty(\Omega;\mathbb{R}^n)} < 3\|L_f\|\},$$
$$B := \{w \in C(\Omega; \mathbb{R}^n) : \Gamma(w) = L_f\}$$

have nonempty intersection. Proceeding toward a contradiction, suppose the intersection $A \cap B$ is empty. Since A is an open set, the Hahn-Banach theorem implies that there are $T \in C(\Omega; \mathbb{R}^n)^*$ and $\gamma \in \mathbb{R}$ such that

$$\langle T, v \rangle < \gamma \leq \langle T, w \rangle$$

for each $v \in A$ and each $w \in B$; see [64, Theorem 3.4, (a)]. As $0 \in A$, we see that $\gamma > 0$.

Claim. $\Gamma^{-1}(0) \subset T^{-1}(0)$.

Proof. Choose $u \in \Gamma^{-1}(0)$ and $w \in B$. Observe that $w + tu$ belongs to B for each $t \in \mathbb{R}$. Thus $T(w) + tT(u) = T(w + tu) \geq \gamma$ for all $t \in \mathbb{R}$. This is impossible unless $T(u) = 0$.

By the claim, there is $S \in CH(\Omega)^*$ such that the diagram

$$
\begin{array}{ccc}
C(\Omega; \mathbb{R}^n) & \xrightarrow{\ \Gamma\ } & CH(\Omega) \\
& {\scriptstyle T}\searrow & \downarrow {\scriptstyle S} \\
& & \mathbb{R}
\end{array}
$$

commutes. As the evaluation map $\mathfrak{E} : BV_c(\Omega) \to CH(\Omega)^*$ is bijective (Theorem 10.5.5), there is $g \in BV_c(\Omega)$ with $\mathfrak{E}(g) = S$. By (10.5.3),

$$-\int_\Omega v \cdot d(Dg) = \langle F_v, g \rangle = \langle \mathfrak{E}(g), F_v \rangle = \langle S, \Gamma(v) \rangle = \langle T, v \rangle \qquad (*)$$

for each $v \in C(\Omega; \mathbb{R}^n)$. If $w \in B$, then

$$\gamma \leq \langle T, w \rangle = \langle S, \Gamma(w) \rangle = \langle S, L_f \rangle$$
$$= \langle \mathfrak{E}(g), L_f \rangle = \langle L_f, g \rangle \leq \|L_f\| \cdot \|Dg\|(\Omega). \tag{$**$}$$

Now select $u \in C_c^1(\Omega; \mathbb{R}^n)$ with $\|u\|_{L^\infty(\Omega;\mathbb{R}^n)} \leq 1$, and observe that the vector field $v := 2\|L_f\| \, u$ belongs to $A \cap C_c^1(\Omega; \mathbb{R}^n)$. According to $(*)$,

$$2\|L_f\| \int_\Omega g \, \operatorname{div} u = \int_\Omega g \, \operatorname{div} v = - \int_\Omega v \cdot d(Dg) = \langle T, v \rangle < \gamma \,.$$

As u is arbitrary, $\|Dg\|(U) \leq \gamma / (2\|L_f\|) < \gamma / \|L_f\|$ contrary to $(**)$. $\square$

Remark 11.3.10. The following are some concluding comments.

(1) Recently P. Bouafia [8] proved that there exists no *uniformly* continuous map

$$F \mapsto v_F : CH(\Omega) \to C(\Omega; \mathbb{R}^n)$$

such that v_F is a weak solution of $\operatorname{div} v = F$ for each $F \in CH(\Omega)$.

(2) If $p > n$ and $f \in L^p(\Omega)$ has compact support, then $\operatorname{div} v = f$ has solutions whose regularity is stronger than continuity. For instance,

$$|v(x) - v(y)| \leq |x - y|^s$$

for all $x, y \in \Omega$ and $0 < s < 1 - (n/p)$; see [45, Theorem 11.2]. On the other hand, D. Preiss [57] constructed a function $f \in C_c(\mathbb{R}^n)$ such that $\operatorname{div} v = f$ has no weak solution that is locally Lipschitz.

(3) For more results related to Theorem 11.3.8 we refer to [10, 23, 24, 25].

Bibliography

I not only use all the brains I have,
but all that I can borrow.

Woodrow Wilson

[1] L. Ambrosio, N. Fusco, and D. Pallara, *Functions of Bounded Variations and Free Discontinuity Problems*, Oxford Univ. Press, Oxford, 2000.

[2] A.S. Besicovitch, *On sufficient conditions for a function to be analytic, and behaviour of analytic functions in the neighbourhood of non-isolated singular points*, Proc. London Math. Soc. **32** (1931), 1–9.

[3] ――――, *A general form of the covering principle and relative differentiation of additive functions*, Proc. Cambridge Philos. Soc. **41** (1945), 103–110.

[4] ――――, *A general form of the covering principle and relative differentiation of additive functions II*, Proc. Cambridge Philos. Soc. **42** (1946), 1–10.

[5] B. Bongiorno, W.F. Pfeffer, and B.S. Thomson, *A full descriptive definition of the gage integral*, Cand. Math. Bull. **39** (1996), 390–401 .

[6] B. Bongiorno, L. Di Piazza, and V. Skvortsov, *A new full descriptive characterization of Denjoy-Perron integral*, Anal. Math. **22** (1996), 3–12.

[7] B. Bongiorno, L. Di Piazza, and D. Preiss, *Infinite variations and derivatives in* $\mathbb{R}^m$, J. Math. Anal. Appl. **224** (1998), 22–33.

[8] P. Bouafia, *Retractions onto the space of continuous divergence-free vector fields*, Ann. Fac. Sci. Touluse. **XX** (2011), 767–779.

[9] N. Bourbaki, *Espaces Vectoriels Topologiques*, chapters III and IV, Hermann at Cie, Paris, 1955.

[10] J. Bourgain and H. Brezis, *On the equation* $\operatorname{div} Y = f$ *and application to control of phases*, J. Amer. Math. Soc. **16** (2003), 393–426.

[11] Z. Buczolich, *Density points and bi-Lipschitz functions in* $\mathbb{R}^m$, Proc. Amer. Math. Soc. **116** (1992), 53–59.

[12] Z. Buczolich and W.F. Pfeffer, *Variations of additive functions*, Czechoslovak Math. J. **47** (1997), 525–555.

[13] ――――, *On absolute continuity*, J. Math. Anal. Appl. **222** (1998), 64–78.

[14] R. Caccioppoli, *Misure e integrazione sugli insiemi dimensionalmente orientati*, Rend. Accad. Naz. Lincei **12** (1953), 3–11, 137–141.

[15] G.-Q. Chen and M. Torres, *Divergence-measure fields, sets of finite perimeter, and conservation laws*, Arch. Ration. Mech. Anal. **175** (2005), 245–267.

[16] G.-Q. Chen, M. Torres, and W.P. Ziemer, *Gauss-Green theorem for weakly differentiable vector fields, sets of finite perimeter, and balance laws*, Comm. Pure Appl. Math. **62** (2009), 242–304.

[17] P. Cousin, *Sur les fonctions de n variables complexes*, Acta Math. **19** (1895), 1–61.

[18] E. De Giorgi, *Su una teoria generale della misura* $(r - 1)$-*dimensionale in uno spazio ad r-dimensioni*, Ann. Mat. Pura Appl. **36** (1954), 191–213.

[19] ———, *Nuovi teoremy relativi alle misure* $(r - 1)$-*dimensionali in uno spazio ad r-dimensioni*, Ricerche Mat. **4** (1955), 95–113.

[20] E. De Giorgi, *Sulla differentiabilità e l'analiticità delle estremali degli integrali multipli regolari*, Mem. Acad. Sci. Torino **III, parte I** (1957), 25–43.

[21] T. De Pauw, *Topologies for the space of BV-integrable functions in* $\mathbb{R}^n$, J. Func. Anal. **144** (1997), 190–231.

[22] T. De Pauw and W.F. Pfeffer, *The Gauss-Green theorem and removable sets for PDEs in divergence form*, Adv. Math. **183** (2004), 155–182.

[23] ———, *Distributions for which* $\operatorname{div} v = F$ *has a continuous solution*, Comm. Pure Appl. Math. **61** (2008), 261–288.

[24] T. De Pauw, L. Moonens, and W.F. Pfeffer, *Charges in middle dimensions*, J. Math. Pures Appl. **92** (2009), 86–112.

[25] T. De Pauw and M. Torres, *On the distributional divergence of vector fields vanishing at infinity*, Proc. Royal Soc. Edingburgh **141A** (2011), 65–76.

[26] R.M. Dudley, *Real Analysis and Probability*, Wadsworth & Brooks/Cole, Pacific Grove, 1989.

[27] R.E. Edwards, *Functional Analysis*, Dover Publications, New York, 1995.

[28] R. Engelking, *General Topology*, PWN, Warsaw, 1977.

[29] L.C. Evans and R.F. Gariepy, *Measure Theory and Fine Properties of Functions*, CRC Press, Boca Raton, 1992.

[30] K.J. Falconer, *The Geometry of Fractal Sets*, Cambridge Univ. Press, Cambridge, 1985.

[31] H. Federer, *The Gauss-Green theorem*, Trans. Amer. Math. Soc. **58** (1945), 44–76.

[32] ———, *A note on the Gauss-Green theorem*, Proc. Amer. Math. Soc. **9** (1958), 447–451.

[33] ———, *Geometric Measure Theory*, Springer-Verlag, New York, 1969.

[34] W.H. Fleming and R. Richel, *An integral formula for total gradient variation*, Arch. Math. **11** (1960), 218–222.

[35] C. Goffman, T. Nishiura, and D. Waterman, *Homeomorphisms in Analysis*, Amer. Math. Soc., Providence, 1997.

[36] E. Giusti, *Minimal Surfaces and Functions of Bounded Variation*, Birkhäuser, Basel, 1984.

[37] R.A. Gordon, *The Integrals of Lebesgue, Denjoy, Perron, and Henstock*, Amer. Math. Soc., Providence, 1994.

[38] J. Harrison, *Stokes' theorem for nonsmooth chains*, Bull. Amer. Math. Soc. **29** (1993), 235–242.

[39] J. Harrison and A. Norton, *The Gauss-Green theorem for fractal boundaries*, Duke J. Math. **67** (1992), 575–588.

[40] K. Hoffman and R. Kunze, *Linear Algebra*, Prentice-Hall, Englewood Cliffs, 1961.

[41] E.J. Howard, *Analyticity of almost everywhere differentiable functions*, Proc. Amer. Math. Soc. **110** (1990), 745–753.

[42] W.B. Jurkat, *The divergence theorem and Perron integration with exceptional sets*, Czechoslovak Math. J. **43** (1993), 27–45.

[43] J.L. Kelley, *General Topology*, Van Nostrand, New York, 1955.

[44] P.Y. Lee and R. Výborný, *An Integral: An Easy Approach after Kurzweil and Henstock*, Australian Math. Soc. Lecture Ser., vol. 14, Cambridge Univ. Press, Cambridge, 2000.

[45] E.H. Lieb and M. Loss, *Analysis*, Amer. Math. Soc., Providence, 1997.

[46] P. Mattila, *Geometry of Sets and Measures in Euclidean Spaces*, Cambridge Univ. Press, Cambridge, 1995.

[47] F. Morgan, *Geometric Measure Theory*, Academic Press, New York, 1988.

[48] W.F. Pfeffer, *The Gauss-Green theorem*, Adv. Math. **87** (1991), 93–147.

[49] _____, *The Riemann Approach to Integration*, Cambridge Univ. Press, New York, 1993.

[50] _____, *The Lebesgue and Denjoy-Perron integrals from a descriptive point of view*, Ricerche Mat. **48** (1999), 211–223.

[51] _____, *Derivation and Integration*, Cambridge Univ. Press, New York, 2001.

[52] _____, *Derivatives and primitives*, Sci. Math. Japonicae **55** (2002), 399–425.

[53] _____, *The Gauss-Green theorem in the context of Lebesgue integration*, Bull. London Math. Soc. **37** (2005), 81–94.

[54] _____, *A devil's platform*, Amer. Math. Monthly **115** (2008), 943–947.

[55] W.F. Pfeffer and B.S. Thomson, *Measures defined by gages*, Canad. J. Math. **44** (1992), 1306–1316.

[56] G. Pisier, *The Volumes of Convex Bodies and Banach Space Geometry*, Cambridge Univ. Press, New York, 1989.

[57] D. Preiss, *Additional regularity for Lipschitz solutions of PDE*, J. Reine Angew. Math. **485** (1997), 197–207.

[58] K.P.S. Bhaskara Rao and M. Bhaskara Rao, *Theory of Charges*, Acad. Press, New York, 1983.

[59] C.A. Rogers, *Hausdorff Measures*, Cambridge Univ. Press, Cambridge, 1970.

[60] P. Romanovski, *Intégrale de Denjoy dans l'espace á n dimensions*, Math. Sbornik **51** (1941), 281–307.

[61] H.L. Royden, *Real Analysis*, Macmillan, New York, 1968.

[62] W. Rudin, *Principles of Mathematical Analysis*, McGraw-Hill, New York, 1976.

[63] _____, *Real and Complex Analysis*, McGraw-Hill, New York, 1987.

[64] _____, *Functional Analysis*, McGraw-Hill, New York, 1991.

[65] S. Saks, *Theory of the Integral*, Dover, New York, 1964.

[66] R. Schneider, *Convex Bodies: the Brunn-Minkowski Theory*, Cambridge Univ. Press, Cambridge, 1993.

[67] V.L. Shapiro, *The divergence theorem for discontinuous vector fields*, Ann. Math. **68** (1958), 604–624.

[68] L. Simon, *On a theorem of De Giorgi and Stampacchia*, Math. Z. **155** (1977), 199–204.

[69] E.H. Spanier, *Algebraic Topology*, McGraw-Hill, New York, 1966.

[70] E.M. Stein, *Singular Integrals and Differentiability Properties of Functions*, Princeton Univ. Press, Princeton, 1970.

[71] _____, *Harmonic Analysis: Real-Variable Methods, Orthogonality, and Oscillatory Integrals*, Princeton Univ. Press, Princeton, 1993.

[72] I. Tamanini and C. Giacomelli, *Un tipo di approssimazione "dall'interno" degli insiemi di perimetro finito*, Rend. Mat. Acc. Lincei **(9) 1** (1990), 181–187.

[73] B.S. Thomson, *Derivatives of Interval Functions*, Mem. Amer. Math. Soc., 452, Providence, 1991.

[74] M. Šilhavý, *Divergence measure fields and Cauchy's stress theorem*, Rend. Sem. Univ. Padova **113** (2005), 15–45.

[75] W.P. Ziemer, *Weakly Differentiable Functions*, Springer-Verlag, New York, 1989.

List of symbols

Euclidean spaces and related concepts

$\mathbb{R}$, $\mathbb{R}_+$, $\overline{\mathbb{R}}$, $\mathbb{R}^m$, 3

$\mathbb{N}$, $\mathbb{Z}$, $\mathbb{Q}$, $\mathbb{Q}_+$, $\mathbb{C}$, 3

$A \triangle B$, 3

$x \cdot y$, $|x|$, 4, 205

$d(E)$, clE, $\operatorname{int} E$, ∂E, 4

$\operatorname{dist}(A, B)$, $\operatorname{dist}(x, B)$, 4

$U(E, r)$, $U(x, r)$ 4

$B(E, r)$, $B(x, r)$ 4

$A \Subset B$ 4

e_i, π_i, Π_i, 6

$A + B$, tA, $-A$, 7

$A \ominus B$, $A \odot B$, 32

$C(x; r, h)$, 50

$B^\bullet$, 53

(xy), $[xy]$, 58

L_u, 59

S^{n-1}, $\Pi_{e,t}, H_{e,t}$, 66

Ω_ε, 77

C^ε, 98

$E_{z,r}$, 131

C^* 194

$\mathbb{R}^\infty$, 211

Functions and maps

$f \equiv c$, $f \upharpoonright B$, 4

$\{f = t\}$, $\{f > t\}$, $\{f \neq t\}$, 4

f^+, f^-, $|f|$, 4

$f(x)$, $f[x]$, $\langle f, x \rangle$, 4

$\mathfrak{u}_{A,B}$, 4

χ_E, $\overline{\phi}$, $\operatorname{spt} \phi$, 6

$\phi^{-1}(B)$, 7

$\Gamma(s)$, $\alpha(s)$, 13

$\operatorname{Lip} \phi$, 17

$\operatorname{Lip} \phi(x)$, $H_s\phi(x)$, 18

L_f, L_μ, F_v, 36

$\mathbf{V}(f, \Omega)$, $\mathbf{V}(f)$, 75

$\eta_\varepsilon * f$, 77

$(f)_E$, 110

$\phi_\# g$, 117

Tf, 168

Tv, 178

$O(1)$, $o(1)$, 185

$\mathfrak{E}$, 212, 216

ϕ_v, 216

Families of sets

$\mathcal{D}_k$, $\mathcal{DC}$, $\mathcal{DF}$, 6

P, $[P]$, 24

$\mathcal{S}t(x, \mathcal{E})$, 24

$\overline{\mathcal{DF}}$, 32

$\mathcal{A} \mathbin{\vdash} E$ 42

$\mathcal{P}(\Omega)$, $\mathcal{P}_{\mathrm{loc}}(\Omega)$, 56

$\mathcal{BV}(\Omega)$, $\mathcal{BV}_{\mathrm{loc}}(\Omega)$, 73

$\mathcal{B}$, 79

$\mathcal{S}t(A, \mathcal{E})$, $\mathcal{C}^\varepsilon$, 98

$\mathcal{BV}_c(\mathbb{R}^n)$, 146

$N(K, r)$, $P(K, r)$, 182

E_∞, 185

$\mathcal{DC}(\Omega)$, $\mathcal{DF}(\Omega)$ 193

$\mathcal{BV}_c(\Omega)$, 216

Measures and related concepts

μ, $\mu \llcorner Y$, $\mathrm{spt}\,\mu$, 10

$\phi \sim \psi$, $A \sim B$, 11

$\mathrm{ess}\,\mathrm{spt}\,\phi$, $\mathrm{ess}\,\mathrm{sup}\,\phi$, 11

$\mathcal{L}^n$, $|E|$, 13

$\int_E f(x)\,dx$, $\int_E f$, $\int_E f\,d\mathcal{L}^n$, 13

$\mathcal{H}^s_\delta$, $\mathcal{H}^s$, 14

$\mathrm{ext}_* E$, $\mathrm{int}_* E$, $\mathrm{cl}_* E$, $\partial_* E$, 49

$s(E)$, 51

$\Theta(E,x)$, 54

$\mathbf{P}(E,\Omega)$, $\mathbf{P}(E)$, 56

$\mathrm{ext}_*^L E$, $\mathrm{int}_*^L E$, $\mathrm{cl}_*^L E$, $\partial_*^L E$, 64

$\mathbf{V}(E,\Omega)$, $\mathbf{V}(E)$, 73

$\mu \llcorner h$, 79

$\|\nu\|$, 81

ν_+, ν_-, 82

$(h)_{B,\mu}$, 84

$\mathrm{w\text{-}lim}\,\mu_k$, 86

μ_E, 126

$\partial^* E$, 127

$H_\pm(E,x)$, $H(E,x)$, 127

$\mathrm{int}_c E$, 158

$\rho(n)$, 159

$\mathcal{M}^{*t}(E)$, $\mathcal{M}^t_*(E)$, 181

$\mathcal{M}^{*t}(K,\partial_* A)$, 184

$\overline{\dim}\,K$, 186

$V_\delta F(E)$, $VF(E)$, 197

Function spaces and norms

$C(E;\mathbb{R}^s)$, $C(E)$, 4

$C^k(\Omega;\mathbb{R}^s)$, $C^\infty(\Omega;\mathbb{R}^s)$, 5

$C_c(\Omega;\mathbb{R}^s)$, 6

$L^0(E,\mu;\mathbb{R}^s)$, 11

$\|\cdot\|_{L^p(E,\mu;\mathbb{R}^s)}$, $1 \le p \le \infty$, 11

$L^p(E,\mu;\mathbb{R}^s)$, 12

$L^p(E,\mu)$, $\|\cdot\|_{L^p(E,\mu)}$, 12

$L^p_{\mathrm{loc}}(\Omega,\mu;\mathbb{R}^s)$, 12

$L^p(E;\mathbb{R}^s)$, $L^p_{\mathrm{loc}}(\Omega;\mathbb{R}^s)$, 13

$Lip(E;\mathbb{R}^m)$, $Lip(E)$, 17

$Lip_c(\Omega)$, $Lip_{\mathrm{loc}}(\Omega)$, 17

$Adm(E;\mathbb{R}^m)$, $Adm(E)$ 27

$\mathcal{D}(\Omega;\mathbb{C})$, $\mathcal{D}(\Omega)$, 35

$\mathcal{D}'(\Omega;\mathbb{C})$, $\mathcal{D}'(\Omega)$, 36

$BV(\Omega)$, $\|\cdot\|_{BV(\Omega)}$, 76

$BV_{\mathrm{loc}}(\Omega)$, 76

$BV(\Omega;\mathbb{R}^n)$, 178

$AC_*(\Omega)$, 200

$CH(\Omega)$, $\|\cdot\|_U$, 206

$BV_c(\Omega)$, 213

Differentiation

$D\phi(x)$, 16

$D_i f(x)$, $\mathrm{div}\,v(x)$, 17

$\det D\phi$, J_ϕ, 19

D^α, 35

Λ, p_Λ, 37

$\triangle$, $\bar{\partial}$, 37

Df, $\|Df\|$, 92

$D_i f$, 93

$\mathrm{adj}\,D\phi$, 117

$D_E \phi(x)$, 155

$D_{E,i} f(x)$, 157

$\mathrm{div}_E v(x)$, 158

$\underline{D}F(x)$, $\overline{D}F(x)$, $F'(x)$, 193

$\eth\mathfrak{iv}\,v(x)$, 201

Topology

$\mathrm{cl}_{\mathcal{J}} E$, G_δ, 7

X^*, $\mathcal{W}^*$, $\mathcal{S}^*$, 9

$\|x^*\|_B$, 9

$\mathrm{inj}\lim(X_\alpha, \mathcal{J} \llcorner X_\alpha)$, 43

$\mathfrak{U}_\mathrm{e}$, 209

$\mathcal{W}$, 212

$\mathcal{J}$, $\overline{\mathcal{J}}$, 214

Index

Washek F. Pfeffer is Professor Emeritus of Mathematics at the University of California in Davis. He was born in Prague, Czech Republic, where he studied mathematics at Charles University (1955–60). He immigrated to the United States in 1965, and in 1966 received his Ph.D. from the University of Maryland in College Park. Dr. Pfeffer has worked at the Czechoslovak Academy of Sciences in Prague, and has taught at the Royal Institute of Technology in Stockholm, George Washington University, University of California in Berkeley, University of Ghana in Accra, and King Fahd University in Dhahran, Saudi Arabia. In 1994–95 he was a Fulbright Lecturer at Charles University. His primary research areas are analysis and topology. Dr. Pfeffer is a member of the American and Swedish Mathematical Societies, and an honorary member of the Academic Board of the Center for Theoretical Study at Charles University. Presently, he is a Research Associate in the Mathematics Department of the University of Arizona. He has written the books *Integrals and Measures* (Marcel Dekker, 1977), *The Riemann Approach to Integration*, and *Derivation and Integration* (Cambridge University Press, 1993 and 2001).